Engineering Materials

This series provides topical information on innovative, structural and functional materials and composites with applications in optical, electrical, mechanical, civil, aeronautical, medical, bio- and nano-engineering. The individual volumes are complete, comprehensive monographs covering the structure, properties, manufacturing process and applications of these materials. This multidisciplinary series is devoted to professionals, students and all those interested in the latest developments in the Materials Science field, that look for a carefully selected collection of high quality review articles on their respective field of expertise.

Indexed at Compendex (2021)

More information about this series at http://www.springer.com/series/4288

Jing Li · Chengbin Shi

Carbide in Special Steel

Formation Mechanism and Control Technology

冶金工业出版社
Metallurgical Industry Press

Springer

Jing Li
State Key Laboratory of Advanced Metallurgy
University of Science and Technology Beijing
Beijing, China

Chengbin Shi
State Key Laboratory of Advanced Metallurgy
University of Science and Technology Beijing
Beijing, China

ISSN 1612-1317 ISSN 1868-1212 (electronic)
Engineering Materials
ISBN 978-981-16-1455-2 ISBN 978-981-16-1456-9 (eBook)
https://doi.org/10.1007/978-981-16-1456-9

Jointly published with Metallurgical Industry Press, China
The print edition is not for sale in China Mainland. Customers from China Mainland please order the print book from: Metallurgical Industry Press.
ISBN of the Metallurgical Industry Press's edition: 978-7-5024-8773-7

This Springer imprint is published by the registered company Springer Nature Singapore Pte Ltd.
The registered company address is: 152 Beach Road, #21-01/04 Gateway East, Singapore 189721, Singapore

Preface

Special steel is the steel with special chemical composition, special structure, and properties, as well as special producing process, aiming to meet special requirements. There are many kinds of special steels with excellent properties, mainly including high-end spring steel, bearing steel, tool steel, heat-resistant steel, and stainless steel. Special steels are key materials for national key project construction and equipment manufacturing. The production and application of special steels represent the levels of industrialization and manufacturing development of a country. Although there is still a gap between the overall quality of our country's special steels and the international advanced level, breakthroughs have been made in some grades of steel.

Carbide is one of the important phases in special steels. Reasonable control of its type, quantity, shape, size, and distribution exerts a great effect on the properties of steel.

With the support of more than 10 scientific projects such as the National Natural Science Foundation of China "Study on Effects of Magnesium on Inclusion Formation and Carbides Precipitation for H13 Hot Work Die Steel" and "Study on the Controlling of Primary Carbides in High-Carbon Martensitic Stainless Steel used as Raw Materials for Superior Knives and Shears" and Guangdong YangFan Innovative and Entrepreneurial Research Team Program "Process Innovation, Product Research and Development for the Preparation of High-quality Knives and Shears", the author deeply studied the carbide control technology in special steels. The primary carbide control technology in the electroslag remelting process, the primary carbide control technology in directional solidification of electroslag remelting, carbide control technology during hot rolling and cold rolling, carbide control technology during heat treatment, and carbide control technology during roll forging-heat treatment are proposed. The mechanism of magnesium and rare earth elements in refining carbides is clarified, the effect of nitrogen on the formation of carbides is preliminarily studied, and the effect of titanium on carbides in special steels and the feasibility of its application in improving carbides are discussed. The research results have been applied to the production of high-carbon martensitic stainless steel for high-quality knives and shears, and the product quality has met the requirements of users.

This book summarizes the research results and process practices from the authors and research team in the control of carbides in special steels in past two decades and strives to form a relatively complete system of carbide control process. This book will be a useful reference and guiding tool for the researchers, producers, and managers engaged in the fields of metallurgy and materials, as well as the teachers and students involved in the related majors in universities.

Beijing, China Jing Li

Acknowledgments

The author wish to express gratitude to the State Key Laboratory of Advanced Metallurgy of USTB (University of Science and Technology Beijing) and the Chinese Society for Metals for their guidance and support for the book. Special thanks are due to the members of our research team, including Dingli Zheng, Qingtian Zhu, Hao Wang, Yongfeng Qi, Wentao Yu, Jie Zhang, and Shouhui Li, for their efforts, strong support, and contributing excellent research work to the book. I also acknowledge Ph.D. Dingli Zheng from the University of Science and Technology Wuhan and Ph.D. Jie Zhang from the University of Science and Technology Beijing for their help rendered in co-ordinating the activities related to the publication of this book. Finally, I would like to thank Metallurgical Industry Press personnel for their attention to the multitude of editing and publication details that led to the quality production published in association with Springer Press.

January 2021 Jing Li

Contents

List of Figures

List of Tables

Chapter 1
Carbides in Special Steel

Abstract Carbide is the binary compound that formed by carbon and the elements with less or similar electronegativity (except hydrogen). Carbide is one of the important precipitated phases in special steels, which plays an important role on the properties of steel. In this chapter, the definition and classification of carbides are presented, and the instruments and technologies for the analysis of carbides in special steels are introduced. Through the thermodynamic analysis and experimental study on the formation of carbides in many typical special steels, it is found that inevitable microsegregation in the solidification process is the key reason for the precipitation of primary carbides. These primary carbides have large size and the high content of alloying elements, which significantly reduce the processing and mechanical properties of special steels. By improving the solidification process, modification and heat treatment, the processing and service properties of special steels could effectively promote the breaking and dissolution of primary carbides. The secondary carbides in special steel mainly generate in the annealing process. Optimizing the heat processing of steel and adding alloying elements could significantly refine the secondary carbides, and consequently improve mechanical properties of steel.

Keywords Carbide · Thermodynamic calculation · Evolution · Special steel

Special steel is the steel with special chemical composition, special structure and properties, as weel as special producing process, which can meet the special needs. Most of the alloying elements in special steel are transition metal elements, which are easy to form carbides with carbon. Carbide is one of the most important phases in special steel, which has great influence on the properties of steel. Therefore, the reasonable control of its type, amount, shape, size and distribution of carbide is one of the primary tasks in the production of special steel.

At present, most of the medium and high carbon alloy die special steel is produced by electroslag remelting technology. As for the medium and high carbon alloy die steel produced by electroslag remelting process, the electroslag remelting solidification process is the source of the formation of primary carbide and the key to control primary carbide. The forging and rolling process can crush and refine the inevitably generated primary carbide, and the subsequent heat treatment process can promote the dissolution and fracture of the primary carbide and reduce the adverse effect of

J. Li and C. Shi, *Carbide in Special Steel*, Engineering Materials,
https://doi.org/10.1007/978-981-16-1456-9_1

the primary carbide on the steel properties. The heat treatment process can control the precipitation of the large amount and dispersed fine secondary carbides, which can strengthen the matrix. The role of rare earth, magnesium, nitrogen, titanium and other alloying elements and their oxides is to act as heterogeneous nucleating agents, which can play an important role in reducing the size of primary carbide and further improving the distribution of carbide. Based on those, this book focus on the evolution and control of carbides in special steels during electroslag remelting, rolling and heat treatment, as well as the influence of alloying treatment on carbides in special steels.

1.1 Carbide and Its Characterization Method

There are many kinds of carbides. Although the researches on carbides are very extensive, the researches on carbides in specific steels are still relatively scarce. Therefore, this section introduces several common analysis methods for commonly carbides in special steels.

1.1.1 Definition of Carbide

Carbides are the compounds formed by carbon element and other elements(except hydrogen) whose electronegativity is smaller or similar to that of carbon element. In steel, partial carbon elements play a solid solution strengthening role in the matrix, and the others combine with alloying elements to form carbides.

Carbides are with the characteristic of brittleness, high melting point and high hardness. The microhardness and melting point of carbides are listed in Table 1.1. The covalent bonds between carbon and alloy elements make the large hardness of carbides. Because of the metal bonds between metal atoms, the carbides have conductive properties and a positive temperature coefficient of resistance. Therefore, carbides have the characteristics of metal bond (dominated) and covalent bond.

Table 1.1 Microhardness and melting point of carbides

Carbide	TiC	ZrC	NbC	VC	WC	Mo_2C	$Cr_{23}C_6$	Cr_7C_3	Fe_3C
H_μ, kg/mm^2	2850~3200	2890	2400	2094	2200	1500	1650	2100	~860
Melting point, °C	3150	3530	3500	2830	2860	2690 (dissociate)	1520 (dissociate)	1780 (dissociate)	1650

1.1.2 Types of Carbide

The type difference of carbides in steel are due to their different alloying elements and precipitation mechanism, and the types of carbides in different steels are different. Carbide in steel can be classified into primary carbide and secondary carbide according to the order of precipitation. During the solidification process, the primary carbides precipitated directly from liquid steel, including primary carbides precipitated before the precipitation of austenite and eutectic carbides precipitated during eutectic reaction in hypereutectic steel, which are mainly MC, M_2C and M_6C-types carbides. The secondary carbides are the carbides precipitated from supersaturated solid phase at elevated temperature or from solid matrix during heat treatment, which are mainly MC, M_2C, M_6C, M_7C_3 and $M_{23}C_6$-types carbides. The primary carbides are generally with large size, shape irregular and high melting point, which can damage the continuity of the steel. The cracks are easy to generate in the matrix around primary carbides under stress, which reduces the processability and service properties of the steel. The secondary carbides are generally with small size. The type, morphology, amount and distribution of secondary carbides can be controlled by hot working and heat treatment. As an important second phase in steel, the control of secondary carbides plays an important role in improving the properties of steel.

Depending on the difference of crystal structure, the carbides in steel include carbides with simple crystal structure and carbides with complex crystal structure. The reference [1] reported that the radius ratio of carbon atom to metal atom is one of the main factors determining the crystal structure of carbides. When the radius ratio of carbon atom to transition elements atom is less than 0.59, the formed carbides are with simple crystal structure. The crystal structure can be the planar centrocube lattice, the volume centrocube lattice, the densely arranged hexagonal lattice, or simple hexagonal lattice. Under this condition, carbon atoms fill in the gap of the metal cubic lattice or hexagonal lattice, which makes the carbide has metal bond. Therefore, the carbide retains obvious metal characteristics, such as TiC, ZrC, VC, NbC, WC, etc. These carbides are with high melting point and high hardness. They are the main components of hard alloy, powder high speed steel and high temperature cermets, as well as important alloy phases of industrial steel. When the radius ratio of carbon atom to transition elements atom is large than 0.59, the gap in the lattice of simple metal atoms is smaller than the diameter of the carbon atom. In order to accommodate carbon atoms, the matrix of metal atoms deforms to form carbides with complex structures, such as M_3C, M_6C, $M_{23}C_6$ and M_7C_3-types carbides. These carbides are with high hardness and are important strengthening phase in alloy steels, but their melting point and hardness are slightly lower than those of the former. Typical carbides with simple structure and complex structure are introduced as follows.

1.1.2.1 MC and M_2C-Types Carbides with Simple Structure

The MC-type carbides are mainly in the steel containing Ti, Nb, V, etc., which has a surface core cubic structure, as shown in Fig. 1.1. Carbon atoms in the crystal lattice occupy the center of the octahedron. MC-type carbides are generally distributed at interdendrite and grain boundary with the morphologies of point [2], point stripe and skeleton. Partical MC-type carbides are with the morphologies of lamellar, dendritic and octahedral, such as TiC [3, 4]. The morphology of carbide is closely related to the mechanical properties of steel. The coarse skeleton MC-type carbide is the origin of the fatigue crack of the superalloy, which greatly reduce the fatigue performance property of the alloy. The distribution of small MC-type carbides in grain boundary and dendrite is beneficial to improve the durability property of the alloy.

Some MC and M_2C-type carbides in the unit cell contains three metal atoms (MC) or six metal atoms (M_2C) and 3 non-metal atoms, such as MoC, WC, Mo_2C and W_2C. The matrix constant(C/A) value of MoC and WC is close to 1, which belongs to simple hexagonal lattice. While the lattice constant(C/A) value of Mo_2C and W_2C is close to 1.6, belonging to a densely arranged hexagonal lattice, which mainly exists in high-speed steel and is generally laminar or feather-like [5].

1.1.2.2 $M_{23}C_6$-Type Carbides with Complex Structure

$M_{23}C_6$-type carbides is the most important carbide in steel without strong carbide forming elements such as Nb, Ti and V. Generally, $M_{23}C_6$-type carbides are mainly Cr-containing carbides, often written as $Cr_{23}C_6$. $M_{23}C_6$-type carbides are with complex face-centered cubic structure, as shown in Fig. 1.2. Each cell $M_{23}C_6$-type carbides contains 116 atoms, including 92 metal atoms and 24 carbon atoms. The lattice constant is about three times of that of the austenite matrix. Due to the different

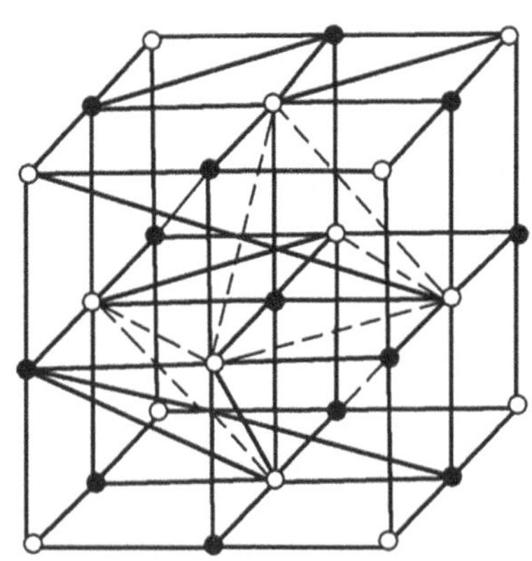

Fig. 1.1 Structue of MC-type carbide

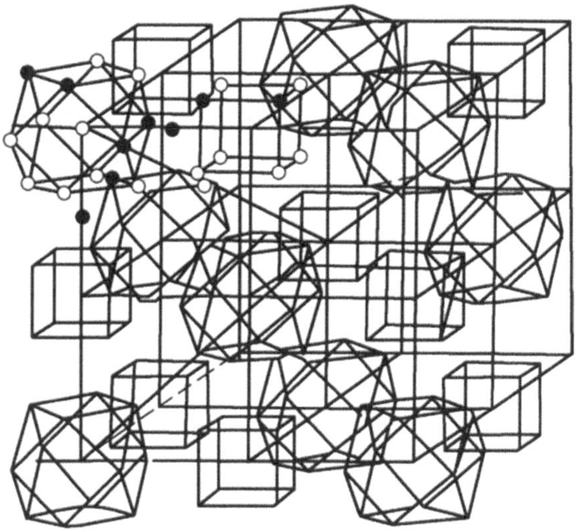

Fig. 1.2 Structue of $M_{23}C_6$-type carbide

alloy composition and heat treatment process, the Cr atom in $M_{23}C_6$-type carbides can be partially replaced by Fe, Mo, W, Ni, etc. The precipitation temperature of $M_{23}C_6$-type carbides mainly ranges from 400 °C to 900 °C.

1.1.2.3 M_6C and $M_{23}C_6$-Types Carbides with Complex Structure

M_6C-type carbides generally exist in steel containing Mo or Nb and are always precipitated near other deposits (carbides or intermetallic compound phases). M_6C-type carbides also has a face-centered cubic structure, as shown in Fig. 1.3. The lattice constant value of M_6C-type carbides is similar to that of $M_{23}C_6$-type carbides. Each cell of M_6C-type carbides contains 96 metal atoms, but the number of carbon atoms in each cell is not fixed. Therefore, it is generally believed that M_6C-type carbides may be an electron compound phase.

The chemical composition of the steel has a great influence on the precipitation of M_6C-type carbides. The N, Nb, Mo and Ni elements can promote the precipitation of M_6C-type carbides. Generally speaking, for alloys with more than 25% Ni content, the higher the Mo, Nb and Ni content, the greater the precipitation tendency of M_6C-type carbides.

M_7C_3-type carbides has a complex hexagonal structure and are the main carbides in high carbon and high chromium steel. The hardness is generally 1200–1800 HV. The main metal elements in M_7C_3-type carbides are Cr and Fe. The formation of M_7C_3-type carbides is that chromium atoms dissolve into the unstable Fe_7C_3 carbide. The Fe_7C_3 crystal is a hexagonal lattice structure, in which each carbon atom is in close contact with six adjacent iron atoms. Cr_7C_3 crystal belongs to the rhomboid

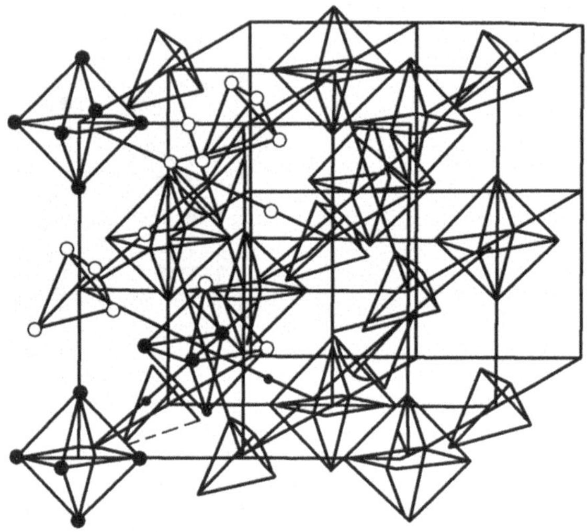

Fig. 1.3 Structue of M_6C-type carbide

system. The chromium atoms in Cr_7C_3 crystal are arranged in a similar way to the iron atoms in Fe_7C_3, and the size of the iron atoms and chromium atoms is also close, which provides conditions for replacing iron atoms by chromium atoms to form $(Fe,Cr)_7C_3$. The hardness of $(Fe,Cr)_7C_3$ crystals is different in different orientations due to the fact that most of the lamellar spreading directions are parallel to the prism plane of hexagonal crystals. The microhardness test of $(Fe,Cr)_7C_3$ crystal shows that the cross section hardness of prism is about 1700–1900 HV, and the side hardness of prism is generally only about 1400 HV. The morphology, type and amount of carbides all affect the wear resistance and mechanical properties of high chromium cast iron [6].

1.1.3 Analysis Method of Carbide

Many methods have been applied to analyze carbides. The following are some of commonly methods and their applicable condition are pointed out.

1.1.3.1 Morphology Observation Method of Carbide

A Optical microscope (OM) analysis

Metallographic analysis is based on the different corrosion potentials of various components in the steel in different chemical reagents. The stable phase in the steel is preserved and the unstable phase is dissolved by corrosion. The depressions formed

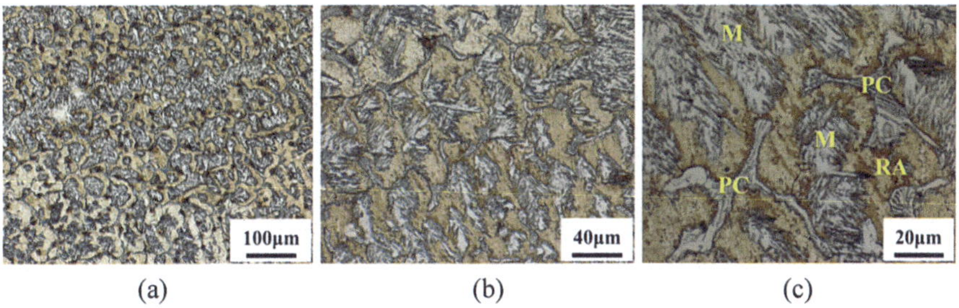

Fig. 1.4 As-cast microstructure of high-carbon martensitic stainless steel 8Cr13MoV: **a** 200×; **b** 500×; **c** 1000× (M: Acicular martensite; RA: Retained austenite; PC: Primary carbide)

by the corrosion outline the preserved phase. Due to the distinguishability limitation of optical microscope, this method can only analyze the morphologies of large-sized carbide, such as primary carbide, eutectic carbide and micron-sized secondary carbides. The optical microscope cannot clearly distinguish the outline and color of too small-sized carbides phase. The as-cast microstructure of high-carbon martensitic stainless steel 8Cr13MoV at different multiples under optical microscope (OM) are shown in Fig. 1.4.

B Scanning electron microscope (SEM) analysis

The crystal structure and surface morphology of carbide can be analyzed by scanning electron microscope (SEM). The morphology and distribution of primary carbides in the high carbon martensitic stainless steel 8Cr13MoV after hot rolling process observed by SEM are shown in Fig. 1.5. It can be seen from Fig. 1.5 that the primary

Fig. 1.5 Primary carbides in the steel after hot rolling process

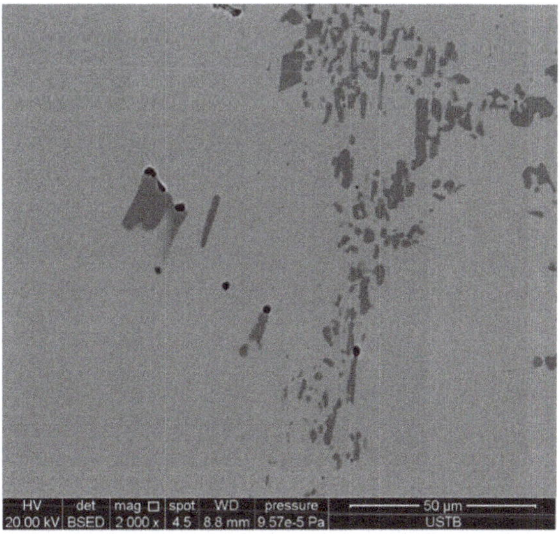

carbides was dark gray and the steel matrix was light gray. The primary carbides in the electroslag ingots are mostly agglomerated rod-like. After hot rolling process, the primary carbides are obviously broken and dispersed.

C Transmission electron microscopy (TEM) analysis

TEM is an electronic optical instrument with high distinguishability and high magnification, which uses the electron beam with very short wavelength as illumination source and focuses imaging with electromagnetic lens. TEM uses electron beam imaging of the penetrating sample and requires that the observed sample be "transparent" to the incident electron beam. The commonly used acceleration voltage for TEM is 100 kV, and the thickness of sample must be between 20 and 200 nm. The thickness of sample for high- distinguishability TEM images is required to be 15 nm. The sample preparation methods include thin film sample and carbon (or gold) extraction. In order to make the film specimen smooth and stable for observation, the specimen is usually placed on the support network. Nanometer carbide particles can be analyzed by TEM, and the microstructure and high distinguishability images can be obtained. But for the large-sized extracted carbides with fluffy structure, the rheology under high distinguishability is more serious, and the results are not ideal. The morphology, diffraction pattern and energy spectrum analysis of $M_{23}C_6$-type carbide under TEM are shown in Fig. 1.6. The morphology of $M_{23}C_6$-type carbide

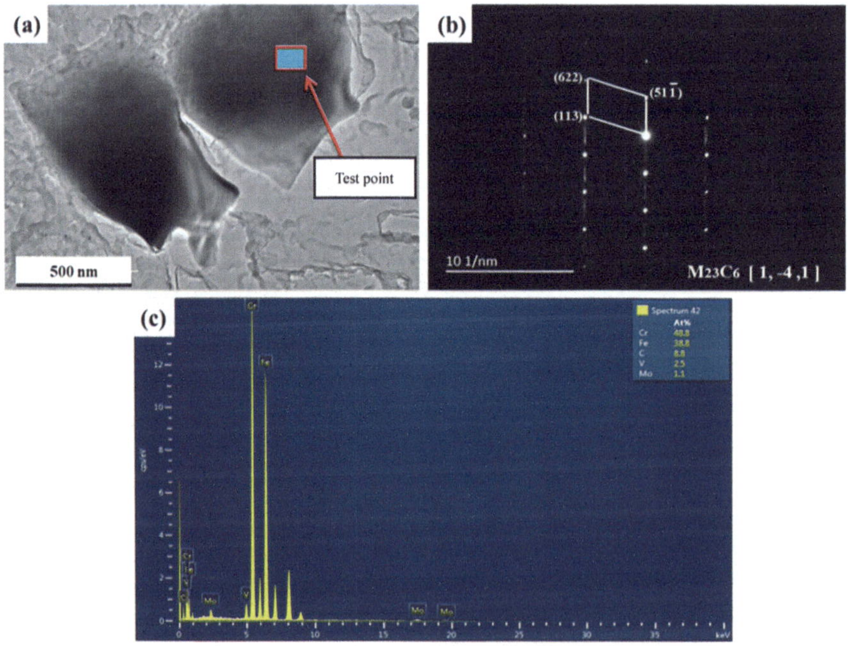

Fig. 1.6 Diffraction pattern and energy spectrum analysis of $M_{23}C_6$-type carbide: **a** TEM image; **b** Calibration results of TEM diffraction spot; **c** TEM energy spectrum analysis

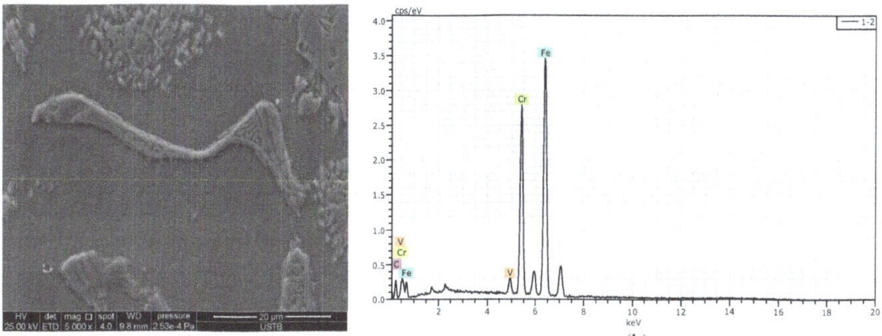

(a) (b)

Fig. 1.7 Carbide in high-carbon martensitic stainless steel: **a** SEM image; **b** EDS results

is irregular block with the size about 1 μm. The atom fraction of chromium in the carbide is 48.8%. In addition, the carbide contains a small amount of molybdenum and vanadium.

D Energy Disperse Spectroscopy (EDS) analysis

The EDS analysis can analyze the types and contents of component elements in the material microregion, which is based on the different quantum energy of X-ray. EDS analysis is usually combined with SEM or TEM, which can scan the point, line and surface of the grain to obtain the content distribution image of each element. The SEM image of carbides in high-carbon martensitic stainless steel and EDS scanning results of corresponding points are shown in Fig. 1.7. It can be seen from Fig. 1.7 that the primary carbide is mainly Cr and Fe-containing carbides.

E Electron Probe Microanalyzer (EPMA) analysis

EPMA is also called microarea X-ray spectrometer and X-ray microanalyzer The principle of EPMA is to bombard the solid surface with focused high-energy electron beam, and the bombarded elements can excite the characteristic X-rays. The qualitative and quantitative chemical analysis of element on the micro-area of solid surface can be carried out according to wavelength and intensity of element. The electron probe can normally be analyzed in a range of diameter and depth not less than 1 m. It is mainly used for quantitative analysis of various elements in primary carbides and large size secondary carbides. The mapping scanning of typical primary carbide and its adjacent areas in 8Cr13MoV remelted ingot was carried out with the electron probe, and the results are shown in Fig. 1.8. The primary carbide grew in the grain boundary in a coiled rod-like structure, in which there was obvious enrichment of carbon, chromium, molybdenum and vanadium.

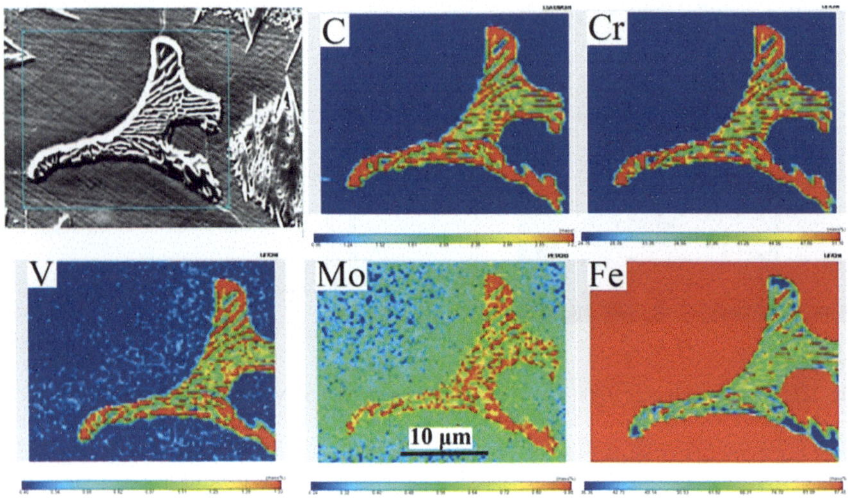

Fig. 1.8 Mapping scanning of typical primary carbide in 8Cr13MoV remelted ingot (Below color bar represents the gradual increase of element content from left to right)

1.1.3.2 Crystal Structure of Carbide Analysis

The crystal structure of carbide can be analyzed by electron backscattering diffraction (EBSD), transmission electron microscopy (TEM), X-ray diffraction (XRD), etc.

A Electron backscattering diffraction (EBSD) analysis

EBSD analysis combined with SEM can be used for phase identification and comparison calculation of carbides in steel, orientation difference analysis of carbides, analysis of size and morphology of carbides, etc. Figure 1.9 shows the morphology and phase distribution of carbide in HP40Nb alloy after holding at 1150 °C for 2 h detected by EBSD [7].

B Transmission electron microscopy (TEM) analysis

TEM can be used to analyze the morphology and structure of nanoscale precipitated phase at grain boundary at the early stage of thermal exposure. The sample preparation are as follows: (1) Sample was sliced into 500 μm in thickness; (2) Sandpaper is used to grind the flakes into 100 m and punch is used to make some small discs with diameter of 3 mm; (3) In the electrolytic double jet instrument, a certain concentration of erosive liquid is used for corrosion at −30 °C, until a small hole appears in the center of the wafer, and then it is quickly removed and rinsed with alcohol; (4) The light field phase and dark field phase of the alloy structure were obtained by means of transmission electron microscope.

The morphology, diffraction pattern and EDS analysis of M_7C_3-type carbide under TEM are shown in Fig. 1.10. According to the diffraction pattern of the carbide, the

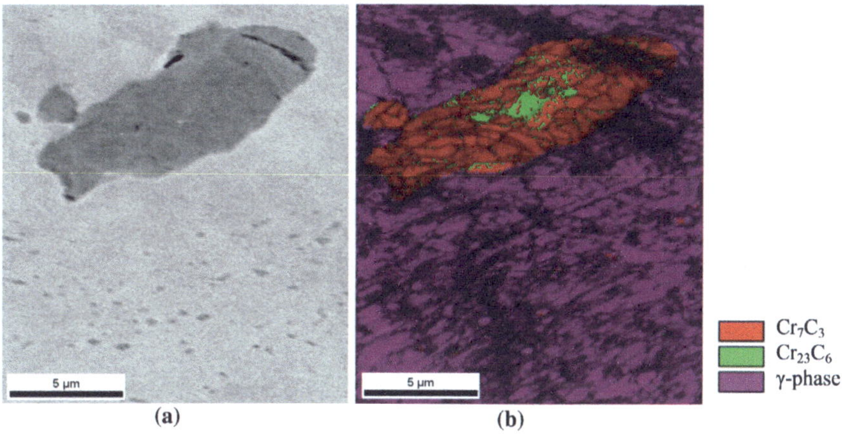

Fig. 1.9 EBSD results of the carbide in the HP40Nb alloy after holding at 1150 °C for 2 h: **a** Morphology; **b** Phase distribution [7]

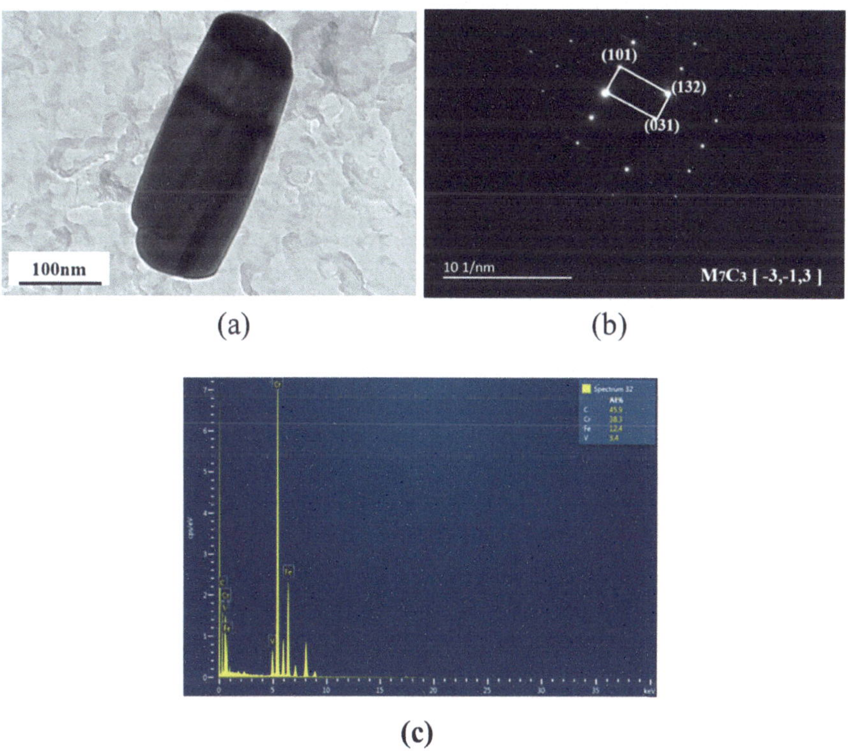

Fig. 1.10 Diffraction pattern and EDS analysis of M_7C_3-type carbide: **a** TEM image; **b** Diffraction spot calibration results of TEM; **c** EDS analysis of TEM

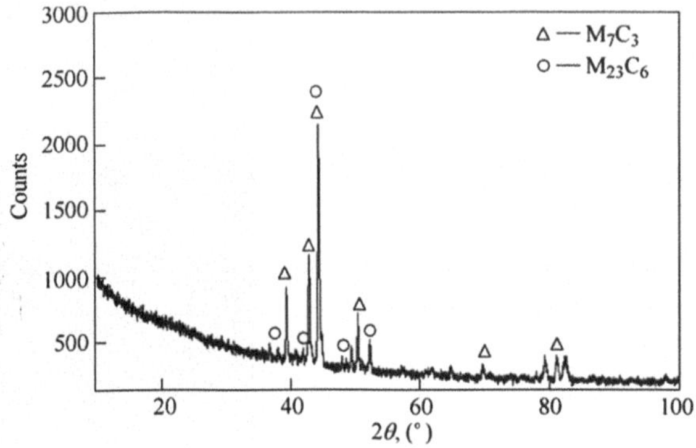

Fig. 1.11 XRD pattern of carbide powder obtained by electrolytic extraction

rod-shaped carbide is M_7C_3-type carbide with orthorhombic system, in which the atom fraction of chromium is 38.3%, and there is a small amount of vanadium in the carbide. It should be noted that there are discontinuous stripes between the diffraction spots of the carbides, which are typical diffraction spots of M_7C_3-type carbide with orthorhombic system [8]. This kind of spot is caused by the high density stacking fault structure of M_7C_3-type carbid [9, 10].

C X-ray diffraction (XRD) analysis

The X-ray diffraction pattern of the material was analyzed to determine the atomic and molecular structure of the crystal. X-ray diffraction (XRD) analysis mainly focuses on the analysis of carbide extracted from alloy steel by chemical or electrolytic methods. There are many kinds of carbides in common alloy steel. The type and amount of carbides can be distinguished from the the specific diffraction peaks of different carbides. XRD pattern of the carbide powder obtained by electrolytic extraction is shown in Fig. 1.11. It can be seen from Fig. 1.11 that the types of carbides are M_7C_3 and $M_{23}C_6$, and the the intensity of diffraction peak indicates that M_7C_3 is the main carbide.

1.1.3.3 Statistical Method for Characteristic Parameters of Carbides

Image Pro Plus (IPP) software is used to analyze the images obtained by OM, SEM or TEM. The size, amount and area ratio of carbides can be obtained by adjusting the contrast of the images, setting the scale, dividing the grid and statistical calculation.

1.1.3.4 Extraction Method of Carbide in Steel

The chemical composition and amount of carbides can be accurately identified by extracting carbide. In this method, carbides are separated by chemical or electrolytic methods. The three-dimensional morphology and size of the extracted carbides can

Fig. 1.12 SEM image of carbide powder

be analyzed by scanning electron microscopy (SEM), and the types of carbides can be identified by X-ray diffraction (XRD).

The SEM image of primary carbides obtained by anodic electrolysis is shown in Fig. 1.12. It can be seen from Fig. 1.12 that the the size of primary carbides is large and with skeletal morphology, which is the typical eutectic carbide morphology. Partial carbides were with flake morphology, which was because that these carbides grew along the grain boundary and were extruded by grains.

1.2 Thermodynamic Analysis of Carbide Formation in Special Steel

The effect of 12% Cr on Fe–C equilibrium diagram is shown in Fig. 1.13 [11].

It can be seen from Fig. 1.13 that the carbide would precipitate from the residual liquid when the carbon content in the residual liquid exceeds the point E. Generally, the carbon content in high carbon steel is not so high, but carbides still precipitate. This is mainly due to the following two factors:

(1) The ferrite forming elements (Cr, Mo, V, etc.) reduce the austenite phase region and make the point E shift to the left, which greatly increases the possibility of primary carbide formation;
(2) Element segregation results to the enrichment of carbon, chromium and other elements in the residual liquid, which makes the primary carbide precipitate from the steel in the solidification process of liquid steel.

The equilibrium phase diagram of 8Cr13MoV high carbon martensitic stainless steel is shown in Fig. 1.14. As shown in Fig. 1.14, the carbon content at eutectic point of M_7C_3-type primary carbide is 0.92%. The carbon content of 8Cr13MoV steel is

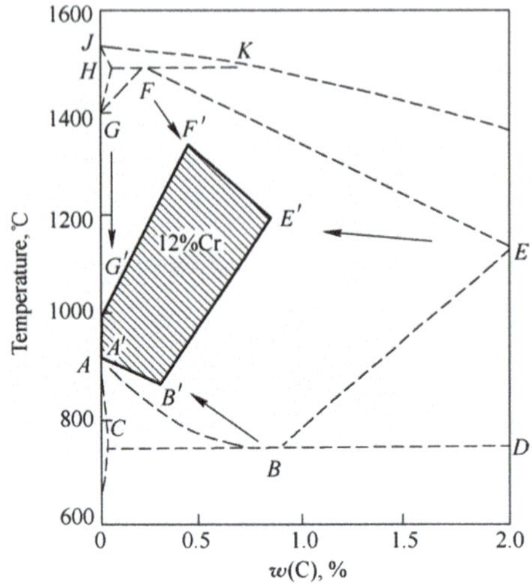

Fig. 1.13 Effect of 12% Cr on Fe–C equilibrium diagram [11]

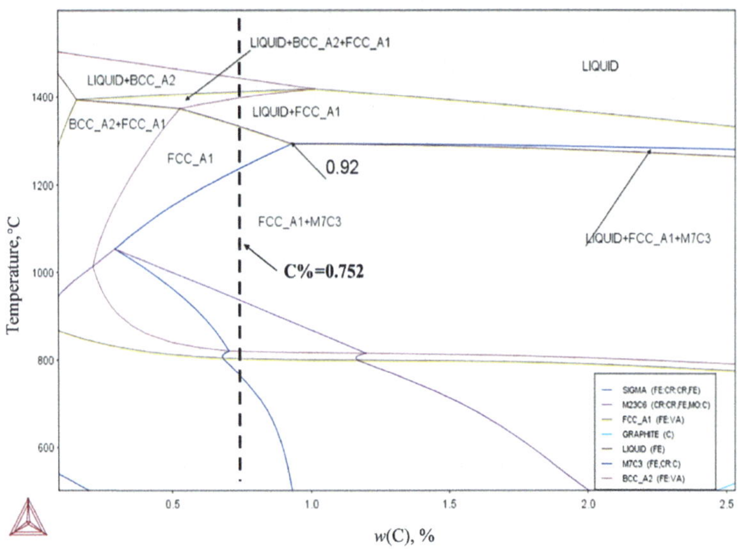

Fig. 1.14 Equilibrium phase diagram of 8Cr13MoV steel

0.752%, which has not met the condition for the precipitation of primary carbide in equilibrium state. However, the actual solidification process is a non-equilibrium process. When the contents of the enriched carbon and alloy elements in the dendrites exceeding the composition of the eutectic point, it achieves the conditions for the precipitation of primary carbide. Therefore, large amount of primary carbides were observed in the solidification structure of 8Cr13MoV steel. The mass fraction of

Cr, Mo and V in H13 die steel are 5%, 1.10–1.75% and 0.80–1.20%, respectively. It is impossible to form primary carbides in H13 steel according to its chemical composition. It is the enrichment of solute atom in the dendrites that promote the formation of primary carbides in H13 steel. are produced. The formation of primary carbides in the two aforementioned steels is the result of chemical composition and selected crystallization.

In order to study the precipitation behavior of carbide in the solidification process, the thermodynamic software Thermo-Calc was applied to calculate the equilibrium and non-equilibrium phase precipitation of austenitic hot work die steel, high carbon martensitic stainless steel, H13 hot work die steel, Cr5 rolled steel and and bearing steel, as well as the precipitation sequence of phases in the steels in non-equilibrium solidification process.

1.2.1 Thermodynamic Analysis of Carbide Formation in Austenitic Hot Work Die Steel

The chemical composition of austenitic hot work die steel is shown in Table 1.2.

1.2.1.1 Equilibrium Phase Diagram

The equilibrium phase diagram of austenitic hot work die steel is shown in Fig. 1.15. Figure 1.15a presents the evolution and phase transformation reaction in austenitic hot work die steel with decreasing, in which the dotted line shows the carbon content corresponding to austenitic hot work die steel. Figure 1.15b presents the precipitation sequence of inclusions and carbides in the steel with the decrease of temperature in the process of equilibrium solidification.

According to the results in Fig. 1.15, the precipitation behavior of phase in austenite hot work die steel in the process of equilibrium phase solidification is as follows:

(1) There are some elevated temperature ferrite phases in the elevated temperature region, but they can be neglected because of their small percentage. When the temperature is lower than 1380 °C, the liquid phase begins to transform into austenite phase. Until the temperature drops to 1210 °C, the liquid phase completely transforms into austenite phase.
(2) When the temperature decreases to 1230 °C, MnS begins to precipitate.

Table 1.2 Chemical composition of austenitic hot work die steel (wt%)

C	Si	Mn	Cr	Mo	V	P	S	Fe
0.70	0.55	14.95	3.45	1.57	1.723	0.0085	0.0023	Bal.

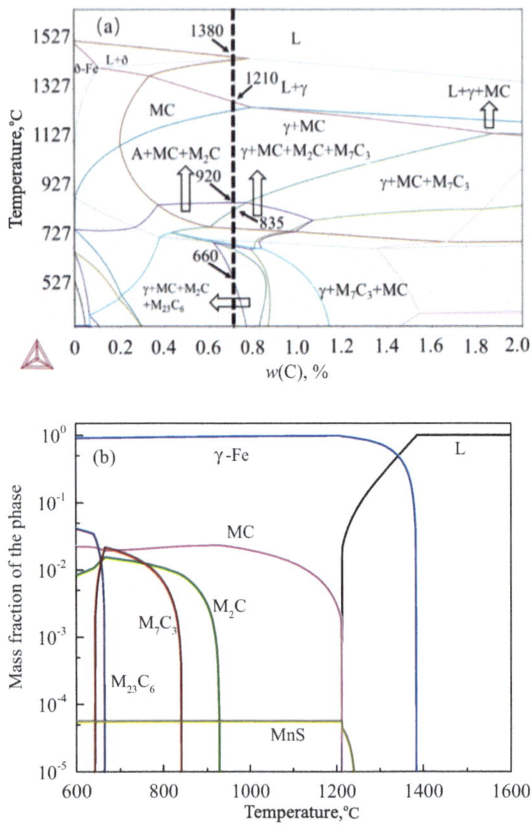

Fig. 1.15 Thermodynamic calculation results of equilibrium phase diagram: **a** Equilibrium phase diagram; **b** Temperature dependence of precipitates

(3) When the temperature decreases to 1210 °C, MC-type carbides begin to precipitate from austenite
(4) When the temperature decreases to 920 °C, M_2C-type carbides begin to precipitate in austenite phase, and no other transformation occurs as MC type carbide.
(5) When the temperature decreases to 835 °C, M_7C_3-type carbides begin to precipitate in austenite phase, which exists in the ranging from 650 °C to 835 °C. M_7C_3-type carbides begin to transform at 680 °C and its mass percentage decreases. $M_{23}C_6$-type carbides begin to precipitate at 680 °C, which can be deduced that M_7C_3-type carbide transforms into $M_{23}C_6$-type carbide.

1.2.1.2 Non-equilibrium Phase Diagram

The segregation of alloy elements is inevitable in the actual solidification process. Therefore, the precipitation sequence and precipitation reaction of carbide and inclusion in austenitic hot work die steel considering segregation behavior during solidification process are simulated and calculated by using Scheil module of Thermo-Calc software. The non-equilibrium phase precipitation diagram in austenitic hot work die steel calculated using Thermo-Calc is shown in Fig. 1.16.

According to Figs. 1.15a and 1.16, it is found that the precipitation temperatures of matrix phase and precipitate phase in non-equilibrium solidification process of austenitic hot work die steel are higher than those in equilibrium solidification process. The precipitation behavior of phase in austenite hot work die steel in the process of non-equilibrium phase solidification is as follows:

(1) When the temperature decreases to 1386 °C, austenite phase begins to precipitate directly from the liquid phase, and the monotectic reaction (L $\rightarrow \gamma + L'$) takes place.
(2) When the temperature decreases to 1212 °C, MC-type carbides begin to form in the liquid phase. At this time, the monotectic reaction (L $\rightarrow \gamma +$L$'$ and L $\rightarrow$ MC + L$'$) or eutectic reaction (L $\rightarrow \gamma + MC$) take place, or accompanied by secondary MC-type carbides precipitating from the supersaturated austenite phase ($\gamma \rightarrow$ MC). The phase composition is austenite, MC-type carbide and residual liquid phase.
(3) When the temperature decreases to 1201 °C, MnS inclusions begin to form in the residual liquid phase.
(4) When the temperature decreases to 1166 °C, primary M_2C-type carbides begin to form in the residual liquid phase. At this time, binary eutectic reaction (L $\rightarrow \gamma + M_2C$) or ternary eutectic reaction (L $\rightarrow \gamma +$ MC + M_2C) may take palce, or accompanied by secondary MC and M_2C-type carbides precipitated from supersaturated austenite phase in the solidification process. The phase

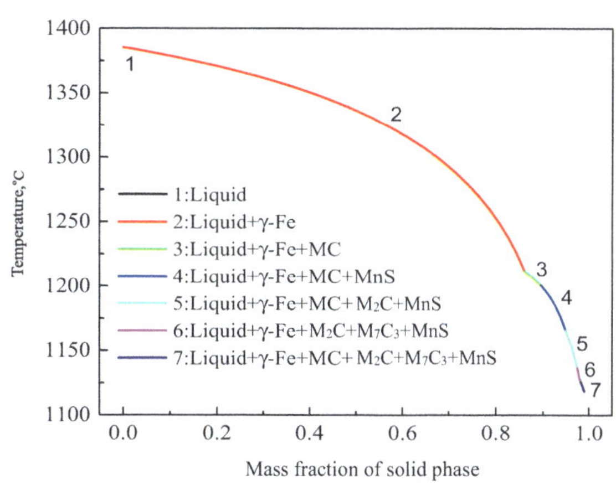

Fig. 1.16 Non-equilibrium phase precipitation diagram in austenitic hot work die steel calculated using Thermo-Calc

compositions are austenite, MC-type carbides, M_2C-type carbides and residual liquid phase.

(5) When the temperature decreases to 1126 °C, M_7C_3-type carbides begin to precipitates, which may be peritectic reaction ($L + \gamma \rightarrow M_7C_3$) or eutectoid reaction ($\gamma \rightarrow MC + M_7C_3$). The phase compositions are austenite, MC-type carbides, M_2C-type carbides, M_7C_3-type carbides and residual liquid phase.

The transformation reaction and microstructure of austenite hot work die steel during non-equilibrium solidification process are shown in Table 1.3. The results show that the phase in austenitic hot work die steel at room temperature is austenite matrix, primary carbide or eutectic carbide precipitated from liquid phase and secondary carbide precipitated from supersaturated austenite solid phase.

1.2.1.3 Composition Analysis of Equilibrium Solidification Phase

Under the condition of equilibrium solidification, it is considered that the solute atoms in solid solution have enough time for diffusion and migration. Therefore, with the decrease of temperature, different kinds of carbides are precipitated or transformed, and the distribution of solute atoms in different kinds of carbides must be different. Therefore, it is necessary to study the distribution of solute atoms in various precipitated carbides during solidification process. The change in alloy element contents of carbides non-equilibrum precipitation from austenitic hot work die steel as a function of temperature is shown in Fig. 1.17.

Figure 1.17a shows MC-type carbide contains 60–70% V element and some Mo element. Figure 1.17b shows that M_2C-type carbide is mainly composed of element Mo and contains about 20% V element. Figure 1.17c shows that M_7C_3-type carbide contains about 50% Cr element and some Fe and Mn elements. Figure 1.17d shows that $M_{23}C_6$-type carbide mainly consists of Cr, Fe and Mn elements. Under the condition of equilibrium solidification process, there is enough time for carbide to nucleate and grow. However, due to the different solid solubility of solute atoms in liquid phase and solid phase in the actual solidification process, there will be element segregation or solute atom segregation behavior in solidification process. Therefore, it is more meaningful to study the distribution behavior of solute atoms in liquid phase in non-equilibrium solidification process.

1.2.1.4 Calculation of Element Segregation and Nonequilibrium Solidification in Actual Solidification Process

A Distribution of alloying elements in phases during solidification process

Due to the inevitable segregation behavior in the actual solidification process, Fig. 1.18 shows the distribution diagram of C, Mo and V elements in various carbides in the solidification process calculated by the thermodynamic software Thermo-Calc.

Table 1.3 Non-equilibrium solidification phase transformation reaction and phase composition

Temperature, °C	Types of carbide	Phase transformation reaction	Transformation behavior	Phase composition
1386	No carbide	$L \rightarrow \gamma + L'$	Austenite phase precipitation	$L + \gamma$
1212	Primary carbide	$L \rightarrow \gamma + L'L \rightarrow MC + L'L \rightarrow \gamma + MC$	Eutectic reaction	$L + \gamma + MC$
	Secondary carbide	$\gamma \rightarrow MC$		
1166	Primary carbide	$L \rightarrow \gamma + M_2CL \rightarrow \gamma + MC + M_2C$	Eutectic reaction	$L + \gamma + MC + M_2C$
	Secondary carbide	$\gamma \rightarrow MC + M_2C$	Eutectoid reaction	
1126	Primary carbide	$L + \gamma \rightarrow M_7C_3$	Peritectic reaction	$L + \gamma + MC + M_2C$ $+M_7C_3$
	Secondary carbide	$\gamma \rightarrow MC + M_7C_3$	Eutectoid reaction	

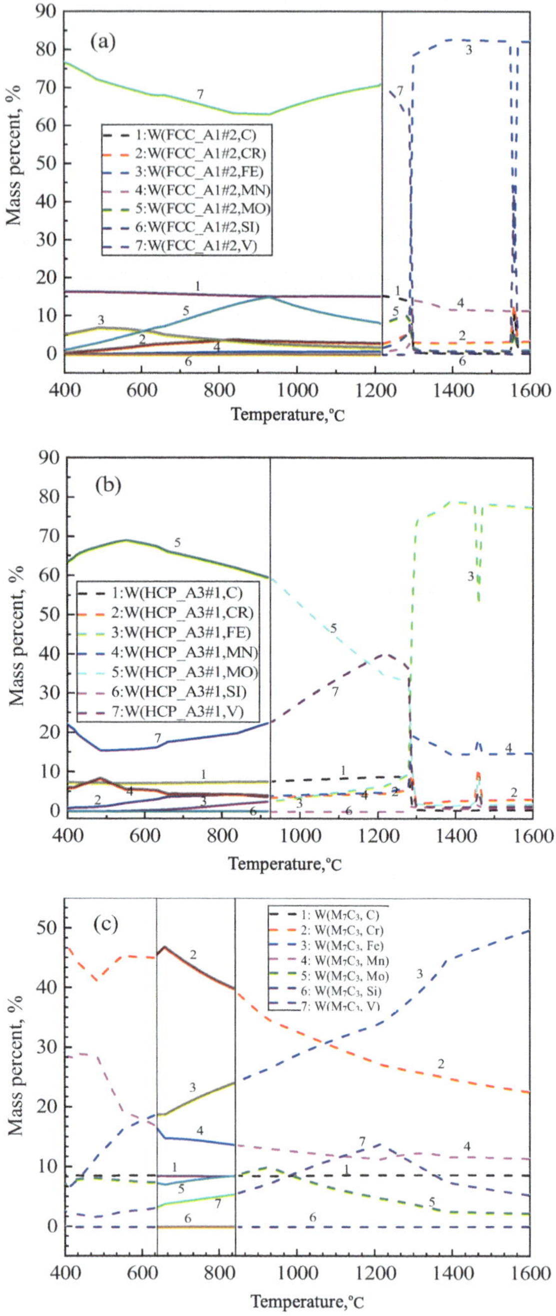

Fig. 1.17 Change in alloy element contents of carbides non-equilibrum precipitation from austenitic hot work die steel as a function of temperature: **a** MC; **b** M_2C; **c** M_7C_3; **d** $M_{23}C_6$

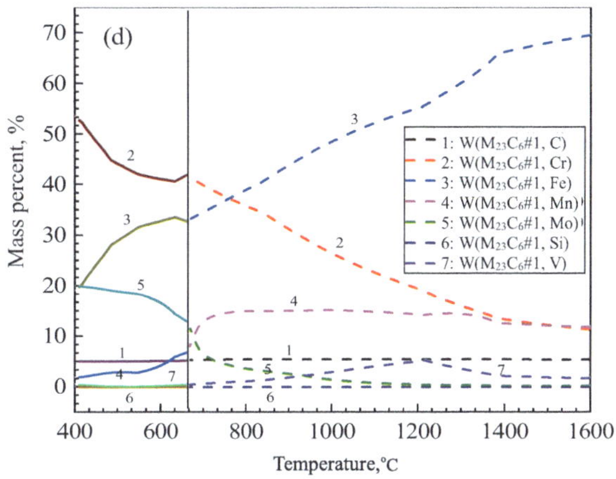

Fig. 1.17 (continued)

As shown in Fig. 1.18a, carbon element is mainly distributed in MC-type carbides and the rest in M_7C_3, M_2C and $M_{23}C_6$-types carbides. According to Fig. 1.18b, c, MC-type carbides are mainly composed of V element and contain 10–18%Mo and 15–16%C elements. M_2C-type carbides are mainly composed of Mo and contain 15–20%V and 7%C elements. In the actual solidification process, the carbide will precipitate on the condition that the contents of the solute atoms of alloy elements in the residual liquid phase reaches the equilibrium concentration product of carbide precipitation.

B Segregation of alloy elements during solidification process

Because of the great difference of the solubility of solute atoms between solid phase and liquid phase, solute atoms are constantly squeezed into the liquid phase at the front of solid-liquid interface during solidification process, which inevitably leads to the uneven distribution of solute atoms between solid phase and liquid phase. According to the chemical composition of austenitic hot work die steel, the relationship between the content of alloy elements in residual liquid phase and solid rate under non-equilibrium solidficiation condition was calculated by using thermodynamic software Thermo-Calc, and the results are shown in Fig. 1.19.

It can be seen from Fig. 1.19 that the alloying elements C, Si, Mn, Cr, Mo and V in austenitic hot work die steel have a certain segregation. Figure 1.19b shows that the mass fraction of V element in the residual liquid phase begins to decrease sharply when the solid phase ratio reaches 0.86. Combined with the results of Fig. 1.18, it can be found that the carbides precipitated from the liquid phase are MC- type carbides riched in V element with a certain content of Mo element. Figure 1.19a, e show that the mass fraction of C and Mo in the residual liquid phase continue to increase, instead of decrease. The main reason is that the content of C and Mo consumed by MC-type carbide precipitation was not increased by the C and Mo segregation in the solidification process. Figure 1.19d shows that the mass fraction of Mo in the residual liquid phase begins to decrease greatly when the solid phase

Fig. 1.18 Dependence of distribution diagram of alloy elements in each phase on temperature: **a** Carbon; **b** Molybdenum; **c** Vanadium

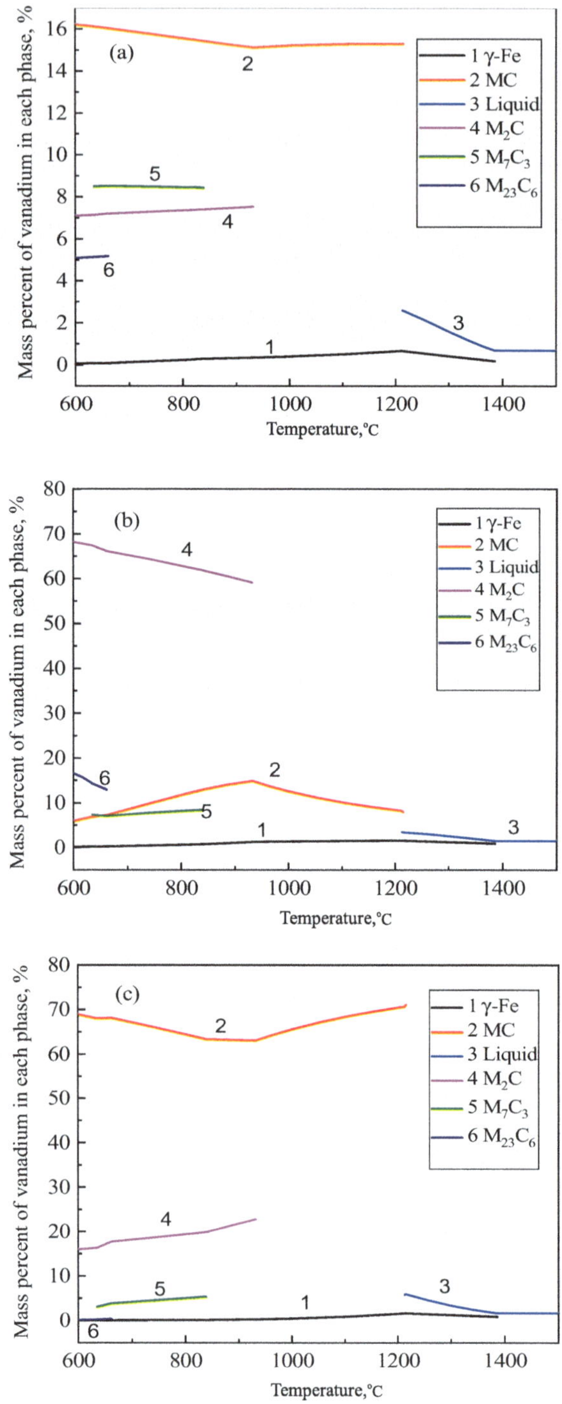

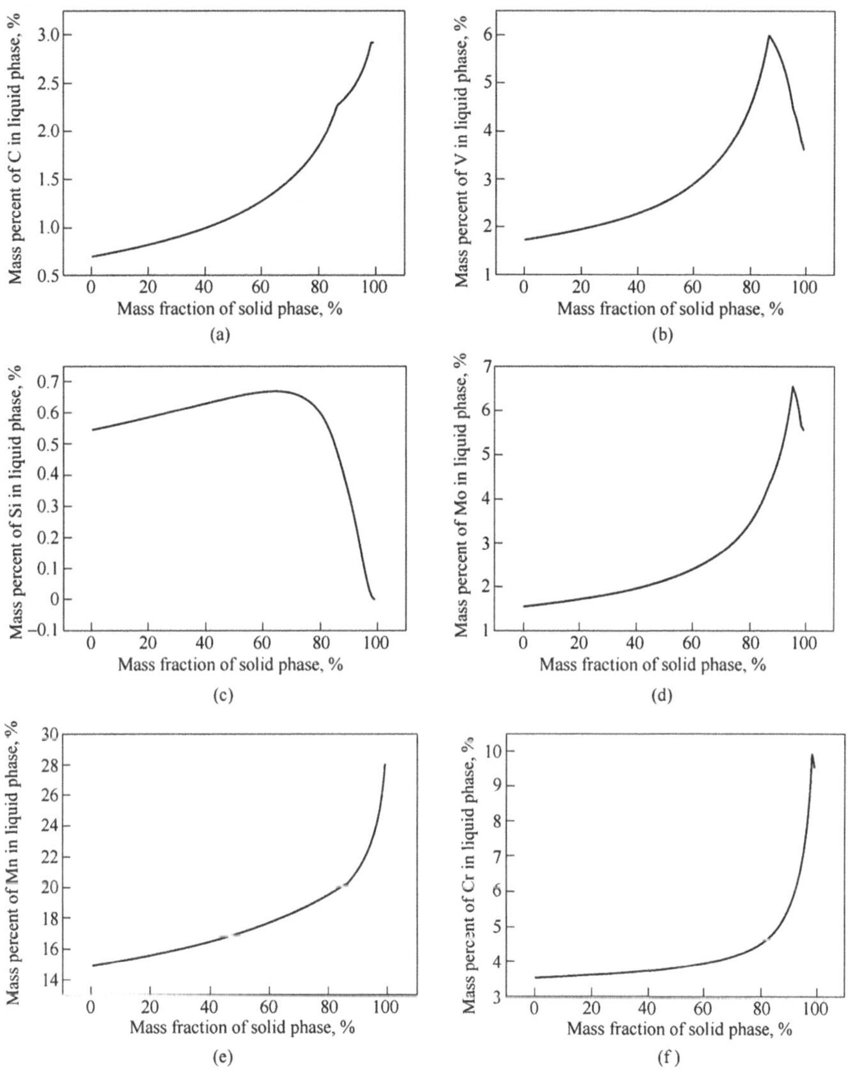

Fig. 1.19 Relationship between the content of alloy elements in residual liquid phase and solid fraction: **a** C; **b** V; **c** Si; **d** Mo; **e** Mn; **f** Cr

ratio reaches 0.95. Combined with the results in Fig. 1.18, it can be seen that the carbides precipitated from the liquid phase are M_2C-type carbides riched in Mo and with a certain content of V element. As the solidification continues, when the solid phase ratio reaches 0.98, Fig. 1.19f shows that the mass fraction of Cr in the residual liquid phase decreases. However, due to the fact that it is near the end of solidification, the content of Cr precipitated from the liquid phase is very small or even not precipitated.

1.2.2 Thermodynamic Analysis of Carbide Formation in 8Cr13MoV Steel

The precipitation behavior of primary carbide in 8Cr13MoV steel during solidification process was calculated by Scheil Gulliver module in Thermo-Calc thermodynamic software. It is assumed that the diffusion rate of each component in the liquid phase is infinite and that in the solid phase is 0. The local equilibrium is established at the solidification interface during the solidification process. The alloy composition at the solidification interface is obviously different from that of the whole steel composition. The composition of liquid phase and solid phase at the interface is determined by the phase diagram of the steel composition. The composition of the solid phase maintains its state at the time of formation, while the liquid composition always is uniform. During the calculation process, with the solidification of molten steel, the new liquid phase composition after each simulation step will be taken as the local global component in the next simulation step. The reverse diffusion of carbon and nitrogen is considered in the calculation, which indicates that the two interstitial elements can be redistributed by diffusion in the solidified solid phase.

High quality cutting tools are generally made of medium and high carbon martensitic stainless steel. The main alloy elements in this kind of martensitic stainless steel include C, Cr, Mo and V [12]. In addition, in order to improve the antibacterial property and surface strength, appropriate alloy elements such as Cu, Ag, Co and W may be added into the steel [13–15]. The chemical composition of 8Cr13MoV steel is shown in Table 1.4, and the equilibrium phase precipitation in 8Cr13MoV steel calculated using Thermo-Calc is shown in Fig. 1.20.

It can be seen from Fig. 1.20 that the steel is liquid when the temperature is above 1442 °C, and the elevated-temperature ferrite (δ- Fe) begins to precipitate when the temperature is lower than 1440 °C. Peritectic reaction (δ-ferrite + L → γ-Fe) take places at 1415 °C. The liquid phase continues to transform to austenite after the peritectic reaction ends at 1400 °C, and it completely solidifies into a single austenite structure at 1310 °C. M_7C_3-type carbides begin to precipitate in austenite when the temperature decreasess to 1240 °C. The content of precipitated M_7C_3-type carbides is maximum at 928 °C, and then M_7C_3-type carbides transform into $M_{23}C_6$-type carbides and completely transforms into $M_{23}C_6$-type carbides at 760 °C. Simultaneously, austenite transforms into α-ferrite when the temperature decreases to 810 °C. The final solidification structures are mainly α-ferrite and $M_{23}C_6$-type carbides, and their mass percentage are 84.67% and 12.97% respectively. The non-equilibrium phase precipitation in 8Cr13MoV steel calculated using Thermo-Calc is shown in Fig. 1.21.

Table 1.4 Chemical composition of 8Cr13MoV steel (wt%)

C	Si	Mn	Cr	Mo	V	S	N	Fe
0.77	0.28	0.45	14.02	0.39	0.45	0.0043	0.011	Bal.

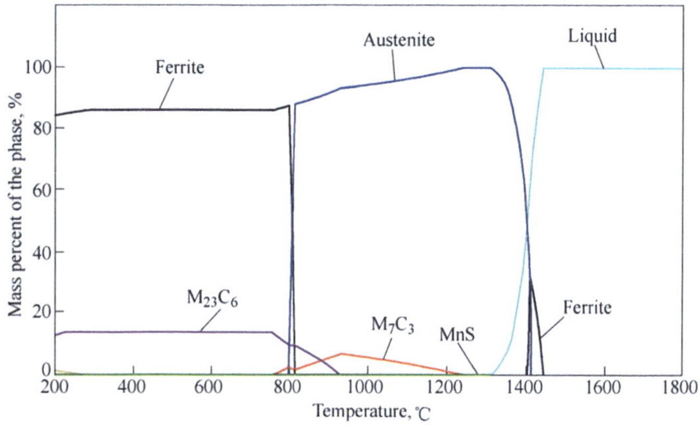

Fig. 1.20 Equilibrium phase precipitation in 8Cr13MoV steel calculated using Thermo-Calc

Fig. 1.21 Non-equilibrium phase precipitation in 8Cr13MoV steel calculated using Thermo-Calc

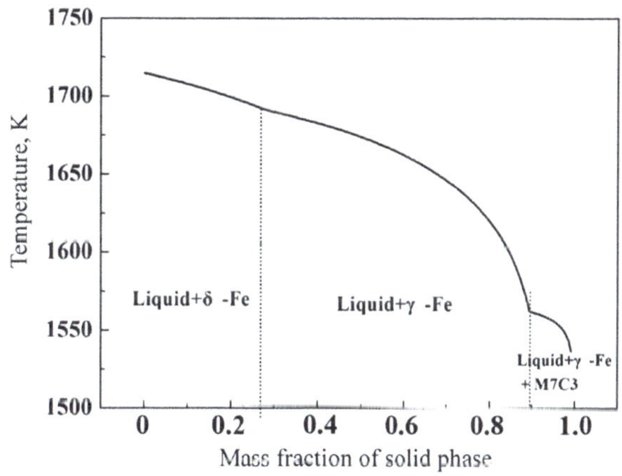

It can be seen from Fig. 1.21 that the liquid steel begins to solidify when the temperature decreases to 1442 °C. When the temperature decreases to 1418 °C, the mass fraction of elevated temperature ferrite reaches 27.9%. Then, peritectic reaction (δ-Fe + liquid → γ-Fe) take places, and the mass fraction of austenite in the system increases gradually. When the temperature decreases to 1289 °C, the mass fraction of austenite reaches 89.4%. Simultaneously, the residual liquid phase reaches eutectic composition and eutectic reaction (liquid → γ-Fe + M_7C_3) take places. When the temperature decreases to 1263 °C, the molten steel is completely solidified (assumes that it is completely solidified when the solid fraction reaches 99%), and M_7C_3-type primary carbides precipitate at the end of solidification process.

1.2.3 *Thermodynamic Analysis of Carbide Formation in High Speed Steel*

The chemical composition of M42 high speed steel was shown in Table 1.5. The evolution behavior of precipitation in M42 high speed steel under equilibrium solidification condition was calculated usingThermo-Calc, as shown in Fig. 1.22.

It can be seen from Fig. 1.22 that austenite phase begins to precipitate in liquid steel when the temperature is lower than 1369 °C, and M_6C-type carbides begin to precipitate when the temperature decreases to 1287 °C. The liquid phase disappears completely and the content of austenite phase reaches the maximum when the temperature decreases to 1201 °C. Simultaneously, the content of M_6C-type carbides begin to decrease. HCP-A3#2 phase begin to precipitate at 1201 °C and dieappear at 505 °C, and it content reached the maximum at 823 °C and 717 °C, respectively. M_7C_3-type carbides begin to precipitate when the temperature is reduced to 960 °C. The content of precipitated M_7C_3-type carbides first increases slowly and increases rapidly in the temperature ranging from 823 °C to 800 °C. Meanwhile, the content of precipitated austenite and ferrite rapidly decreases and increases rapidly, respectively, which results from the transformation between austenite and ferrite. When the

Table 1.5 Chemical composition of M42 high speed steel (wt%)

C	Si	Mn	Cr	V	W	Mo	Co
1.10	0.40	0.275	3.875	1.15	1.50	9.50	8.25

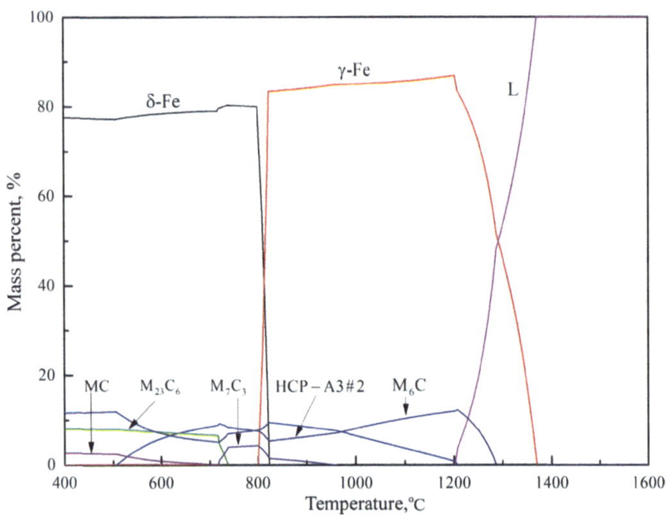

Fig. 1.22 Equilibrium phase precipitation in M42 high speed steel calculated using Thermo-Calc

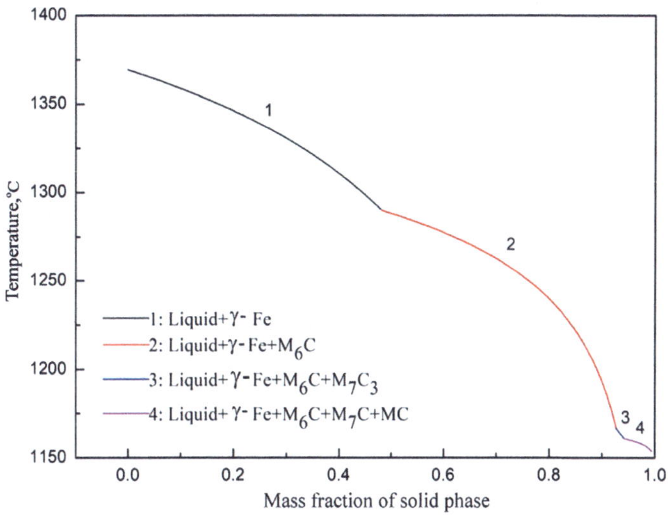

Fig. 1.23 Non-equilibrium phase precipitation in M42 high speed steel calculated using Thermo-Calc

temperature decreases to 738 °C, $M_{23}C_6$-type carbides begin to precipitate. Meanwhile, M_7C_3-type carbide content begins to decrease and disappears at 717 °C. When the temperature decreases to 717 °C, MC-type carbides begin to precipitate, and the trend of precipitated content is the similar to that of $M_{23}C_6$-type carbides.

The Non-equilibrium phase precipitation in M42 high speed steel calculated using Scheil Gulliver module in Thermo-Calc thermodynamic software is shown in Fig. 1.23.

It can be seen from Fig. 1.23 that when the temperature decreases to 1369 °C, the liquid steel begins to solidify and the austenite phase (γ-Fe) precipitates directly from the liquid phase, which is consistent with the temperature at which the solid phase begins to precipitate in the equilibrium solidification process. When the temperature is 1290 °C, the mass fraction of austenite is 48%. The liquid phase begins to form M_6C-type carbides and eutectic reaction (L $\rightarrow \gamma + M_6C$) takes places When the temperature is reduced to 1167 °C, the mass fraction of austenite and M_6C-type carbide is 92%, and M_7C_3-type carbide begins to form in the residual liquid phase, and the phase group is austenite + M_6C + M_7C_3 carbide; when the temperature is reduced to 1161 °C, the mass fraction of solid phase reaches 94%. MC-type carbide is formed in the residual liquid phase, and the phase composition is austenite + M_6C + M_7C_3 + MC. When the temperature is 1153 °C, the mass fraction of solid phase in the steel is 99%. It can be seen from Fig. 1.23 that when the temperature drops to 1369 °C, the liquid steel begins to solidify, and the austenite phase (γ-Fe) precipitates directly from the liquid phase, which is consistent with the temperature at which the solid phase begins to precipitate in the equilibrium solidification process. When the temperature is 1290 °C, the mass fraction of austenite is 48%. The liquid phase begins to form M_6C-type carbide, and eutectic reaction L $\rightarrow \gamma + M_6C$ occurs.

When the temperature drops to 1167 °C, the mass percent of austenite and M_6C-type carbide is 92%. M_7C_3-type carbide begins to form in the residual liquid phase, and the phase composition is austenite + M_6C + M_7C_3. When the temperature decreased to 1161 °C, the mass fraction of solid phase reaches 94%. MC-type carbides are formed in the residual liquid phase, and the phase composition is austenite + M_6C + M_7C_3 + MC. When the temperature is 1153 °C, the mass fraction of solid phase in the steel is 99%, which can be considered that the liquid steel is completely solidified.

1.2.4 Thermodynamic Analysis of Carbide Formation in H13 Hot Working Die Steel

The chemical composition of H13 hot working die steel was shown in Table 1.6. The evolution behavior of precipitation in H13 hot working die steel was calculated by thermodynamic software Thermo-Calc. The non-equilibrium phase precipitation in H13 steel calculated using Thermo-Calc is shown in Fig. 1.24.

It can be seen from Fig. 1.24 that MC, M_2C and M_7C_3-type carbides are the primary carbides precipitated successively in H13 molten steel. The initial precipitation temperatures of the three kinds of carbides are 1285 °C, 1216 °C and 1216 °C, respectively.

Table 1.6 Chemical composition of H13 steel (wt%)

C	Si	Mn	Cr	Mo	V	Ni	Al	N	Ti	Fe
0.41	0.90	0.39	5.24	1.45	0.92	0.21	0.035	0.034	0.011	Bal.

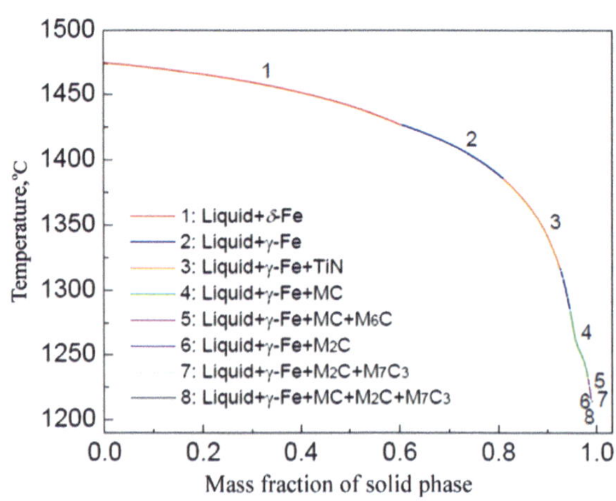

Fig. 1.24 Non-equilibrium phase precipitation in H13 steel calculated using Thermo-Calc

1.2.5 Thermodynamic Analysis of Carbide Formation in Cr5 Rolling Steel

The chemical composition of Cr5 cold rolling steel is shown in Table 1.7.

The Equilibrium phase precipitation in Cr5 cold rolling steel is calculated, that is, the solidification process of Cr5 cold rolling steel under ideal diffusion conditions (elements in solid and liquid phases can be completely diffused), as shown in Fig. 1.25. Under the condition of equilibrium solidification, the liquidus temperature of Cr5 cold rolling steel is 1450 °C. With the decrease of the temperature, austenite precipitates from the liquid phase, and the liquid phase completely changes into austenite at about 1330 °C. At 1067 °C, M_7C_3-type carbide precipitates from austenite. Eutectoid reaction occurs at 790 °C, and austenite transforms into ferrite and cementite. The eutectoid transformation is completed at 753 °C. In this process, more M_7C_3-type carbides are generated.

In the actual solidification process, the segregation of alloy elements is inevitable. Therefore, the Scheil model in Thermo-Calc is used to calculate the non-equilibrium phase precipitation in Cr5 cold rolling steel.

It can be seen from Fig. 1.26 that under the condition of non-equilibrium solidification, the liquidus temperature of Cr5 cold rolling steel is 1450 °C. When the temperature is lower than 1450 °C, and the liquid phase transforms into austenite. When the temperature was 1292 °C and the mass fraction of residual liquid was 3.11%, MnS was precipitated from the liquid phase. When the temperature is 1246 °C and

Table 1.7 Chemical composition of Cr5 cold rolling steel (wt%)

C	Si	Mn	S	Cr	Ni	Mo	V	Fe
0.858	0.658	0.375	0.006	4.9	0.393	0.218	0.148	Bal.

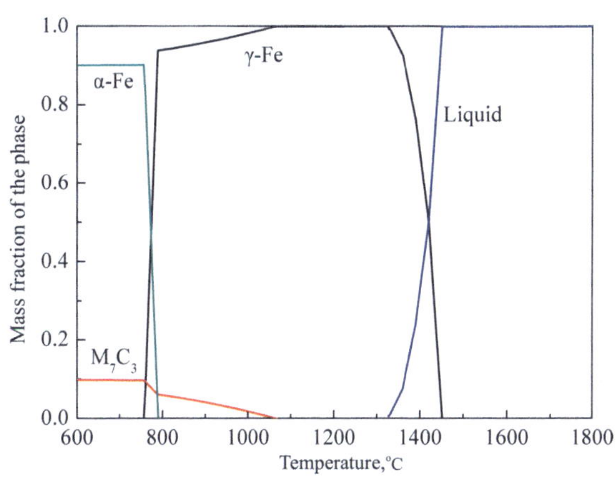

Fig. 1.25 Equilibrium phase precipitation in Cr5 cold rolling steel calculated using Thermo-Calc

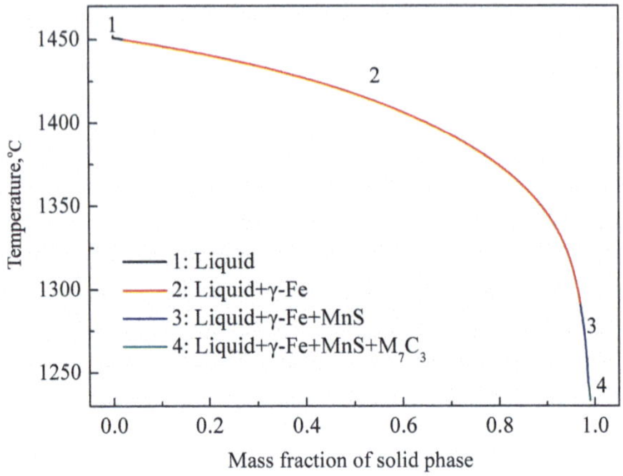

Fig. 1.26 Non-equilibrium phase precipitation in Cr5 cold rolling steel calculated using Thermo-Calc

the residual liquid phase is 1.24%, austenite and M_7C_3-type carbides are precipitated from the liquid phase.

According to the non-equilibrium solidification property diagram, the calculated solidus temperature is 1234 °C, which is different from the solidus temperature of 1330 °C calculated previously. According to the theory of metal solidification, with the decrease of temperature, when the liquid steel solidifies into austenite and liquid phase regions, the carbon content in austenite moves along the solid phase line, and the carbon content in the liquid phase moves along the liquidus. Therefore, during the solidification process, the amount of austenite increases, the amount of liquid phase decreases, and carbon atoms can fully diffuse. In the non-equilibrium solidification, due to the rapid cooling rate, the carbon atoms in liquid phase and austenite can not be fully diffused. With the solidification process, the carbon elements in liquid phase are continuously enriched and the content is increasing, resulting in element segregation. When the carbon content in liquid phase exceeds the maximum amount of carbon dissolved in austenite, the liquid phase composition will enter the hypoeutectic region. With the decrease of temperature, eutectic reaction will occur and eutectic carbides will be produced. The carbon content in liquid phase increases at the end of non-equilibrium solidification, while carbon reduces the solidification temperature of liquid steel, which is also the reason for the decrease of solidus temperature.

1.2.6 Thermodynamic Analysis of Carbide Formation in GCr15 Bearing Steel

Thermodynamic software Thermo-Calc was used to calculate the precipitates and their contents in the solidification process of GCr15 bearing steel under equilibrium conditions, as shown in Fig. 1.27. The chemical composition of GCr15 bearing steel is shown in Table 1.8.

It can be seen from Fig. 1.27 that there are three types of carbides in GCr15 bearing steel, namely M_3C_2, M_7C_3 and M_3C. M_3C-type carbides precipitate from austenite at 910 °C and the content reachs the maximum at 747 °C. At the same time, austenite begins to transform into α—ferrite. When the temperature reaches 490 °C, M_3C-type carbides begin to transform into M_7C_3-type carbides. When the temperature was further reduced to 422 °C, M_3C_2-type carbides began to precipitate, and the final microstructure of bearing steel transformed into α-ferrite, M_7C_3 and M_3C_2-type carbides.

Considering the limited diffusion of carbide forming elements in the solid steel, the non-equilibrium phase precipitation in GCr15 bearing steel wae calculated using Scheil Gulliver model in Thermo-Calc, and the results are shown in Fig. 1.28. It can be seen that when the solid fraction of liquid steel is higher than 0.902 (corresponding to temperature 1162 °C), the eutectic reaction will cause the primary M_3C-type carbide to precipitate from the liquid steel.

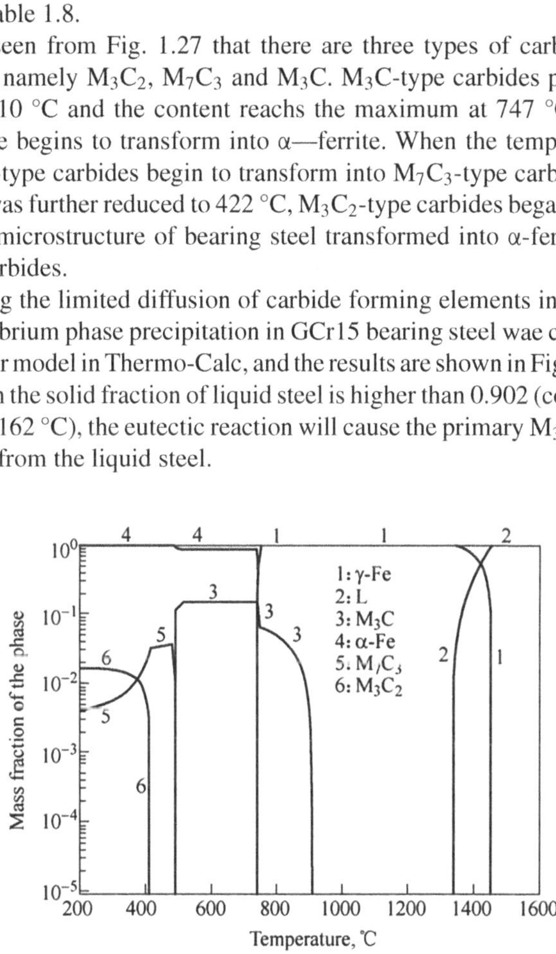

Fig. 1.27 Equilibrium phase precipitation in Cr15 bearing steel calculated using Thermo-Calc

Table 1.8 Chemical composition of GCr15 bearing steel (wt%)

C	Si	Mn	Cr	S	P	O	N	Ti	Fe
1.00	0.21	0.31	1.47	0.0018	0.0068	0.0008	0.0046	0.0061	Bal.

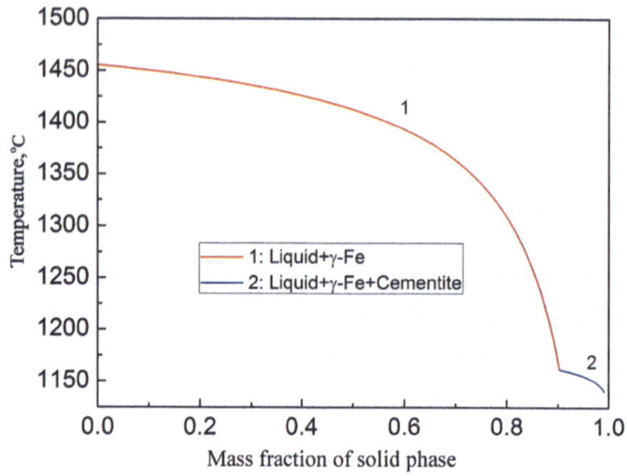

Fig. 1.28 Non-equilibrium precipitates in the solidification process of GCr15 bearing steel

1.3 Growth Characteristics and Morphology Analysis of Carbide

1.3.1 Morphology Analysis of Carbide

The carbides in 8Cr13MoV steel were extracted by deep corrosion technology and electrolytic extraction technology. The growth mode of primary carbides was observed by SEM secondary electron and backscatter diffraction, and the results are shown in Fig. 1.29. Figure 1.29a shows typical rod-shaped carbide (RC) growing from one side of grain boundary to the other side. Figure 1.29b shows typical coiled primary carbide (TC). Figure 1.29c contains both rod-shaped primary carbide, spherical carbide (GC) and block carbide (NC). Figure 1.29d shows the primary carbide extracted by electrolysis. The morphology of rod-shaped carbide and massive carbide can be seen more clearly.

The nucleation and growth of M_7C_3-type primary carbides in 8Cr13MoV steel are mainly affected by the concentration of carbon and chromium. On the condition that the concentration of carbon and chromium reaches a certain level, M_7C_3-type primary carbides nucleate and continue to grow under the driving of concentration gradient and temperature gradient. The main factors affecting the morphology include nucleation conditions, temperature gradient and concentration gradient of related solute atoms (chromium atoms). In the solidification process of liquid steel, the solidification front is the place where the elements are enriched, where the energy is high and there are nucleation spots, which is the best position for nucleation. During the solidification process, austenite grows in the form of dendrite. Some residual liquid phase in the gap between secondary dendrites may reach hypereutectic composition, and massive primary carbide (NC) will be precipitated directly at the solidification front. With the precipitation of massive primary carbide, the concentration of solute atoms in molten steel gradually decreases, and eutectic carbides will be precipitated

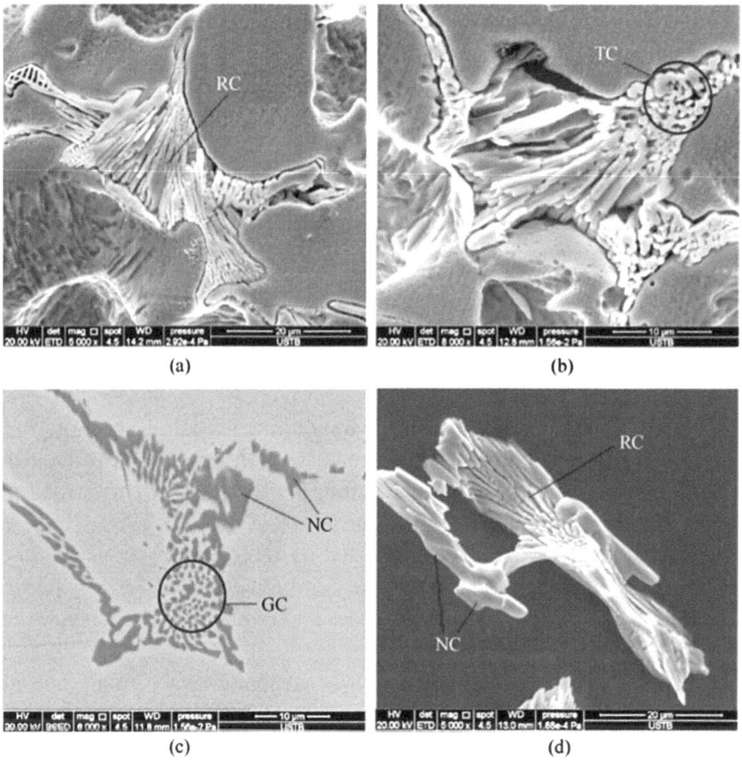

Fig. 1.29 Typical growth mode of primary carbide in as-cast 8Cr13MoV steel (RC represents rod-shaped carbide, TC represents coiled carbide, GC represents spherical carbide, and NC represents massive carbide)

when the eutectic point is reached. The morphology of eutectic carbides is affected by both temperature gradient and concentration gradient. When the temperature gradient plays a leading role, eutectic carbides nucleate at the interface of adjacent secondary dendrites and grow towards the middle, which is easy to form rod-shaped carbide (RC), as shown in Fig. 1.30a. When the concentration gradient and the temperature gradient act together, eutectic carbides will grow to the chromium enriched area in the residual liquid phase driven by the temperature gradient, and then it is easy to form the coiled primary carbide (TC), as shown in Fig. 1.30b. When the concentration gradient plays a dominant role, the primary carbides nucleate in the topographical or heterogeneous nucleation at the solute atom enrichment point in the secondary dendrite gap. There is no obvious growth trend after nucleation and forming spherical carbide (GC), as shown in Fig. 1.30c.

According to the precipitation and growth mechanism of primary carbide, the following measures can be taken to control the primary carbide reasonably:

(1) The necessary condition for primary carbide precipitation is that the content of carbon and chromium in the residual liquid phase reaches a certain level, and the

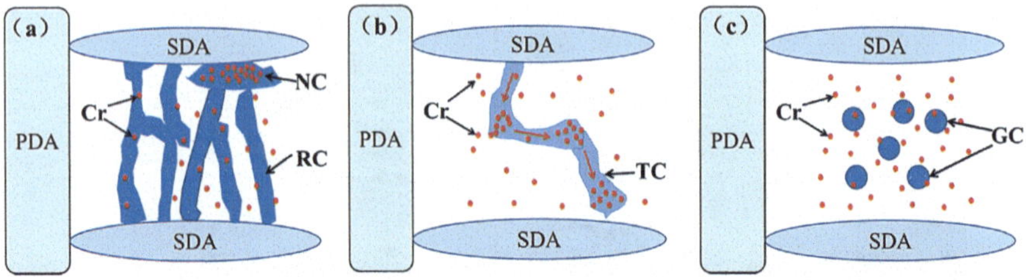

Fig. 1.30 Schematic view of primary carbide growth principle (PDA represents primary dendrite, SDA represents secondary dendrite, NC represents massive primary carbide, RC represents rod-shaped primary carbide, TC represents coiled primary carbide, and GC represents spherical primary carbide)

main alloy element required for primary carbide growth is chromium. Therefore, controlling the uniform distribution of carbon and chromium and reducing the segregation of carbon and chromium are the most effective measures to reduce the precipitation of primary carbides.

(2) The size of primary carbides is affected by the distance between secondary dendrites. Therefore, the size of primary carbide can be effectively reduced by adjusting the parameters of ESR process and reducing the secondary dendrite spacing.

(3) The content of carbon and chromium in austenite near primary carbide has reached the maximum, with carbon content of 0.86% and chromium content of 15.90%. It is difficult for carbon and chromium in primary carbides to diffuse into the surrounding matrix if high temperature annealing is carried out directly at this time. The carbon content of primary austenite is only half of the average carbon content of the matrix (0.40%), and the chromium content is only 11.35%. Therefore, the eutectic carbides can be broken and dispersed around the primary austenite by hot working such as forging and rolling, and then the high temperature diffusion annealing can promote the dissolution of primary carbides.

1.3.2 Formation Characteristics of Primary Carbide

The composition of 8Cr13MoV high carbon martensitic stainless steel is shown in Table 1.4, and the microstructure of remelted ingot is shown in Fig. 1.31.

It can be seen from Fig. 1.31 that the as-cast microstructure of 8Cr13MoV steel is mainly composed of acicular martensite, lath martensite, retained austenite and primary carbide. In the Fig. 1.31, the light yellow area near the grain boundary is retained austenite. The white coiled primary carbide at the grain boundary, and the acicular martensite and the white strip structure are lath martensite. The formation mechanism of the solidified structure can be explained by the concentration distribution characteristics of carbon and alloy elements in the right Figure of Fig. 1.31.

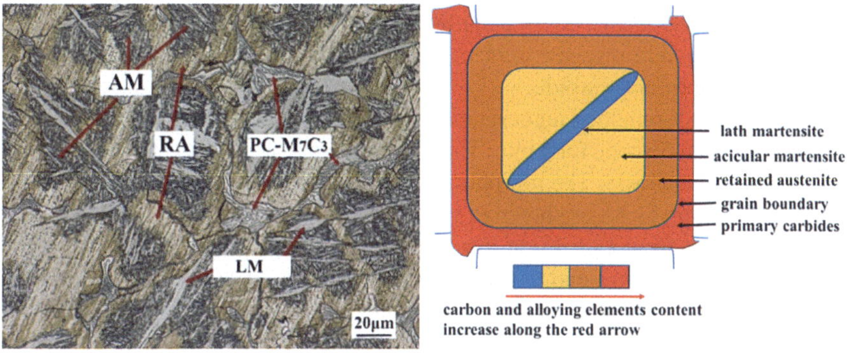

Fig. 1.31 Solidification structure and formation principle of 8Cr13MoV steel (RA, PC-M_7C_3, LM and AM represent retained austenite, M_7C_3-Type primary carbide, lath martensite and acicular martensite respectively)

Fig. 1.32 XRD resluts of 8Cr13MoV remelted ingot

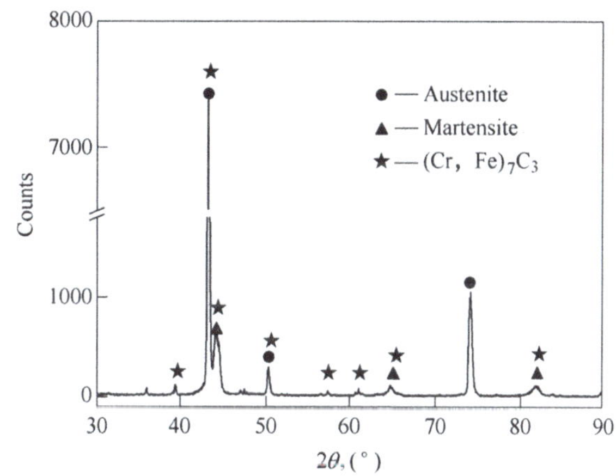

During the solidification process of molten steel, carbon and alloying elements atoms are continuously discharged from the solidification front to the remaining liquid phase, resulting in the enrichment of solute atoms in the residual liquid phase, and the carbon content gradually increases from the grain center to the grain boundary. In the final stage of solidification of liquid steel, the composition of residual liquid phase reaches eutectic point. Eutectic reaction occurs, and primary carbide precipitates at the grain boundary. Due to the high content of carbon and alloy elements and high stability of austenite near the grain boundary, it becomes retained austenite during the cooling process. Because of the lowest carbon content in the center of grain, lath martensite is formed during the cooling process the center of grain is from the center to the center of grain. In the grain boundary region, acicular martensite is formed at the place with relatively high carbon content.

The results of XRD analysis of 8Cr13MoV remelted ingot are shown in Fig. 1.32. XRD results show that austenite is the main phase in the steel, followed by martensite and M_7C_3-type primary carbide.

The 8Cr13MoV remelted ingot sample was heated to 1400 °C under high temperature confocal microscope for 5 min. The surface of steel was completely melted and then cooled down slowly. The morphology of dendrite growth during solidification can be observed by SEM, as shown in Fig. 1.33.

It can be seen from Fig. 1.33a that the primary dendrites are coarse and grow along different directions, and the secondary dendrites grow symmetrically on both sides of the primary dendrites. Figure 1.33b shows the microstructure of 8Cr13MoV electroslag ingot observed under the condition of SEM backscattering. The primary carbides are mainly distributed in the secondary dendrite gap as a network, which indicates that the primary carbides are mainly precipitated at the end of solidification, and the morphology of primary carbides is affected by the growth state of secondary dendrites. After etching 8Cr13MoV as-cast ingot with $FeCl_3$ solution, the morphology of primary carbides can be observed more clearly under SEM. As shown in Fig. 1.33c, the primary carbides are mostly rod-shaped with typical eutectic morphology, and a small amount of massive primary carbides can also be observed.

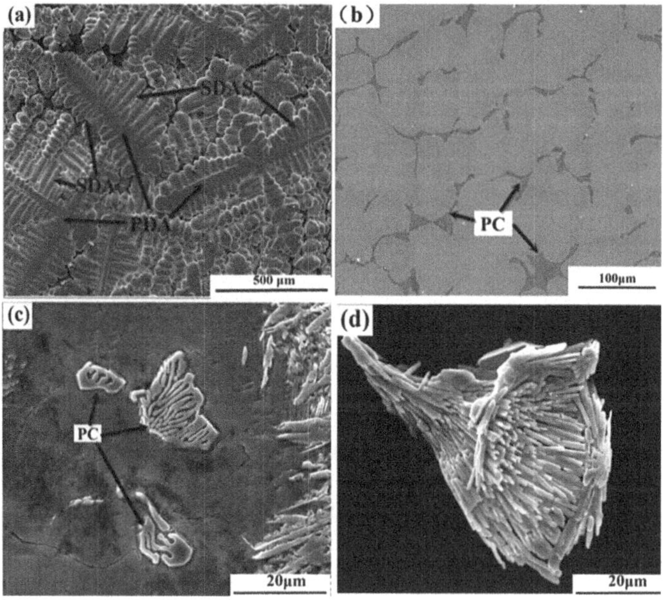

Fig. 1.33 Dendrite morphology and primary carbide morphology in as-cast 8Cr13MoV steel: **a** Dendrite morphology; **b** Primary carbide distribution; **c** Two-dimensional morphology of primary carbide; **d** Stereoscopic morphology of primary carbide (where PDA, SDA, SDAs, PC successively represent primary dendrite arm, secondary dendrite arm, secondary dendrite gap and primary carbide)

Table 1.9 Atomic fraction of elements in primary carbide (at.%)

Element	C	Mo	Cr	V	Fe
1#	30.07	0.38	47.76	1.52	20.28
2#	29.93	0.40	47.75	1.50	20.42
3#	29.19	0.29	44.01	1.08	25.43

The primary carbides were extracted by electrolytic extraction, and the cluster like primary carbides could be seen more clearly under SEM, as shown in Fig. 1.33d.

In conclusion, there are a lot of primary carbides in 8Cr13MoV as-cast ingot, which are mainly distributed at the grain boundary. Most of the primary carbides are eutectic morphology, and the size is large, generally tens of microns or even hundreds of microns. During the solidification process of liquid steel, almost all the carbon and chromium atoms discharged from the solidification front participate in the formation of primary carbide, while some of the enriched molybdenum and vanadium atoms remain in the steel matrix, forming a concentration transition zone between the primary carbide and austenite. In addition, carbon, chromium, molybdenum and vanadium are also enriched in austenite precipitated together with primary carbide during eutectic reaction, which mainly comes from the enrichment of elements in residual liquid phase before eutectic reaction.

The composition of primary carbide in 8Cr13MoV steel was analyzed by EPMA. The results are shown in Table 1.9. The main alloying elements in the primary carbide are chromium and carbon, and a small amount of molybdenum and vanadium. The content of each element in the three parts of primary carbide is determined, in which the carbon atom fraction is about 30%, and the total atomic fraction of Cr + Mo + V + Fe is 69.94%, 70.07% and 70.81% respectively. Namely, the ratio of carbon atom to the sum of other alloy elements is 3:7. According to XRD analysis in Fig. 1.32, the type of primary carbide is M_7C_3, in which M contains Cr, Mo, V, Fe and other elements.

Under the condition of SEM, the primary carbide is dark gray, while the steel matrix is light gray. Ten fields of view (1 mm^2) were randomly selected from the remelted ingot samples. The area fraction of primary carbides in the samples was counted by Image Pro Plus (IPP) image analysis software. After adjusting the relevant parameters, the image analysis software has a high recognition of primary carbides, as shown in Fig. 1.34. Figure 1.34a, c show the primary carbides observed under the condition of SEM backscattering. Figure 1.34b, d show the corresponding primary carbides in Fig. 1.34a, c identified by IPP respectively. The white area in the figure corresponds to the primary carbide.

The area fraction statistics of primary carbides are shown in Table 1.10. The average area fraction of primary carbide in 8Cr13MoV remelted ingot is 2.29%

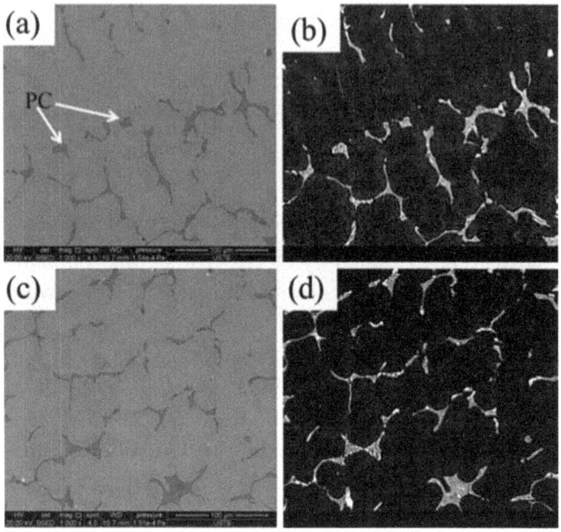

Fig. 1.34 Primary carbides in as-cast 8Cr13MoV steel and the image analysis software identification diagram: **a** and **c** are the primary carbides under the condition of SEM backscattering; **b** and **d** are the primary carbides identified by image analysis software (PC represents M_7C_3-type primary carbide)

Table 1.10 Area fraction statistics of primary carbides in 8Cr13MoV remelted ingot

No.	1#	2#	3#	4#	5#	6#	7#	8#	9#	10#
Area fraction, %	2.63	2.18	2.26	1.80	3.22	1.97	2.25	1.98	2.21	3.68

1.3.3 Precipitation and Growth Behavior of Secondary Carbides

During the production of 8Cr13MoV steel by electroslag remelting, the precipitation state of secondary carbide in consumable electrode and remelted ingot is shown in Fig. 1.35. The consumable electrode is obtained by casting, and its microstructure is mainly composed of martensite, retained austenite and primary carbide. It is difficult to observe the precipitation of secondary carbide. After electroslag remelting, there is no secondary carbide precipitation in the microstructure. After annealing treatment, white bright primary carbides and a large number of spherical secondary carbides can be observed in the microstructure of remelted ingots. Due to the high carbon and carbide forming elements in 8Cr13MoV steel, a large number of secondary carbides will be precipitated under the condition of equilibrium solidification. During solidification of consumable electrode and electroslag ingot, the cooling rate is fast. The kinetic conditions of carbide precipitation are insufficient, resulting in the solid solution of carbon atoms in austenite or martensite and in supersaturation state. After annealing, the supersaturated carbon atoms in austenite or martensite will recombine with chromium, molybdenum, vanadium and other alloy atoms and precipitate

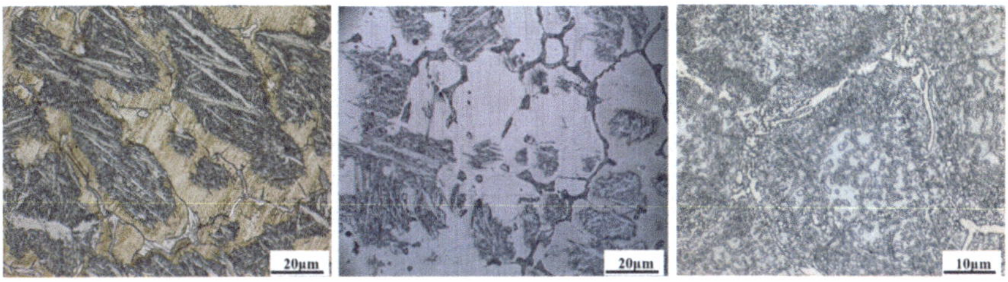

(a) (b) (c)

Fig. 1.35 Precipitation state of secondary carbide in consumable electrode and remelted ingot: **a** Consumable electrode; **b** Remelted ingot; **c** Remelted ingot after annealing treatment

from the matrix, which makes 8Cr13MoV steel system develop to thermodynamic equilibrium state.

The production process of cold rolling 8Cr13MoV sheet steel is: hot rolling, spheroidizing annealing, quenching and tempering. The evolution behavior of secondary carbide is shown in Fig. 1.36.

It can be seen from Fig. 1.36 that the microstructure of hot rolling slab is mainly acicular martensite and retained austenite, and no secondary carbide is found, as shown in Fig. 1.36a. After spheroidizing annealing, a large number of uniformly

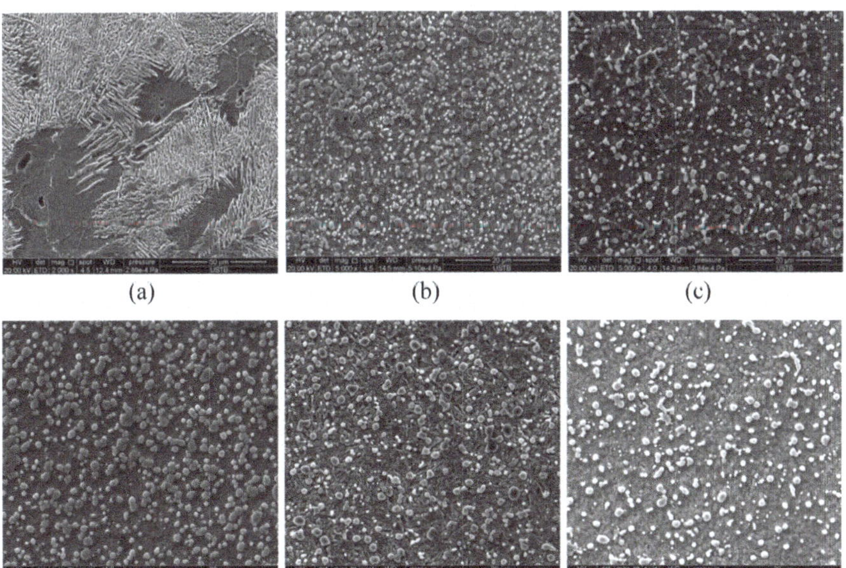

(a) (b) (c)

(d) (e) (f)

Fig. 1.36 Evolution of secondary carbide during hot rolling, spheroidizing annealing, quenching and tempering: **a** Hot rolling; **b** Spheroidizing annealing; **c** Cold rolling; **d** Recrystallization annealing; **e** Quenching; **f** Tempering distributed secondary carbides precipitated from the hot rolling slab, as shown in Fig. 1.36b. The cold rolling process has no obvious effect on the secondary carbide, as shown in Fig. 1.36c. After recrystallization annealing, the original secondary carbides grow slightly, as shown in Fig. 1.36d. After quenching, the microstructure of the steel is mainly martensite and secondary carbide, and the amount of carbide is reduced compared with that of cold rolling slab, as shown in Fig. 1.36e. After tempering, the microstructure of the steel is mainly tempered martensite and secondary carbide. Compared with the quenched state, the number of nano secondary carbides increased, as shown in Fig. 1.36f. The precipitation and evolution of secondary carbides are closely related to the heating, heat preservation and cooling systems of steel. It is necessary to heat to 1180 °C for 2 h before hot rolling and air cooling after rolling. According to thermodynamic calculation, after 8Cr13MoV steel is heated to 1180 °C and held for 2 h, almost all secondary carbides dissolve into austenite matrix. The fast cooling rate after rolling results to the lack of sufficient kinetic conditions for carbide precipitation, so there is almost no secondary carbide in hot rolling slab.

The spheroidizing annealing process is as follows: the hot rolling slab is heated to 750 °C for holding for 2 h, then rapidly heated to 860 °C for 2 h and cooled with the furnace. After 2 days, the temperature drops to 450 °C and is discharged into the furnace for air cooling. According to the thermodynamic calculation, when the temperature is 860 °C, a large number of M_7C_3 and $M_{23}C_6$-type carbides will precipitate. After spheroidizing annealing for 2 days, the carbides has enough time for precipitation, so the steel after spheroidizing annealing contains a lot of secondary carbides.

After spheroidizing annealing, the hot rolling slab should be cold rolling. The cold rolling process may have mechanical crushing effect on secondary carbides, but it can not affect the precipitation, dissolution and growth of carbides. After cold rolling, the steel needs recrystallization annealing at 800 °C for 5 h, and the secondary carbides will precipitate and grow during the annealing process.

In the process of quenching, the annealed cold rolling was heated to 1050 °C for 5 min and then air coolling. According to thermodynamic calculation, the number of secondary carbides begins to decrease when the temperature exceeds 920 °C. Therefore, some secondary carbides will dissolve into the matrix during the quenching heating and holding process The secondary carbides will not precipitate again in the rapid cooling process, which leads to the reduction of the number of secondary carbides after quenching process.

In the tempering process, the quenched cold-rolling slab is heated to 180 °C for 3 h and then air coolling, which is mainly to reduce the internal stress. The precipitation process of carbides is mainly controlled by the diffusion of alloying elements. Although $M_{23}C_6$-type carbides tend to precipitate during tempering process, Cr element is difficult to diffuse due to the low tempering temperature, so nano secondary carbides cannot be precipitated.

The main characteristic phases in quenched microstructure include martensite, primary carbide and secondary carbide. The element composition of martensite, primary carbide and secondary carbide in quenched sample was analyzed by EPMA. The results are shown in Fig. 1.37. The components of martensite are basically similar

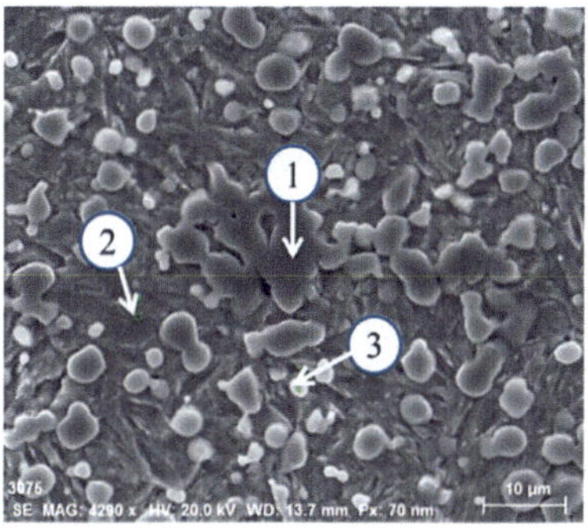

The content of each element of the test point (wt%)

	Fe	Cr	C	Mo	V
1 (PC)	8.8	60.6	8.8	0.9	1.9
2 (M)	81.7	15.4	1.4	0.1	0.1
3 (SC)	52.6	26.3	7.2	0.3	0.2

Fig. 1.37 Analysis of element content of characteristic phase in quenching structure (PC, M and SC represent primary carbide, martensite and secondary carbide respectively)

to those of 8Cr13MoV steel. is closely related to the parent phase of precipitated carbide. The content of Cr and C in primary carbide is 60.6% and 8.8%, which is close to the calculated value. The content of alloying elements in the primary carbide in the quenched sample is basically the same as that in the remelted ingot, which indicates that the spheroidizing annealing, quenching and tempering processes after hot rolling have little effect on the composition of primary carbide. The contents of carbon, chromium, molybdenum and vanadium in secondary carbides are lower than those in primary carbides and higher than those in martensite. It should be noted that the carbon content of secondary carbides basically reaches the thermodynamic equilibrium state, while the chromium content is far lower than the thermodynamic calculation value, which is mainly due to the fast diffusion of carbon atoms and the slower diffusion speed of chromium atoms. Therefore, it is difficult to reach the thermodynamic equilibrium state in the actual production process.

The XRD results of phase analysis of annealed and quenched samples were shown in Fig. 1.38. Ferrite and $M_{23}C_6$-type secondary carbides are the main structures in annealed samples. The microstructure of quenched samples is mainly martensite and $M_{23}C_6$-type secondary carbides.

According to the thermodynamic calculation, $M_{23}C_6$ and M_7C_3-type carbides may precipitate in 8Cr13MoV steel, while $M_{23}C_6$-type carbides is the main secondary carbide in steel under the existing process conditions, and the characteristic diffraction peak of M_7C_3-type carbides is not found. The reasons are as follows: the annealing temperature of 8Cr13MoV remelted ingot in sand pit is low, and some $M_{23}C_6$-type carbides may precipitate during the annealing process. Almost all secondary carbides dissolve into the matrix in the heating furnace before hot rolling. The temperatures of spheroidizing annealing after hot rolling and recrystallization annealing after cold rolling are between 500 °C and 800 °C for a long time, which is the temperature ranging of $M_{23}C_6$-type carbides precipitation. The original M_7C_3-type carbides also transform to $M_{23}C_6$-type carbides in the steel During the quenching process, the transformation from $M_{23}C_6$-type carbides to M_7C_3-type carbides will take place theoretically at the austenitizing temperature (1050 °C).

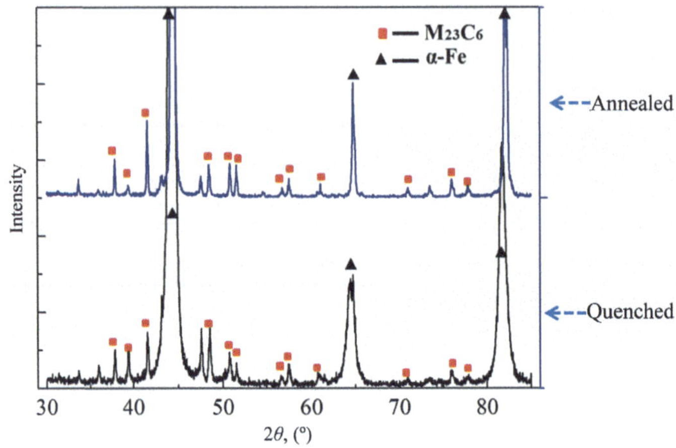

Fig. 1.38 XRD results of phase analysis of annealed and quenched samples

However, due to the short austenitizing time, there is no transformation or less transformation of carbides, resulting in $M_{23}C_6$-type carbides as the main carbide in the steel. Furthermore, it can be seen from Fig. 1.38 that the peak value of α-Fe (martensite) in quenched steel shifts to the left compared with that of annealed α-Fe (ferrite), which indicates that $M_{23}C_6$-type carbides dissolves in the process of austenitizing temperature holding, and carbides cannot precipitate during quenching.

The microstructure of the steel is shown in Fig. 1.39. The main microstructure of the steel is acicular martensite, a small amount of long strip retained austenite, a small amount of primary carbides and a large number of dispersed spherical secondary

Fig. 1.39 Initial microstructure of high temperature confocal specimen (PC represents primary carbide, RA represents retained austenite)

carbides. The kinetic transformation of carbide and microstructure in 8Cr13MoV steel during heating and cooling process was observed by high temperature confocal microscope. The process parameters of the high temperature confocal experiment are as follows: heat to 1350 °C at the rate at 5 °C/s and then cool at the speed of 5 °C/s. The observation results are shown in Fig. 1.40.

It can be seen from Fig. 1.40 that when the temperature rises to 323 °C, the surface of the steel will bulge. It is mainly because that the steel contains martensite and retained austenite. During the heating process, the martensite decomposes and the volume decreases, and the retained austenite remains unchanged. Therefore, the retained austenite floats on the upper surface, and the original position of martensite exists and sinks. When the temperature increases to 1030 °C, the surface relief is more obvious, and the martensite has completely transformed into austenite. In addition,

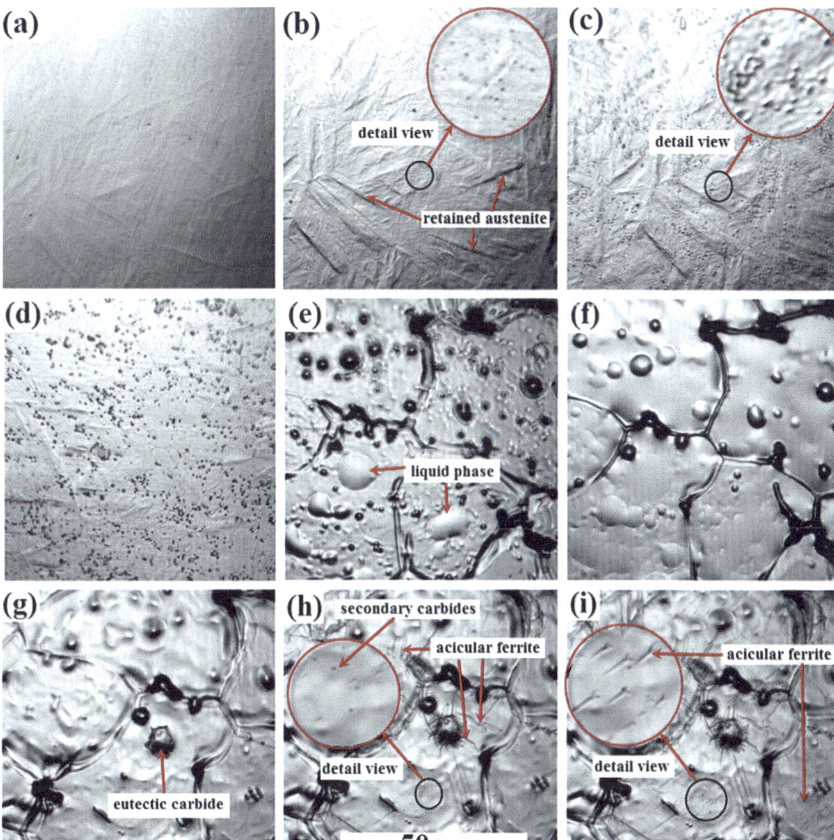

Fig. 1.40 Dynamic observation of microstructure and carbide evolution of 8Cr13MoV steel during heating and cooling: **a** 323 °C; **b** 1030 °C; **c**1202 °C; **d** 1243 °C; **e** 1320 °C; **f** 1350 °C; **g** 1058 °C; **h** 905 °C; **i** 864 °C

there are many small pits in the matrix, where are the traces of the original carbide dissolved. When the temperature rises to 1202 °C, a large number of secondary carbides are precipitated in the steel. Combined with thermodynamic calculation, it can be inferred that $M_{23}C_6$-type carbides is dissolved in Fig. 1.40b, and M_7C_3-type carbide is reprecipitated in Fig. 1.40c. Due to the high heating rate, the transformation temperature of carbide and microstructure in steel may be higher than the calculated theoretical temperature. When the temperature increases to 1243 °C, the content of carbon and alloy elements around the carbide increases due to the gradual dissolution of carbide, which reduces the solidus temperature of steel, and the steel matrix melts at the edge of carbide. When the temperature rises to 1320 °C, most of the carbides have been dissolved, and liquid phase appears at the grain boundary and other carbide enriched areas. When the temperature rises to 1350 °C, the liquid phase ratio is over 80%. When the temperature decreases, the primary carbides preferentially reprecipitate in the region with high carbon and alloy elements, as shown in Fig. 1.40g. When the temperature decreases to 905 °C, acicular ferrite precipitates at the edge of primary carbide and around grain boundary, and secondary carbide precipitates in austenite. When the temperature decreases to 864 °C, acicular ferrite also precipitates around the secondary carbide. The principle of acicular ferrite precipitation is as follows: due to the rapid heating speed and no heat preservation stage, the diffusion of carbon and other alloy elements at the original positions of primary carbide, grain boundary and secondary carbide is not uniform. When the temperature decreases, the carbide directly nucleates in situ, while the content of chromium around the carbide is high. Chromium is a typical ferrite forming element. It is easy to form acicular ferrite with carbide as the core during cooling.

In conclusion, the dissolution, transformation, dissolution and reprecipitation of carbides will occur in 8Cr13MoV steel during heating and cooling. With the increase of temperature, $M_{23}C_6$-type carbides first dissolves, then transforms to M_7C_3-type carbides. When the temperature continues to rise, M_7C_3-type carbides will also dissolve into the matrix and precipitate again in the subsequent cooling process.

1.4 Influence of Carbide on Properties of Steel and Its Control Method

1.4.1 Effect of Electroslag Remelting on Primary Carbide in Steel

Electroslag remelting (ESR) is a process of remelting the steel made by common smelting methods (mold casting or continuous casting), and the molten metal droplets pass through the slag pool and then solidify again. It is a special metallurgical technology which integrates the refining of molten steel and the control of solidification structure. In the process of electroslag remelting, with the melting, dropping and solidification of the electrode, the primary carbides in the steel also undergo the process of dissolution and precipitation.

The carbides in 8Cr13MoV remelted ingot is shown in Fig. 1.41. The amount of carbides was counted by Image Pro Plus (IPP) image analysis software, and the results are shown in Table 1.11.

It can be seen from Table 1.11 that the amount, area and size of carbides are significantly reduced after electroslag remelting, The formation of primary carbide is mainly related to the segregation of alloy elements. The cooling intensity of electroslag remelting process is better than that of cast iron mold smelted by vacuum induction furnace. The element segregation can be improved and the amount of primary carbide can be reduced by increasing the cooling intensity.

Samples are taken at the center, 1/2 radius and edge of the 8Cr13MoV remelted ingot after annealling, as shown in Fig. 1.42. The effect of ESR on the morphology, distribution, type and amount of primary carbides in remelted ingots was analyzed by SEM.

The distribution and morphology of primary carbide at 1/2 radius of remelted ingot are shown in Fig. 1.43. A large number of secondary carbides were precipitated in the remelted ingot after annealling in the sand pit for a long time. The primary and secondary carbides are bright white under optical microscope. Under SEM, the primary carbide is dark gray, and the secondary carbide is light gray. The size of

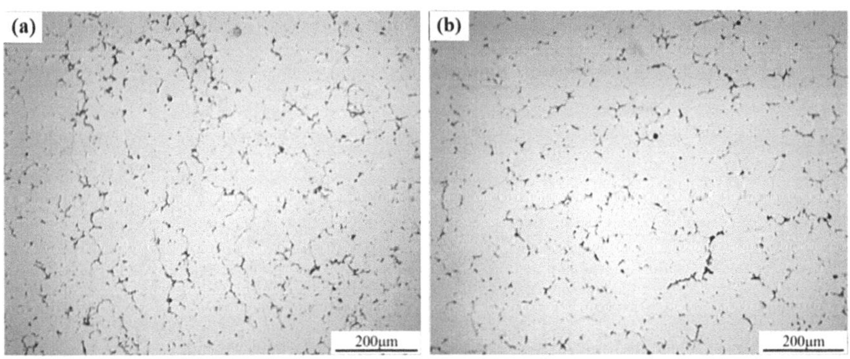

Fig. 1.41 Microstructure of carbide in the 8Cr13MoV steel after erosion: **a** Before electroslag remelting; b After electroslag remelting

Table 1.11 Statistical results of carbides

Sample	Amount	Area, μm^2	Average length, μm	Average width, μm
Before electroslag remelting	1112	39031.48	6.65	3.11
After electroslag remelting	896	19376.91	4.73	2.04

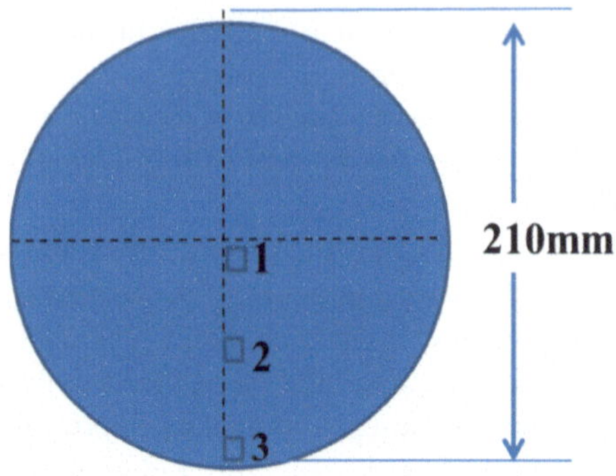

Fig. 1.42 Schematic view of 8Cr13MoV remelted ingot

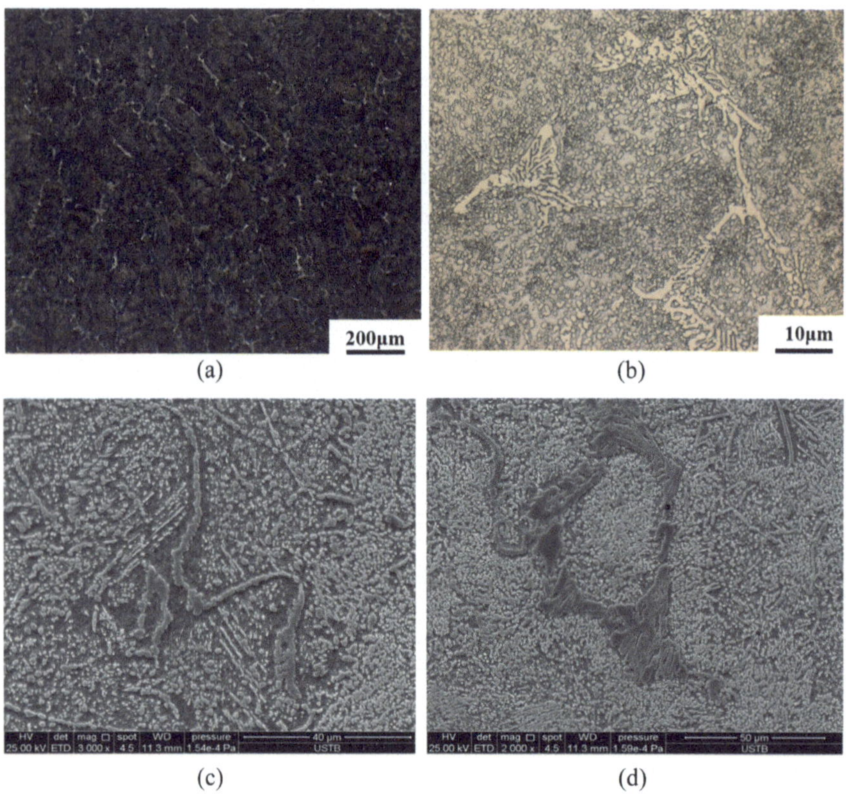

Fig. 1.43 Carbide distribution and morphology in remelted ingot: **a**, **b** are carbides under OM; **c** and **d** are carbides under SEM

primary carbides is large, generally tens of microns or even hundreds of microns, which is irregular block, coiled rod or spherical, and the secondary carbides are mainly small balls.

It can be seen from Fig. 1.43a that the primary carbides are mainly distributed in the dendrite gap, some are continuous network, and some exist independently. From Fig. 1.43b, the typical morphology of primary carbides are in the form of block, strip and coil, and a large number of small spherical secondary carbides precipitated from the matrix. In Fig. 1.43c, fine primary carbides and a large number of small particles of secondary carbides can be seen along the grain boundary. In Fig. 1.43d, it can be seen that the primary carbides grow along the grain boundary in a mixture of coiled and massive carbides. Therefore, there are a lot of primary carbides in electroslag ingot, and the morphology and distribution of primary carbides are similar to those in the consumable electrode.

The area fraction of primary carbides in the center, 1/2 radius and edge of remelted ingot was counted by using image Pro Plus (IPP) image analysis software. Each sample was randomly selected with 10 fields in the view of 1 mm^2. The results are shown in Table 1.12.

It can be seen from Table 1.12 that the area fraction of primary carbide in remelted ingot is less than that in consumable electrode by 2.29%, and the amount of primary carbide precipitation gradually decreases from the center to the edge of remelted ingot. The average area fraction of primary carbide in remelted ingot is 1.37%. The ESR process can effectively reduce the area fraction of primary carbide in steel. Because tha the edge of remelted ingot is close to the mold, the cooling intensity is high. The solidification of molten steel is only carried out in a very small volume, so that the full diffusion of solid and liquid phase is restrained and the composition segregation is reduced. Therefore, the area fraction of primary carbide in remelted ingot gradually decreases from the center to the edge. The primary carbides from the center to the edge of remelted ingot were observed by scanning electron microscope backscatter technique. The results are shown in Fig. 1.44. The bright white area in the figure is primary carbide.

It can be seen from Fig. 1.44 that the primary carbide at the center of remelted ingot is coarse and has strong continuity. The continuity of primary carbide at 1/2 radius is weakened, and the morphology becomes finer. The size of primary carbide at the edge is smaller and the distribution tends to be discrete. It demonstrates that with the increase of cooling intensity, not only the total amount of primary carbides decreases, but also the growth and distribution of primary carbides change.

The primary carbides are mainly precipitated in the secondary dendrite gap. The growth and distribution of primary carbides are affected by the growth state of dendrites, and the size of primary carbides is affected by the distance between

Table 1.12 Area fraction of primary carbide at different positions of remelted ingot

Sample no.	1	2	3
Area fraction, %	1.73	1.44	0.95

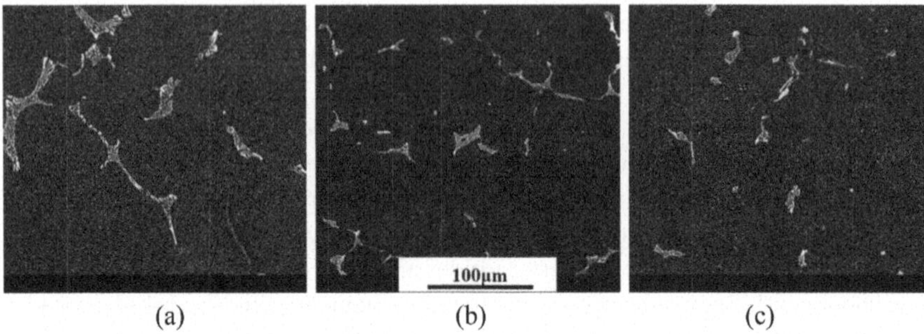

Fig. 1.44 Morphology and distribution of primary carbide from center to edge of remelted ingot: **a** Center; **b** 1/2 radius; **c** Edge

secondary dendrites. During electroslag remelting process, the molten steel in the mold is forced to water-cooled at the bottom and side, and the cooling intensity is the largest near the edge of the mold, which is easy to form fine-grained area. The dendrite gap is the smallest to form fine primary carbides. In addition, the dendrite which nucleates and grows rapidly at the edge of the mold may form a dendrite closed area, which prevents the continuous flow of the remaining molten steel in the dendrite. It restrains the enrichment of carbon and alloy elements and makes the primary carbide dispersed. From the edge to the center of remelted ingot, the cooling intensity and secondary dendrite gradually decreases and increases, respectively. The dendrite closed area gradually decreases, which makes the primary carbide size larger and the growth continuity stronger.

The carbide in 8Cr13MoV remelted ingot was extracted by electrolysis with anhydrous organic solution. The XRD result of the obtained carbide powder is shown in Fig. 1.45.

Fig. 1.45 XRD result of the carbide in 8Cr13MoV remelted ingot

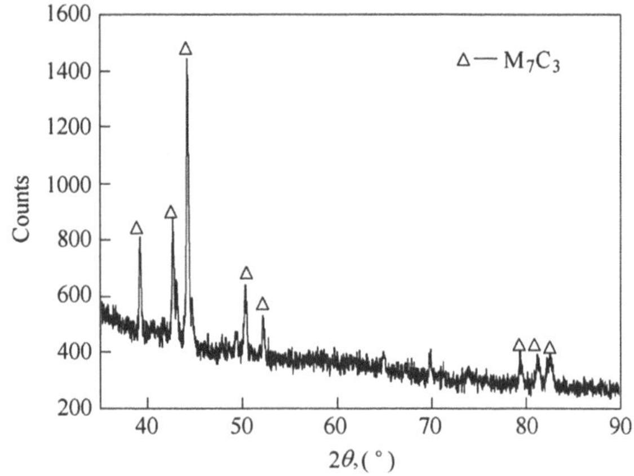

It can be seen from Fig. 1.45 that the primary carbide in remelted ingot is still M_7C_3-type. The ESR process does not change the type of primary carbide in 8Cr13MoV steel.

In conclusion, the ESR process can change the dendrite growth mode, improve the morphology and distribution of primary carbides, and not affect the type of primary carbides. But it can effectively reduce the total amount of primary carbide precipitation.

1.4.2 Effect of Carbides on the Properties of Steel

As an important second phase in steel materials, the morphology, size and distribution of carbides have an important influence on the workability and serviceability of steel. Chromium in stainless steel is used to increase the electrode potential of the steel matrix to reduce the corrosion rate, but the carbide generated by it often reduces the corrosion resistance of the matrix [16]. Primary and secondary carbides have great differences in precipitation temperature, alloying element content, morphology, size, thermophysics properties, which have different effects on the properties of steel. Therefore, the effects of primary and secondary carbides on the properties of steel are discussed:

(1) Effect of primary carbide on the properties of steel

The effect of primary carbides on the properties of steel is diverse. The primary carbides with high hardness in high speed steel can effectively protect the matrix and improve the wear resistance of steel [17, 18]. The coarse primary carbides in high chromium cast iron are easy to fracture and fall off from the matrix due to their high hardness and brittleness, which worsens the wear resistance and fracture toughness of the cast iron. In the low chromium cast iron, the hardness of the eutectic carbide with network structure is relatively low, but the wear resistance and toughness of the cast iron are improved. The addition of molybdenum element can produce Mo_2C-type primary carbide, which can improve the wear resistance of cast iron without reducing the toughness [19]. Large size carbides are easy to cause stress concentration and reduce fatigue strength and toughness of steel. There are a lot of alloy elements in the large carbides, which reduce the content of alloy elements in the matrix structure. It reduces the secondary hardening effect after tempering, and decreases the toughness and hardness of materials [20, 21]. The coarse primary carbides segregated at grain boundaries can improve the hardness of the roll, but obviously worsen the fracture toughness [22]. Large primary carbides significantly reduce the fatigue life of GCr15 bearing steel [23]. The coarse primary carbide in high chromium cast iron reduces its impact toughness [24].

The influence of primary carbide on the impact toughness of H13 shows that the higher the primary carbide level, the lower the impact energy of the material [25]. Table 1.13 presents the relationship between primary carbide and mechanical

Table 1.13 Relationship between primary carbide and mechanical properties

Mechanical properties	$R_{p0.2}$, N/mm^2	R_m, N/mm^2	A, %	Z, %	A_{KV}, J	Carbide level
1#	910	1090	11.0	26.0	7.0	4
2#	925	1120	12.0	24.5	9.0	3
3#	920	1090	13.0	39.0	12.0	2
4#	925	1100	12.0	40.5	15.0	1
5#	925	1110	14.5	48.0	21.0	0

properties. The primary carbides in the steel are in the form of point and chain, hard and brittle, and seriously isolate the matrix, which is the stress concentration point and the source of fatigue crack propagation.

(2) Effect of secondary carbide on the properties of steel

Secondary carbide is an important strengthening phase in steel materials. The size, type, quantity, distribution and morphology of secondary carbide can be controlled in the process of hot working and heat treatment. The dissolution and precipitation behavior of secondary carbides has an important influence on the workability and serviceability of steel. In the process of steel processing, the secondary carbides are generally controlled to precipitate uniformly and grow moderately by annealing process, resulting in the good processability of steel. The final heat treatment can strengthen the steel materials by controlling the fine, uniform and dispersed precipitation of secondary carbides. By reasonably controlling the type and quantity of secondary carbides, the requirements of wear resistance and corrosion resistance of steel can be achieved.

Fine and uniform $M_{23}C_6$-type carbide precipitates in the 12% Cr-containing steel after high temperature austenitizing and tempering, which can significantly improve the hardness and strength of the steel [26]. After isothermal pretreatment of high carbon martensitic stainless steel, the obtained uniform and fine $M_{23}C_6$-type not only plays the role of precipitation strengthening, but also inhibits the growth of austenite grains, which can reduce the ductile brittle transition temperature and improve the strength of steel [27]. For high CoNi alloy steel, when tempering temperature is 482 °C, cementite carbide completely transforms into M_2C-type acicular carbide, which is coherent with martensite matrix and can effectively improve the hardness of steel [28]. The precipitation of intergranular carbide in TWIP steel can reduce the delayed fracture time of steel [29].

Improper control of secondary carbides may also cause adverse influence on steel. It is easy to precipitate network carbide along the grain in boundary hypereutectoid steel during the cooling process. The higher the level of network carbide, the lower the impact toughness of steel [30]. As shown in Fig. 1.46, excessive secondary carbides in the metallographic structure of the steak knife reduces the chromium content in the matrix and its corrosion resistance [31].

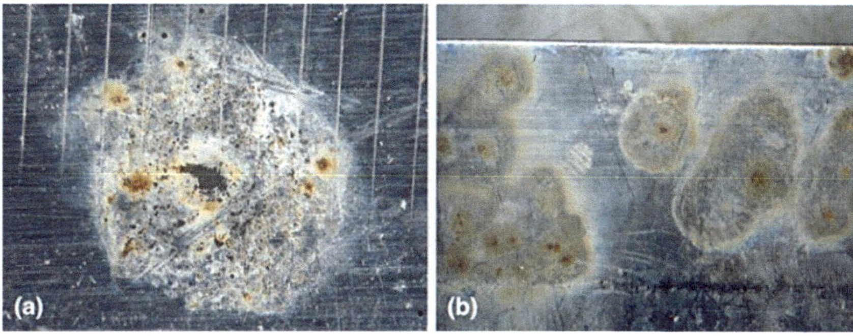

Fig. 1.46 Effect of secondary carbide precipitation on the corrosion resistance of bullrow cutters: **a** Tool corrosion morphology with high carbide content; **b** Tool corrosion morphology with low carbide content

1.4.3 Controlling Method of Carbide

1.4.3.1 Controlling Method of Primary Carbide

Primary carbides are formed in the solidification process of molten steel. Due to selective crystallization, carbon and many carbide forming elements gradually accumulate in the residual molten steel during cooling process. When the element concentration reaches a certain level, primary carbides will precipitate from the molten steel, mainly including the primary eutectic carbides precipitated prior to austenite and the eutectic carbides precipitated simultaneously with austenite. Due to the high melting point and large size of primary carbides, it is difficult to remove them by ordinary solid phase transformation. Therefore, the key to control the primary carbide is the solidification process of molten steel, followed by the subsequent hot working and heat treatment process.

In the study of M2 high speed steel, it is found that control the melting rate and depth of liquid metal pool of ESR process can promote the primary carbides in flat and flat layers transform into isolated rod-shaped and granular carbides, and the size of primary carbides can be reduced [32]. In the field of electroslag remelting, conductive mold technology [33], electroslag remelting continuous directional solidification technology [34], electroslag remelting liquid casting technology [35] can reduce element segregation by reducing the depth of liquid metal pool and local solidification time, so as to control the formation of primary carbide in steel. The effects of continuous casting parameters and electromagnetic stirring on the central carbon segregation of 82B billet show that the reduction of superheat and casting speed and the increase of cooling intensity are beneficial to the reduction of central carbon segregation. The carbon segregation index of 82B billet is the lowest at the optimal end electromagnetic stirring parameters [36]. Other studies have shown that the electromagnetic stirring process with appropriate parameters can effectively reduce carbon segregation [37–40]. Carbon segregation is one of the important reasons for the formation of primary carbides.

High temperature deformation treatment can break the primary carbide in GCr15 bearing steel and further promote the dissolution of primary carbide [41]. According to the research on the fracture behavior of eutectic carbide in as cast M2 high speed steel during hot compression, it is found that hot deformation can break rod-shaped or irregular eutectic carbides into small particles, and processing stress is the main driving force for eutectic carbide crushing, while high temperature or low strain rate lead to the reduction of processing stress [42].

When Ti, V, Nb and other strong carbide forming elements are added in the production process of high chromium cast iron, MC-type carbide formed preferentially during solidification can act as the heterogeneous nucleation point, which can improve the nucleation rate of M_7C_3-type carbide and refine the primary M_7C_3-type carbide [43–46]. The morphology of primary carbides can be improved and primary carbides can be refined by multiple rare earth modification of high chromium cast iron [47–49]. The addition of rare earth cerium can obviously refine the primary carbide in high chromium cast iron. As shown in Fig. 1.47, with the increase of cerium content, the isotropy of carbide morphology increases [48].

The addition of Mg element in steel can modify Al_2O_3 to $MgO \cdot Al_2O_3$, which can act as the nucleation core of (Ti, V)N [50]. Because the formation temperature of TiN is higher than that of VC, it can act as the nucleation core to refine primary carbide and improve the distribution of primary carbide [51].

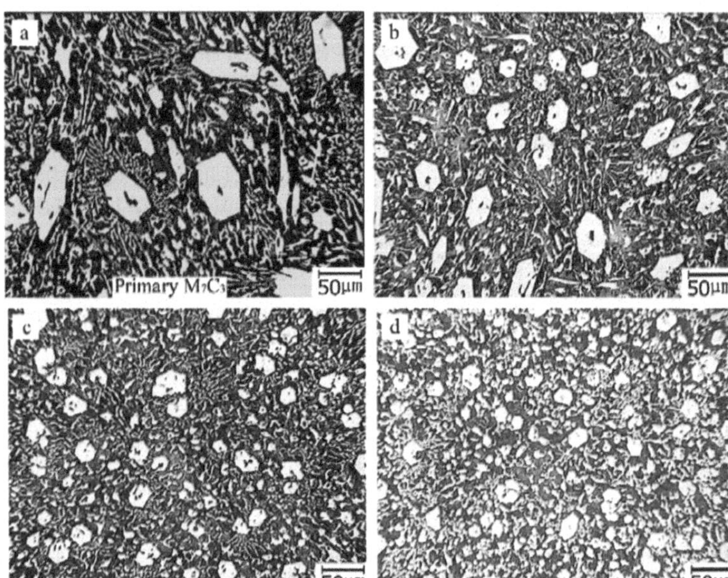

Fig. 1.47 Effect of rare earth cerium on the size and morphology of primary carbide in high chromium cast iron: **a** 0% cerium; **b** 0.5% cerium; **c** 1% cerium; **d** 1.5% cerium

1.4.3.2 Controlling Method of Secondary Carbide

The precipitation and dissolution of secondary carbides are mainly affected by heat processing, heat treatment and alloy elements. In general, the size of secondary carbides is small and the dissolution temperature is not high. Most of the secondary carbides can dissolve into the matrix in the heat treatment process, and they can be orderly precipitated in the appropriate heat preservation conditions and cooling process. The precipitation of secondary carbides is not only affected by temperature, but also related to physical and chemical properties of carbides, steel matrix structure and interface energy. The thermophysics properties of carbides mainly refer to its crystal structure, alloy element composition, size and morphology.

(1) Effect of heat treatment process on secondary carbide

After tempering at 560 °C, the precipitation of secondary carbide of M2 high speed steel is affected by the austenitizing temperature during quenching. When the austenitizing temperature is 1180 °C, only $M_{23}C_6$-type carbides is precipitated during tempering. When the austenitizing temperature is 1220 °C, there are $M_{23}C_6$ and M_6C-type carbides after tempering. When the temperature is 1260 °C, only M_6C-type secondary carbides are found after tempering [52]. The research on the transformation behavior of carbides in 8% Cr-containing rolling steel showed that the dissolution of carbides was mainly affected by heating temperature and had a certain order. As shown in Fig. 1.48, $M_{23}C_6$-type carbides could be completely dissolved at 850 °C and M_7C_3-type carbides could be completely dissolved at 1150 °C [53].

(2) Effect of hot working process on secondary carbide

The controlled rolling and cooling process is an effective way to control the precipitation of network carbide in hypereutectoid steel. By controlling the finish rolling temperature, the network carbide can be changed into fine particles, as shown in Fig. 1.49 [54].

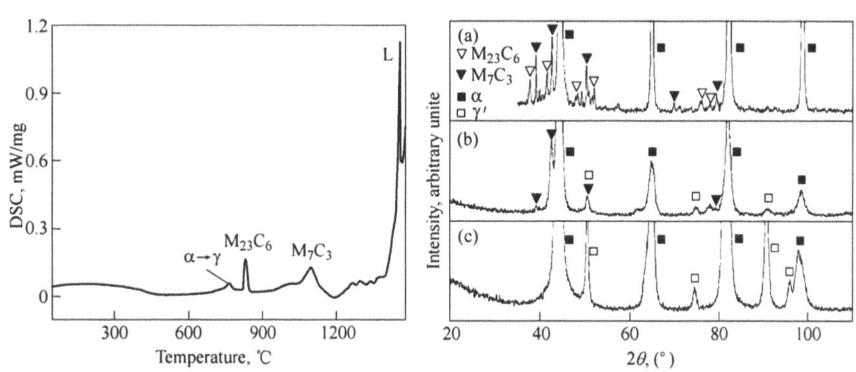

Fig. 1.48 Effect of heating temperature on carbide transformation behavior of $M_{23}C_6$ and M_7C_3-type carbides: **a** Before heating; **b** Holding at 850 °C for 10 h; **c** Holding at 1150 °C for 10 h

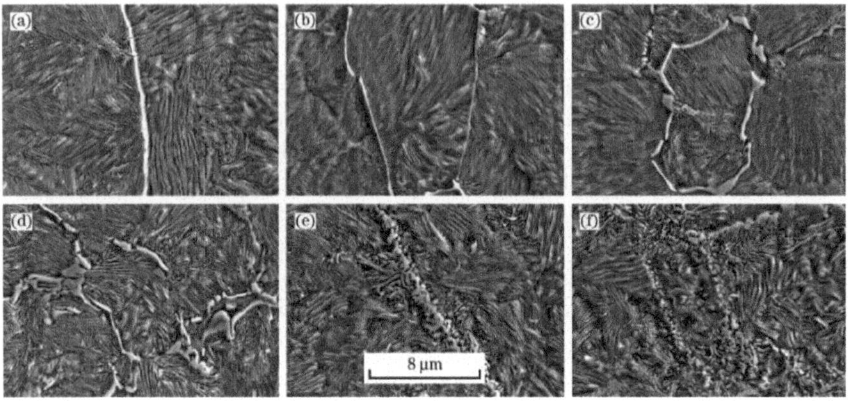

Fig. 1.49 Effect of finish rolling temperature on network carbide in GCr15SiMn steel: **a** 980 °C; **b** 900 °C; **c** 850 °C; **d** 800 °C; **e** 750 °C; **f** 700 °C

(3) Effect of alloying elements and rare earth elements on secondary carbides

The study on the coarsening behavior of $M_{23}C_6$-type carbides in 12% Cr-containing martensitic steel showed that adding Mn and reducing V and Ta can increase the coarsening rate of $M_{23}C_6$-type carbides. Furthermore, the coarsening rate of $M_{23}C_6$-type carbides increases exponentially every 50 °C when the temperature is 600–750 °C [55]. The study on Fe–Cr–Ni–Mo high strength steel showed that the V content can significantly affect the type and size of carbides in the steel due to the competition between V and other alloying elements in the steel. The carbides in the steel without V are mainly $M_{23}C_6$-type carbides. When the V content reaches 0.03%, M_2C and M_6C-type carbides were found in the steel. With the continuous increase of V content, MC-type carbides were found in the steel, and the size of carbides in the steel gradually decreases [56]. The addition of appropriate amount of Cerium in stainless steel can make the secondary carbides more dispersed and fine [57]. Rare earth lanthanum can also change the morphology of carbides and reduce the segregation of carbides in 4Cr13 steel [58].

The high magnetic field can change the Gibbs free energy of carbide precipitation and the precipitation behavior of carbide. Strengthening the magnetic field during the annealing process of Fe–C–Mo alloy can promote the precipitation of M_6C-type carbides and inhibit the formation of M_2C and M_3C-type carbides [59]. The application of enhanced magnetic field can also make the eutectoid point of hypereutectoid steel move to the upper right, which reduces the number of pre eutectoid cementite and increases the lamellar spacing of pearlite, as shown in Fig. 1.50 [60].

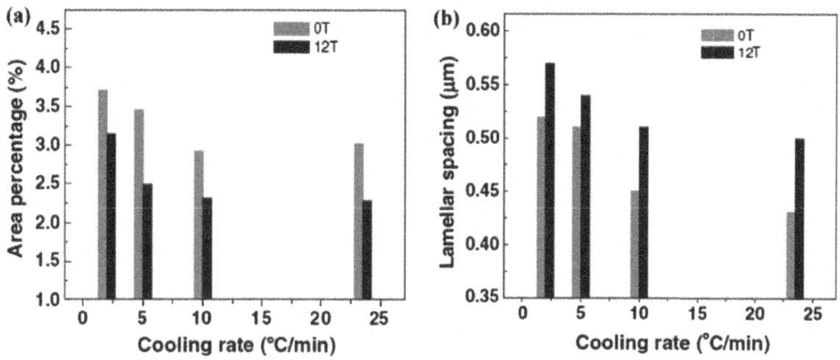

Fig. 1.50 Effect of high magnetic field on proeutectoid cementite number and pearlite lamellar spacing in hypereutectoid steel: **a** Proeutectoid cementite; **b** Pearlite lamellar spacing

References

1. Guo KX (1957) Researches on carbides in alloy steels. Acta Metallurgic Sinica 2(3):305–321
2. Xu L, Xing J, Wei S et al (2006) Investigation on wear behaviors of high-vanadium high-speed steel compared with high-chromium cast iron under rolling contact condition. Mater Sci Eng, A 434(1):63–70
3. Liu L, Fu HZ (1989) Theoretical morphology of MC carbides precipitated in Ni-base superalloy melt. Mater Sci Progress 3(5):396–400
4. Wang ZT, Chen HH, Sun JF et al (2006) Microstructure and properties of In-Situ synthesis TiC particle refinforced metal matrix composite coating. Heat Treat Met 31(6):57–60
5. Zhou X-F, Fang F et al (2010) Effect of pouring temperature on microstructure and eutectic carbides of high-speed steel. In: Proceedings of the annual conference of china foundry association, Bei Jing2010
6. Jun-peng DANG (2012) High-chromium cast ball heat treatment process research. Foundry Technol 33(1):83–84
7. Kondrat'ev SY, Kraposhin VS, Anastasiadi GP et al (2015) Experimental observation and crystallographic description of M7C3 carbide transformation in Fe–Cr–Ni–C HP type alloy. Acta Materialia 100: 275–281
8. Wang HC (2018) Ex situ and in situ TEM investigations of carbide precipitation in a 10Cr martensitic steel. Journal of Materials Science 53(10):7845–7856
9. Jack DH, Jack KH (1973) Invited review: carbides and nitrides in steel. Mater Sci Eng 11(1):1–27
10. Morniroli JP, Khachfi M, Courtois A et al (1987) Observations of non-periodic and periodic defect structures in M7C3 carbides. Philos Mag A 56(1):93–113
11. Monypenny JHG (1951) Stainless iron and steel
12. Zi-li SONG, Xiao-dong DU et al (2011) Microstructure and impact toughness of 7 Cr17 Mo martensitic stainless steel. Trans Mater Heat Treatment 32(5):95–99
13. Tao MA, Yun-gang LI (2015) Development situation and application prospect of antibacterial stainless steels. Mater Rev 29(13):98–101
14. Li NAN, Yong-qian LIU et al (2007) Study on antibacterial properties of coppercontaining antibacterial stainless steels. Acta Metall Sin 43(10):1065–1070
15. Xuan Y, Zhang C, Fan N et al (2014) Antibacterial property and precipitation behavior of Ag-added 304 austenitic stainless steel. Acta Metallurgica Sinica (English Lett) 27(3):539–545
16. Hall EL, Briant CL (1984) Chromium depletion in the vicinity of carbides in sensitized austenitic stainless steels. Metall Mater Trans A 15(15):793–811
17. Fu HG, Xiao Q, Fu HF (2005) Heat treatment of multi-element low alloy wear-resistant steel. Mater Sci Eng, A 396(1–2):206–212

18. Badisch E, Mitterer C (2003) Abrasive wear of high speed steels: influence of abrasive particles and primary carbides on wear resistance. Tribol Int 36(10):765–770
19. Oh H, Lee S, Jung J et al (2001) Correlation of microstructure with the wear resistance and fracture toughness of duo-cast materials composed of high-chromium white cast iron and low-chromium steel. Metall Mater Trans A 32(3):515–524
20. Rong-Bin WANG (1990) Influence of high-speed steel carbide appearance to hot-working technology and die & mold life. Die Mould Technol 7(4):106–110
21. Bu-qing ZHAO, Zhen-zhong GON et al (2011) Influence of carbide to high-speed steel cutting tool life. Heat Treatment Technol Equip 32(3):60–64
22. Hwang KC, Lee S, Lee HC (1998) Effects of alloying elements on microstructure and fracture properties of cast high speed steel rolls Part I: microstructural analysis. Mater Sci Eng, A 254(1):282–295
23. Bao-Ping FENG, Ya-Jun QIU et al (2003) Effect of carbide on contact fatigue life of GCr15 bearing steel. Bearing 10:30–32
24. Zhi XH, Xing JD, Gao YM et al (2008) Effect of heat treatment on microstructure and mechanical properties of a Ti-bearing hypereutecitc high chromium white cast iron. Mater Sci Eng 487(1–2):171–179
25. Wang H et al (2012) Effect of liquid carbide analysis on impact properties of 4 Cr5MoSiV1 mandrel. In: Eighth national academic conference on materials science and image technology, Rongcheng, Shandong
26. Kim HD, Kim IS (1994) Effect of austenitizing temperature on microstructure and mechanical properties of 12%Cr steel. ISIJ Int 34(2):198–204
27. Tsuchiyama T, Ono Y, Takaki S (2000) Microstructure control for toughening a high carbon martensitic stainless steel. ISIJ Int 40(Supplement):184–188
28. Zheng-Fei HU, Xing-Fang WU et al (2001) Precipitation and transformation of the second carbides in isothermal tempered high CoNi alloy steel. Acta Metall Sin 34(4):381–385
29. Hong S, Lee J, Lee B et al (2013) Effects of intergranular carbide precipitation on delayed fracture behavior in three twinning induced plasticity (TWIP) steels. Mat Sci Eng A 587:85–99
30. Chou YJ, Ye JY, Zhang YY (2008) Influence of network carbide on impact property of steel GCr15. Bear 4:28–31
31. Banuta M, Tarquini I (2010) Premature corrosion of steak knives due to extensive precipitation of chromium carbides. J Fai Ana Pre 10(6):458–462
32. Chu W, Xie C, Wu XC (2013) Research on controlling eutectic carbides in M2 high speed steel of ESR process. Shanghai Metals 35(5):23–26
33. Jiang ZH (2014) Recent progress on electroslag remelting technology. In: Conference of Special Steel, Tianjin
34. Zhan LC, Chi HX, Ma DS et al (2013) The as-cast microstructure of esr-cds m2 high speed steel. J Mater Eng 7:29–34
35. Jiang ZH, Li ZB (2009) Recent progress on electroslag remelting technology. Special Steel 30(6):10–13
36. Wang T, Chen WQ, Wang HB et al (2013) Effect of concasting parameter and f-ems on center carbon segregation of 82b steel billet. Special Steel 34(1):49–51
37. Jia YL, Miao F, Zhao WC (2012) Effects of f-ems on carbon content segregation in φ400 mm round billets. Hot Work Tech 41(7):61–65
38. Hu Y, Chen WQ, Han HB et al (2016) Study on improvement of center carbon segregation in 60Si2MnA spring steel billet. Shanghai Metal 38(4):41–49
39. Liu Y, Wang XH (2007) Effect of electromagnetic stirring at secondary cooling area on central segregation of a continuously cast slab. J Uni Sci Tech Beijing 29(6):582–590
40. Jiang D, Zhu M (2017) Center segregation with final electromagnetic stirring in billet continuous casting process. Metall Mater Trans B 48(1):444–455
41. Kong XH, Liu JZ, Liu ZL et al (2014) Effect of high temperature deformation on liquation carbide of GCr15 bearing steel. Trans Mater Heat Treat 35(7):173–176
42. Bin Z, Yu S, Jun C et al (2011) Breakdown behavior of eutectic carbide in high speed steel during hot compresion. J Iron Steel Res Int 18(1):41–48

43. Lin C, Chang C, Chen J et al (2010) The effects of additive elements on the microstructure characteristics and mechanical properties of Cr–Fe–C hard-facing alloys. J Alloy Compo 498(1):30–36
44. Chung RJ, Tang X, Li DY et al (2009) Effects of titanium addition on microstructure and wear resistance of hypereutectic high chromium cast iron Fe-25 wt%Cr-4 wt%C. Wear 267(1–4):356–361
45. Zhou Y, Yang Y, Yang J et al (2012) Effect of Ti addition on (Cr, Fe)7C3 carbide in arc surfacing layer and its refined mechanism. Appl Surf Sci 258(17):6653–6659
46. Chung RJ, Tang X, Li DY et al (2013) Microstructure refinement of hypereutecitc high Cr cast irons using hard carbide-forming elements for improved wear resistance. Wear 301(1):695–706
47. Qu Y, Xing J, Zhi X et al (2008) Effect of cerium on the as-cast microstructure of a hypereutectic high chromium cast iron. Mater Lett 62(17–18):3024–3027
48. Hou Y, Wang Y, Pan Z et al (2012) Influence of rare earth nanoparticles and inoculants on performance and microstructure of high chromium cast iron. J Rare Earth 30(3):283–288
49. Chen ZX, Xu XE, Qu X et al (2011) Effect of modification and heat treatment on microstructure and properties of high-chromium cast iron. Heat Treatment 26(4):35–39
50. Shi CB, Chen XC, Guo HJ et al (2013) Control of MgO·Al2O3 spinel inclusions during protective gas electroslag remelting of die steel. Metall Mater Trans B 44(2):378–389
51. Ma DS, Zhou J, Chen ZZ et al (2009) Influence of thermal homogenization treatment on structure and impact toughness of H13 ESR steel. J Iron Steel Res Int 16(5):56–60
52. Chaus AS, Domankova M (2013) Precipitation of secondary carbides in M2 high-speed steel modified with Titanium diboride. J Mater Eng Perform 22(5):1412–1420
53. Wang Z, Lv Z, Bai X et al (2012) Study on transformation characteristics of carbides in an 8% Cr roller steel. J Mater Sci 47(20):7132–7137
54. Zhou WS, Hua JS, Yu F et al (2015) Effects of deformation temperature on the evolution behaviour of carbide network in GCr15SiMn bearing steel. Iron Steel 6:87–93
55. Xiao X, Liu G, Hu B et al (2013) Coarsening behavior for M23C6 carbide in 12%Cr-reduced activation ferrite/martensite steel: experimental study combined with DICTRA simulation. J Mater Sci 48(16):5410–5419
56. Wen T, Hu X, Song Y et al (2013) Carbides and mechanical properties in a Fe–Cr–Ni–Mo high-strength steel with different V contents. Mater Sci Eng, A 588(12):201–207
57. Li YB, Wang FM, Li CR (2009) Effect of Cerium on Grain and Carbide in Low Chromium Ferritic Stainless Steels. J Chinese Rare Earth Society 27(1):123–127
58. Zhang HM, Cui CY, Zhao LP et al (2011) Effects on Corrosion Resistance and Mechanical Property of Lanthanum-Contained 4Cr13 Steel. J Chinese Rare Earth Soc 29(1):100–104
59. Hou TP, Li Y, Zhang YD et al (2014) Magnetic field-induced precipitation behaviors of alloy carbides M2C, M3C, and M6C in a molybdenum-containing steel. Metall Mater Trans A 45(5):2553–2561
60. Zhang YD, Esling C, Gong ML et al (2006) Microstructural features induced by a high magnetic field in a hypereutectoid steel during austenitic decomposition. Scripta Mater 54(11):1897–1900

Chapter 2
Carbides Control in Electroslag Remelting Process

Abstract Liquid steel solidification process is the source of the formation of the primary carbides in special steel. This chapter takes electroslag remelting (ESR) as an example to study the control of primary carbide in different ESR solidification processes. It is found that the precipitation of primary carbides can be effectively restrained and the size of primary carbides in special steel can be reduced by appropriately reducing the melting rate, and increasing the filling ratio and cooling intensity of ESR. When reducing the melting rate of ESR from 150 kg/h to 133 kg/h, increasing the filling ratio from 0.23 to 0.33, the area fraction of primary carbide in ESR ingot is reduced by 23%. The directional solidification of ESR process gives a shallow U-shaped molten metal pool, which improve the dendrite growth direction, refine the dendrite structure and reduce the segregation of alloying elements. Meanwhile, the directional solidification of ESR process can effectively reduce the size and amount of carbides, and improve the morphology and distribution of carbides in the solidification structure, whereas it cannot change the types of carbides.

Keywords Carbide · Electroslag remelting · Melting rate · Filling ratio · Cooling intensity · Directional solidification

Generally, the characteristics of carbides in tool and die steels, such as the size, quantity, type, morphology and distribution of carbides, are the key factors affecting hardness, wear resistance and toughness. Once coarse primary carbides or eutectic carbides are formed during solidification, these carbides are difficult to remove by hot forging or heat treatment. Therefore, the size, quantity, type and morphology of primary or eutectic carbides in ESR ingot are the key controlled points of ESR process.

The parameters of ESR process include geometric parameters, control parameters and target parameters. Geometric parameters refer to the specific size parameters reflected in macroscopic, including mold diameter and height, consumable electrode diameter and length. Among them, the ratio of consumable electrode cross-sectional area to mold cross-sectional area is defined as filling ratio, which has a great influence on metallurgical quality to a certain extent. The control parameters include slag system, power supply system, deoxidation system and cooling water system. The target parameters mainly include shape and size of liquid metal pool, electrode

J. Li and C. Shi, *Carbide in Special Steel*, Engineering Materials,
https://doi.org/10.1007/978-981-16-1456-9_2

melting speed, slag thickness, local solidification time, secondary dendrite spacing, power consumption, etc. In terms of electroslag remelting process, the main measures to control the shape of liquid metal pool are to reduce the melting speed of consumable electrode [1–3], increase the filling ratio of consumable electrode to mold [4], and appropriately increase the amount of slag [5, 6]. The shallower the liquid metal pool is, the shorter the local solidification time is, the smaller the secondary dendrite spacing is and the smaller the microsegregation degree of elements is, which are beneficial to ensure the good solidification quality of ESR ingot [7, 8] and reduce quantity the primary carbide. The melting rate of electroslag remelting is the key factor to determine the depth of liquid metal pool. The filling ratio of ESR has an important influence on remelting process, including the interface heat transfer between liquid slag pool and liquid metal pool, and the temperature field distribution of liquid metal pool [9], and also affects the size and shape of liquid metal pool. Therefore, the researches on the influence of melting rate and filling ratio on liquid metal pool, temperature field, two-phase zone and element segregation in ESR process have great signification, these can provide theoretical guidance for reducing element segregation and characteristic control of primary carbide in ESR ingot.

Directional solidification electroslag remelting process can also affect the shape of the liquid metal pool. The V-shape of the traditional electroslag remelting process is changed to the shallow flat U-shape. The growth direction of dendrite formed by directional solidification electroslag remelting process is approximately parallel to the central axis of ingot, which reduces the thickness of liquid-solid two-phase region, increases the distribution of temperature gradient, and increases the solidification rate of liquid metal. It is beneficial to refine the dendrite structure, reduce the size and amount of carbides, and make the distribution of carbides more uniform and dispersed. Meanwhile, the morphology of primary carbide is improved to make it easier to be decomposed by the later heat treatment. Therefore, it is necessary to research the effect of directional solidification ESR on primary carbides.

2.1 Effects of Parameter of ESR on the Primary Carbides

ESR is a black box process, it is difficult to grasp the heat transfer, flow field, temperature field, two-phase distribution in the smelting process, and these invisible target parameters are the important factors affecting the quality of ESR ingot. Therefore, the influence of different ESR process parameters on the solidification quality of 8Cr13MoV steel ESR ingot was simulated by meltflow ESR software. Meltflow ESR software is a set of numerical simulation software developed and upgraded by Prof. Alec Mitchell (the University of British Columbia, Canada) and IRI company for ESR smelting process of high performance alloy [10]. Meltflow ESR has changed the traditional trial and error process exploration method, using computer simulation to simulate the heat, electricity, flow, magnetism, phase transformation, chemical composition, dendrite spacing, macro segregation in the ESR process [11, 12].

2.1.1 *Effect of Melting Rate of ESR on Primary Carbides in Steel*

Melt-Flow software was applied to invertigate the effect of parameters of electroslag remelting process on solidification quality of 8Cr13MoV steel. The related operating parameters of ESR are shown in Table 2.1. The physical parameters of 8Cr13MoV steel and slag are listed in Table 2.2.

The slag used in ESR experiments consists of 30wt% Al_2O_3 and 70wt% CaF_2. The melting rate of ESR was 150 kg/h, and the fill ratio was 0.23.

The heat transfer, temperature and electric field in ESR process can be simulated by means of coupled-field technology, metallic material data and experience data. Melting rates of 133 kg/h, 150 kg/h, 165 kg/h and fill ratios of 0.23, 0.33, 0.50, 0.75 were employed to study their effect on the primary carbides in ESR ingot, respectively.

From the point of view of heat balance, the shape of liquid metal pool depends on the heating and cooling methods. Under the same cooling conditions, the shape of the liquid metal pool is mainly affected by the heat supply conditions of the slag metal interface, and the heat supply of the slag metal interface is mainly determined by the melting rate of electroslag remelting. The distribution of temperature field and solid-liquid two-phase region in liquid metal pool and ESR ingot in stable melting stage under three kinds of ESR melting rates (133 kg/h, 150 kg/h and 165 kg/h) were simulated by meltflow software. The results are shown in Fig. 2.1.

Table 2.1 Operating parameters of ESR

Parameter	Value
Diameter of electrode, mm	110
Diameter of mold, mm	228
Depth of slag, mm	13
Current frequency, Hz	50
Cooling intensity, W/(m^2·K)	4000

Table 2.2 Physical properties of 8Cr13MoV steel and slag

Parameter	Value
Density of slag, kg/m^3	2569 [13]
Viscosity of slag, Pa·s	0.25 (1673 K), 0.033 (1773 K) [14]
Density of ESR ingot, kg·m^3	7444
Electrical conductivity of slag, S/m	320 (1873 K) [15]
Electrical conductivity of steel, S/m	8.8 × 105 [16]
Liquidus temperature of steel, K	1715
Solidus temperature of steel, K	1536

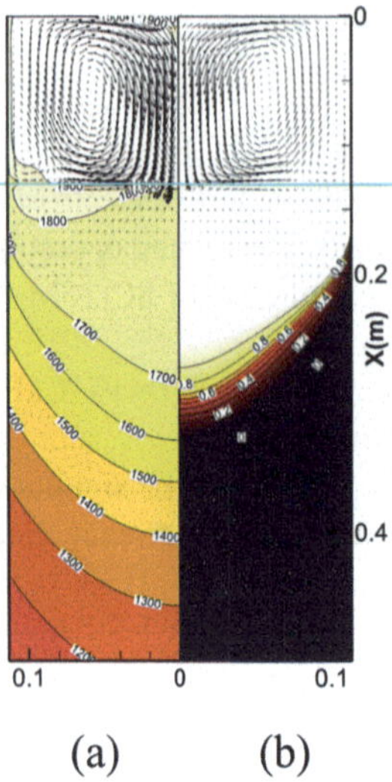

Fig. 2.1 Temperature field and mushy zone at different melting rates: **a**, **c**, **e** The temperature field under the melting rate of 133 kg/h, 150 kg/h, 165 kg/h, respectively. **b**, **d**, **f** The mushy zone under the melting rate of 133 kg/h, 150 kg/h, 165 kg/h, respectively

The depth of liquid metal pool and width of two-phase zone at the center of ESR ingot are shown in Table 2.3. It can be seen from Fig. 2.1 and Table 2.3 that with increasing the melting rate of ESR, the liquid metal pool becomes deeper and the solid-liquid two-phase zone widens. Hoyle, a British scholar, summarized the previous work and believed that the ideal depth of the liquid metal pool should be 1/2 of the diameter of the ESR ingot, and its front shape should be parabolic. At this time, the dendrites are solidified in an appropriate angle sequence, and the crystallization quality is good [17]. According to the above theory, the diameter of simulated ESR ingot is 228 mm, and the depth of ideal liquid metal pool is 114 mm. It can be seen from Table 2.3 that when the melting rate of ESR is reduced to 133 kg/h, the depth of liquid metal pool is reduced to 112 mm, which is close to the ideal depth of liquid metal pool, which is conducive to the improvement of internal quality of ESR ingot.

The two-phase zone refers to the area between the complete solidification of ESR ingot (solid rate is 100%) and the parabolic front of liquid metal pool (solid rate is 0). The main advantage of electroslag remelting is that only a small volume of molten

Table 2.3 Depth of liquid metal pool and width of two-phase zone corresponding to different melting rate of ESR

Melting rate, kg/h	133	150	165
Depth of liquid metal pool, m	0.112	0.142	0.146
Width of two-phase zone, m	0.097	0.106	0.119

steel gradually solidfies from bottom to top at any stage, which greatly reduces the macro segregation. The width of the two-phase zone represents the size of the solidification zone during ESR. Generally, the larger the width of the two-phase zone, the longer the local solidification time and the larger the tendency of element segregation. According to the precipitation and growth mechanism of primary carbide, it can be inferred that the content of primary carbide increases with the increase of element enrichment. It can be seen from Table 2.3 that the width of the two-phase zone decreases with the decrease of the melting rate. Therefore, reducing the melting rate is conducive to reducing the segregation of elements and reducing the precipitation of primary carbides.

The cooling intensity is higher near the mold wall and bottom water tank. The molten steel solidifies first, and its carbon content is lower than the average carbon content of the system. In the process of electroslag remelting, the solidification of molten steel is affected by the direction of heat flow, and dendrites grow along the normal direction of the curve at the bottom of the liquid metal pool, resulting in the continuous enrichment of carbon atoms along the growth direction of dendrites. In addition, due to the cooling intensity of the upper part of the ESR ingot is lower than that of the edge and bottom, and is continuously affected by the thermal radiation of the electrode and slag pool, the solidification rate of the molten steel is slowed down, which provides sufficient time for the enrichment of carbon atoms. Therefore, the enrichment degree of carbon in ESR ingot increases gradually from edge to center and from bottom to top. The distribution of carbon in ESR ingot at different melting rates is shown in Fig. 2.2. The initial mass fraction of carbon in ESR ingot is 0.77%. It can be seen from Fig. 2.2 that with increasing the melting rate of ESR, the area where the carbon content exceeds the average content of the liquid metal pool increases, and the carbon enrichment is intensified. Therefore, reducing the melting rate has a positive effect on reducing carbon segregation.

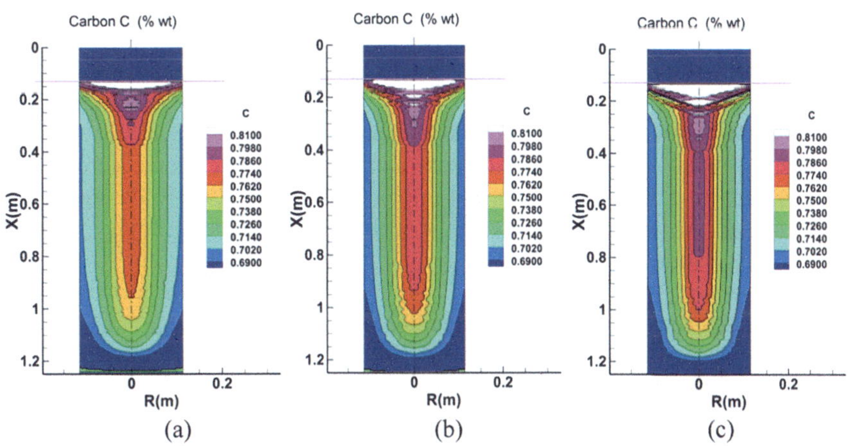

Fig. 2.2 Distribution of carbon at different ESR melting rates: **a** 133 kg/h, **b** 150 kg/h, **c** 165 kg/h

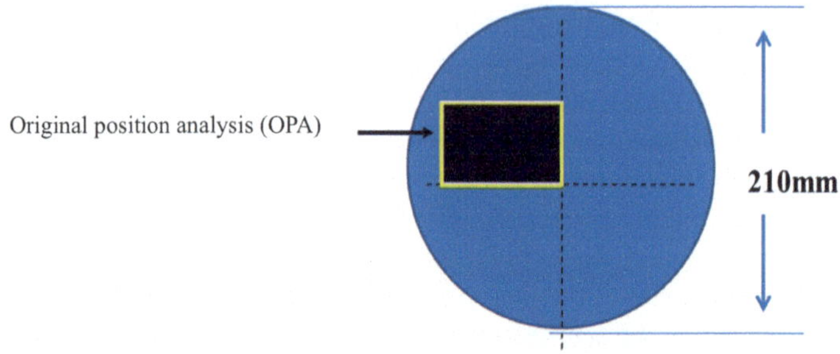

Fig. 2.3 Schematic view for original position analysis (OPA) of ESR ingot

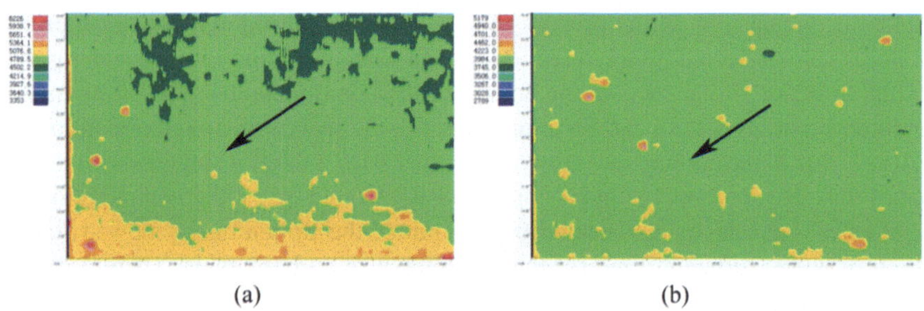

Fig. 2.4 OPA of carbon at different melting rate of ESR (arrow from ESR ingot edge to center): **a** 150 kg/h, **b** 133 kg/h

The segregation of elements in ESR ingots produced at 133 kg/h and 150 kg/h was compared with other ESR process parameters. The 1/4 area of the upper and middle section of ESR ingot was cut for, and the schematic view is shown in Fig. 2.3.

The OPA results of carbon element at different rate are shown in Fig. 2.4. The green area represents the average carbon concentration of the system. The blue area represents the area where the carbon concentration is lower than the average concentration, and the yellow to red area represents the carbon enrichment area. The degree of carbon enrichment increases with the deepening of color.

It is clear that both the carbon-rich zone and carbon-depleted zone decreased at lower melting rate. Carbon-rich zone distributed randomly in the ESR ingot with lower melting rate, while it enriched in the center of ESR ingot at a higher melting rate. The distribution of carbon can be more uniform with the melting rate of 133 kg/h. Therefore, reducing the melting rate of ESR can reduce the enrichment of carbon in ESR ingot and make the distribution of carbon more uniform.

Reducing the remelting rate of electroslag is helpful to reduce the carbon segregation in the ESR ingot, which can be analyzed by dendrite growth in ESR ingot. During the solidification process of electroslag remelting, dendrites nucleate at the solidified metal and grow along the temperature gradient direction (normal direction of liquid metal pool). The solute atoms can be gradually discharged from the residual liquid phase with the growth of dendrites, as shown in Fig. 2.5.

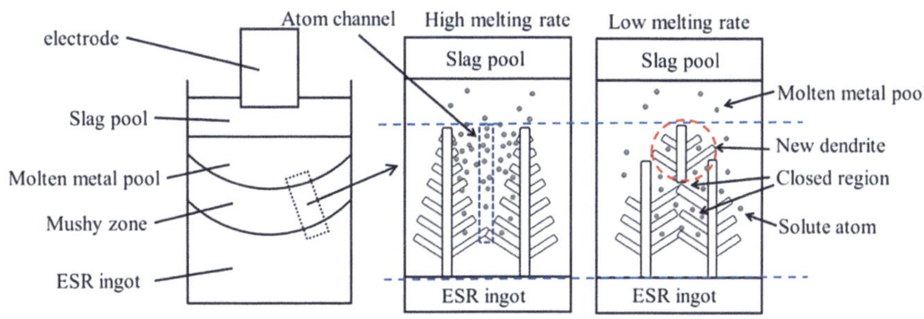

Fig. 2.5 Schematic view for the effect of melting rate on element segregation

It can be seen from Fig. 2.5 that when the melting rate of ESR is high, the liquid metal pool is deep. The heat transferred from the liquid metal pool to the solid-liquid two-phase region is more, and the solidification rate of liquid metal pool is slow. Dendrites nucleate at the solidified metal and grow towards the liquid metal pool gradually. The solute atoms form an atomic channel in the dendrite gap. With the growth of dendrites, solute atoms gradually enrich from the edge to the center of ESR ingot. When the melting rate of ESR is low, the liquid metal pool is shallow, the relative cooling intensity of solid-liquid two-phase region is relatively high, and the secondary dendrite grows fast, which is easy to produce dendrite sealing phenomenon and prevent solute atoms from further enriching to the center of ESR ingot. In addition, the cooling intensity will lead to dendrite nucleation in other parts of the solid-liquid two-phase region. At this time, the new dendrite will meet with the growing dendrite, which will also block the enrichment of solute atoms. After solidification of the closed region produced by the rapid growth of secondary dendrites, a solute rich micro region will be formed. As the dendrite closed area may appear in any part of the ESR ingot, the element enrichment micro area will also be randomly distributed in the ESR ingot, as shown in the yellow area in Fig. 2.4b.

The effect of the melting rate of ESR on the distribution of chromium in ESR ingot is shown in Fig. 2.6. For 8Cr13MoV steel, chromium is the most difficult element to segregate. Because chromium is not easy to segregate, the effect of ESR melting rate

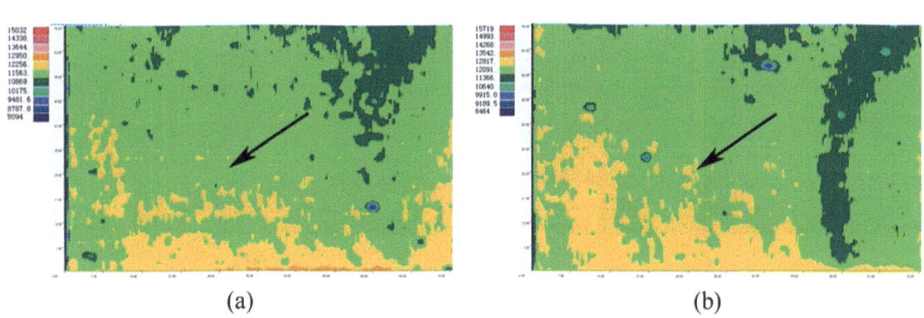

Fig. 2.6 OPA of chromium at different melting rate of ESR (arrow from ESR ingot edge to center): **a** 150 kg/h, **b** 133 kg/h

on the distribution of chromium is not obvious. As a whole, chromium is enriched in the center of ESR ingot. According to the principle of primary carbide precipitation and growth, the enrichment of chromium will cause the increase of primary carbide content in the center of ESR ingot.

The morphology and distribution of dendrites and primary carbides at 1/2 radius of ESR ingot are shown in Fig. 2.7. The white areas in Fig. 2.7c, d are primary carbides.

It can be seen from Fig. 2.7 that the size of secondary dendrite in ESR ingot decreases, the number increases, and the growth becomes fuller, and the dendrite gap decreases after decreasing the melting rate. By observing the morphology and distribution of primary carbides, it can be seen that the morphology of primary carbides is more fine and the network structure of primary carbides is reduced and the distribution is more uniform after reducing the melting rate. The primary carbides nucleate and grow along the dendrite gap. The main reason for the primary carbides to become thinner is the decrease of dendrite gap, and the reduction of primary carbide network structure also indicates the decrease of dendrite size.

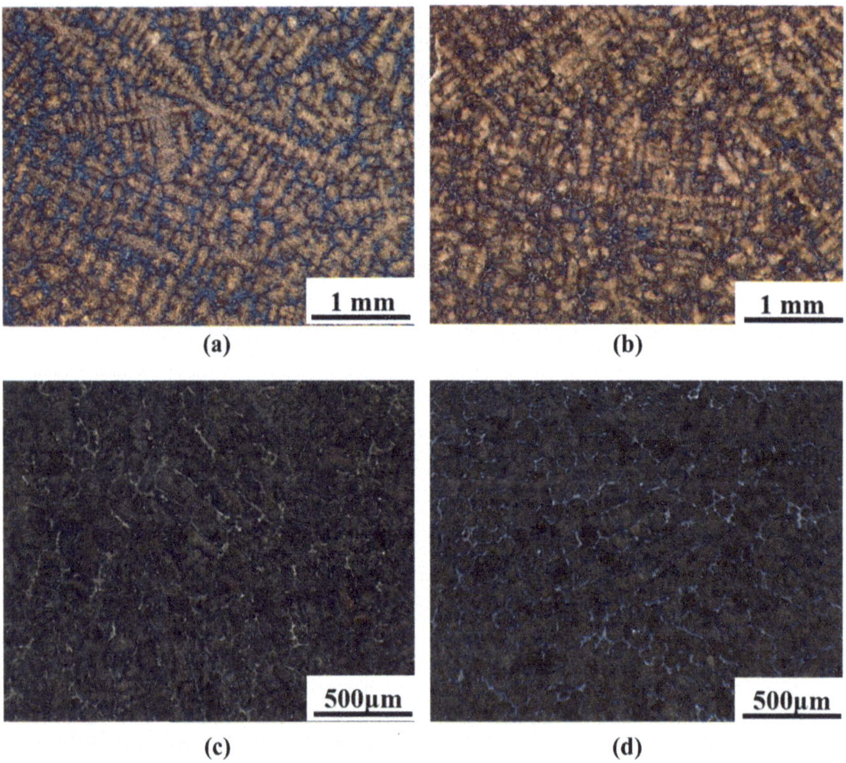

Fig. 2.7 Morphology and distribution of dendrites and primary carbides at 1/2 radius of ESR ingot: **a** and **c** 150 kg/h, **b** and **d** 133 kg/h

It can be seen from Figs. 2.2 and 2.4 that decreasing the melting rate of ESR can increase the solid fraction before eutectic reaction, reduce the dendrite gap and reduce the size of primary carbide. In addition, decreasing the melting rate of ESR can make the liquid metal pool shallower, improve the relative cooling intensity of the solid-liquid two-phase region, increase the dendrite nucleation rate and reduce the dendrite size. Therefore, the dendrite growth state is improved by reducing the melting rate of ESR, thus promoting the fine and uniform distribution of primary carbides.

The area fraction of primary carbide from the center to the edge of ESR ingot was counted by image analysis software under different ESR melting rates. The schematic view is shown in Fig. 2.8.

The statistical results of area fraction primary carbide are shown in Fig. 2.9.

It can be seen from Fig. 2.9 that at a fixed melting rate, the volume fraction of primary carbides decreased from the center to the edge of ESR ingot. The main reason is that the relative cooling intensity from the center to the edge gradually increases, the local solidification time of molten steel decreases gradually, and the element segregation at the edge of ESR ingot is light, which leads to the decrease of

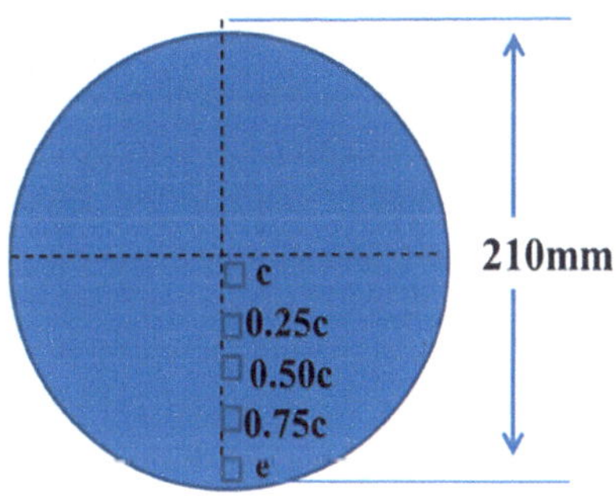

Fig. 2.8 Schematic view for determination of primary carbide area fraction in ESR ingot

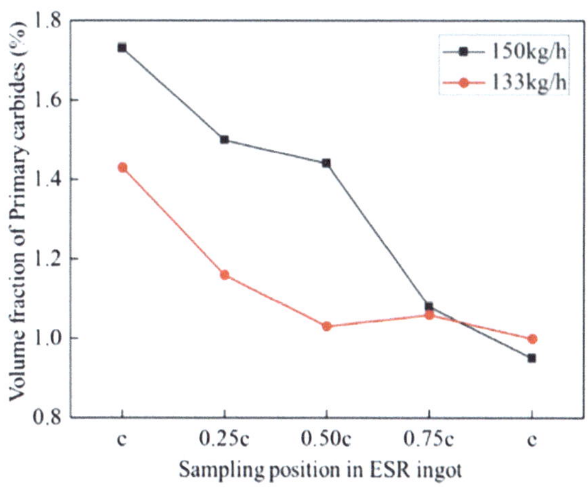

Fig. 2.9 Effect of melting rate on the volume fraction of primary carbides. *Note* 'c' represents center of the ingot, while 'e' represents edge of the ingot

primary carbide area fraction. By decreasing the melting rate, the volume of primary carbides decreased obviously at the positions of c, 2.25c, 0.50c in ESR ingot, and the volume fraction of primary carbides in the position of 0.75c and edge was not sensitive to melting rate. This is because that during the solidification process at the center, 0.25c and 0.50c of ESR ingot, the heat transfer mode mainly includes the heat dissipation from the bottom of ESR ingot and the heat supplied by the liquid metal pool. Reducing the melting rate can make the liquid metal pool shallower, reduce the heat input in the solid-liquid two-phase region, improve the cooling intensity of the solid-liquid two-phase region, reduce the local solidification time, reduce the element segregation, and make the total amount of primary carbide decrease. The ESR ingots near the edge of 0.75c and the edge are located at the upper edge of the liquid metal pool with the shape of parabola. The depth of the liquid metal pool has little influence on its heat transfer, and the edge is subjected to the forced water-cooling effect of the mold wall. Therefore, the element segregation at the edge is not affected by the depth of the liquid metal pool, and the primary carbide area fraction is generally low.

The average volume fraction of primary carbides at melting rates of 133 kg/h and 150 kg/h was 1.14% and 1.37%, respectively. As the melting rate decreased by 11%, the amount of primary carbides reduced by 16.8%, which demonstrated that reducing the melting rate in ESR process played an important role on reducing primary carbides.

Although reducing the melting rate can improve the quality of crystallization and reduce the precipitation of primary carbide, the melting rate should be controlled within a reasonable range in the actual electroslag remelting process. In general, reducing the electroslag remelting rate will reduce the depth of electrode embedded in the slag pool, lead to excessive current fluctuation and even open arc phenomenon, and make the electroslag remelting process unstable. In addition, after reducing the melting rate, the liquid metal pool will become shallow, which will weaken the secondary slag capacity of the liquid metal pool, lead to uneven thickness of slag skin, easy to reduce the surface quality of ESR ingots.

2.1.2 Effect of Fill Ratio in ESR Process on Primary Carbides in 8Cr13MoV Steel

As one of the important geometric parameters in electroslag remelting process, filling ratio affects the temperature field distribution and solute transport in slag pool and liquid metal pool to a certain extent [18]. Melflow-ESR software was used to simulate and calculate the influence of filling ratio on depth of liquid metal pool and width of two-phase zone, carbon segregation, heat conduction law in smelting process. The geometric parameters used in calculation process are shown in Tables 2.1 and 2.2. The corresponding filling ratios are 0.23, 0.33, 0.50 and 0.75 when the melting rate of ESR is 133 kg/h. The diameter of electrodes are 110 mm, 130 mm, 161 mm and 197 mm, and the mold diameter is 228 mm. The distribution of temperature field, solid-liquid two-phase region and slag pool flow field in ESR ingot with different filling ratios are shown in Fig. 2.10.

It can be seen from Fig. 2.10 that when the filling ratio increases from 0.23 to 0.33, the depth of liquid metal pool becomes shallower obviously. With the increase of filling ratio, the depth of liquid matel pool does not change much. With the increase of filling ratio, the depth of liquid metal pool and width of two-phase zone change as shown in Fig. 2.11.

It can be seen from Fig. 2.11 that when the filling ratio increases from 0.23 to 0.33, the depth of liquid metal pool decreases from 142 mm to 121 mm. When the filling ratio is greater than 0.33, the depth of the liquid metal pool becoming shallower with the continuous increase of the filling ratio. When the filling ratio is 0.75, the depth of liquid metal pool decreases to 115 mm. When the filling ratio is 0.33, the minimum

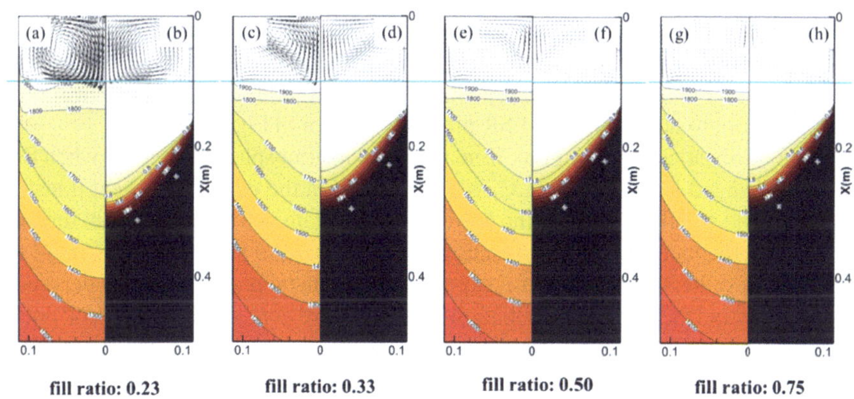

Fig. 2.10 Temperature field and mushy zone under different fill ratios: **a**, **c**, **e**, **g** were the temperature field at fill ratio of 0.23, 0.33, 0.50, 0.75, respectively. **b**, **d**, **f**, **h** were the mushy zone at fill ratio of 0.23, 0.33, 0.50, 0.75, respectively. *Note* the short black lines on the top of each image represent heat flux in slag pool

Fig. 2.11 Relationship between the fill ratio and the depth of liquid metal pool and the width of mushy zone

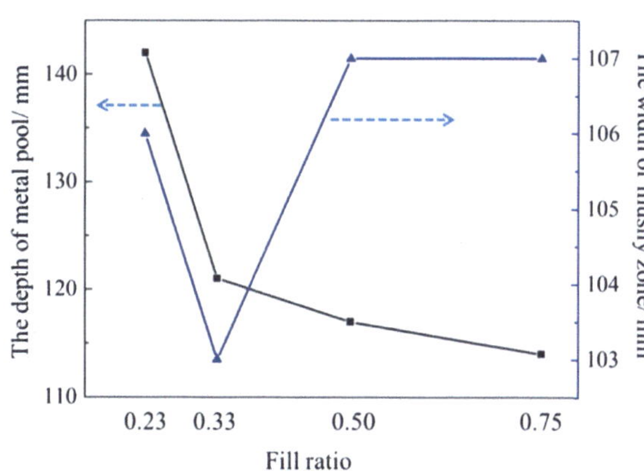

width of the solid-liquid two-phase zone is 103 mm. When the filling ratio increases from 0.23 to 0.75, the width of solid-liquid two-phase zone fluctuates from 103 mm to 107 mm, and the change is not obvious.

In the process of ESR, the increase of filling ratio weaken the skin effect of alternating current and make the temperature field of slag pool more uniform. In Fig. 2.10, the length of the black line in the slag pool above represents the rate of slag flow. It can be seen that with the increase of filling ratio, the flow rate of slag slows down, and the thermal convection in the slag pool weakens. The weakening of thermal convection in the slag pool will weaken the scouring effect of slag on the electrode end, which will make the temperature field distribution along the radius direction of the electrode end tend to be uniform, resulting in the shape of the electrode end being flat, as shown in Fig. 2.12. Compared with the end face of the inverted cone electrode, the flat end of the electrode can make the metal droplet enter the liquid metal pool more evenly, which makes the temperature field of the liquid metal pool more uniform, and finally obtain a shallow and flat liquid metal pool.

The heat resistance generated in slag pool mainly transferred to three directions: atmosphere, mold and liquid metal pool. By using different fill ratios, the heat flux from slag pool to atmosphere in ESR process is shown in Fig. 2.13.

It can be observed that heat flux from slag pool to atmosphere increased when the fill ratio changed from 0.23 to 0.33, which resulted in the decrease of heat transfer into the liquid metal pool. When the power supply system and cooling system remain unchanged, the increase of heat radiation from slag pool to atmosphere will lead to the decrease of heat transfer from slag pool to liquid metal pool. Due to the decrease of heat input and the uniform distribution of temperature field in the liquid metal pool, the liquid metal pool becomes shallower, and the liquid metal pool decreases from 142 mm to 121 mm, the decrease is obvious. When the filling ratio is greater than 0.33, with the increase of filling ratio, the heat radiation from slag pool to atmosphere decreases, and the heat transfer from slag pool to liquid metal pool increases, which leads to the deepening of liquid metal pool. However, with the increase of filling ratio, the temperature field becomes more uniform and the depth of liquid metal pool

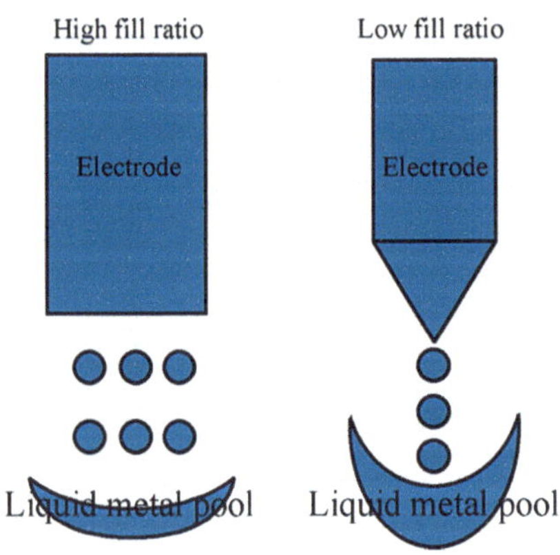

Fig. 2.12 Effect of filling ratio on the shape of electrode and liquid metal pool

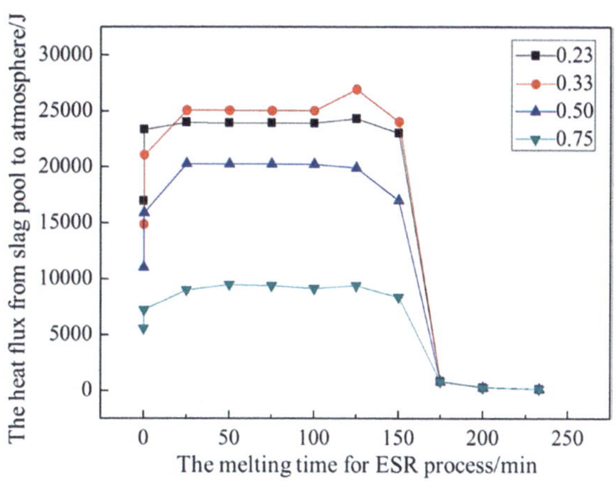

Fig. 2.13 Heat flux from slag pool to atmosphere with different fill ratio in ESR process

becomes shallower. Under the joint action of these two opposite factors, the depth of liquid metal pool decreases generally, but the decrease trend of liquid metal pool depth slows down due to the increase of heat transfer from slag pool to liquid metal pool.

Although the depth of liquid metal pool decreases with the increase of filling ratio, the width of solid-liquid two-phase zone decreases first and then increases. When the filling ratio is 0.33, the width of solid-liquid two-phase zone is the smallest. The heat in the solid-liquid two-phase region mainly comes from the liquid metal pool, and the decrease of the heat input from the liquid metal pool will directly lead to the decrease of the heat accepted by the solid-liquid two-phase region. Therefore, there is an obvious corresponding relationship between the effect of filling ratio on the width of solid-liquid two-phase zone and the influence of filling ratio on heat flow radiated from slag pool to atmosphere, as shown in Figs. 2.11 and 2.13. When the filling ratio is 0.33, the heat transfer from slag pool to liquid metal pool is the minimum, and the width of solid-liquid two-phase zone reaches the minimum.

The solid-liquid two-phase region is the final region of liquid steel solidification and the primary carbide formation area. The ideal ESR process is to make the molten steel solidify in a small range in order to obtain uniform and dense solidification structure and reduce the segregation of elements. The width of solid-liquid two-phase zone directly affects the degree of element segregation, thus affecting the precipitation and growth of primary carbides. The effect of the filling ratio of electroslag remelting on carbon segregation in ESR ingot is shown in Fig. 2.14. The white area in the figure is the area with the most serious carbon enrichment.

It can be seen from Fig. 2.14 that with the increase of filling ratio, the degree of carbon segregation in the center of ESR ingot gradually decreases. When the filling ratio increases from 0.23 to 0.33, the degree of carbon segregation and the depth of liquid metal pool decrease obviously. When the filling ratio continues to increase, there is little change between them.

In the actual production process, the upper limit diameter of electrode is restricted by the safety clearance between the electrode and the mold wall, that is, the gap

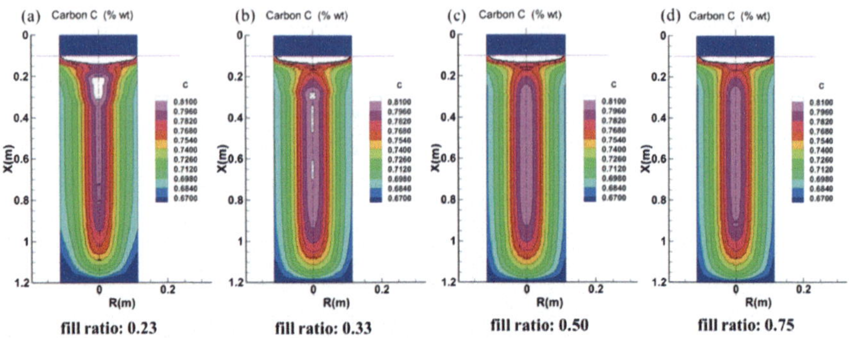

Fig. 2.14 Effect of fill ratio on the macrosegregation of carbon. **a**, **b**, **c**, **d** demonstrate the carbon distribution when the fill ratio is 0.23, 0.33, 0.50 and 0.75, respectively

between the two must be greater than the minimum safety distance. Generally, for 1–3t electroslag furnace, the safety value should be greater than 30–60 mm. When the distance between electrode and inner wall of mould is too small, the bypass current ratio from the side of slag pool to the mold wall increases, which increases the heat loss of slag pool and leads to the increase of power consumption. According to the simulation results, when the filling ratio increases from 0.33 to 0.50, the depth of liquid metal pool decreases by 4 mm, but the distance between the crystallizer and the electrode is 34 mm, which is very close to the limit of safety clearance. The probability of electrode collision caused by electrode wall collision increases greatly during electroslag remelting. Therefore, the best filling ratio of 8Cr13MoV steel is set to 0.33. At this time, the distance between the mold and the electrode is 49 mm, which can not only ensure the safety of production, but also obtain the ESR ingot with good solidification quality.

The effect of filling ratio of 0.23 and 0.33 on primary carbide in ESR ingot was studied. The schematic view is shown in Fig. 2.8.

The area fraction of primary carbides in ESR ingots with filling ratios of 0.23 and 0.33 were calculated by image analysis software. The results are shown in Fig. 2.15.

It can be seen from Fig. 2.15 that the area fraction of primary carbide decreases gradually from the center to the edge of ESR ingot under the two filling ratios. With the increase of filling ratio, the area fraction of primary carbide in ESR ingot generally decreases. When the filling ratio is increased from 0.23 to 0.33, the area fraction of primary carbide in ESR ingot is reduced from 1.14% to 1.05%, and the reduction ratio is 7.9%. Compared with the original ESR process, the area fraction of primary carbide is reduced by 23%. The decrease of primary carbide area fraction is attributed to the shallowing of liquid metal pool, the decrease of width of two-phase zone and the reduction of carbon segregation.

Based on the above research results, with the melting rate of ESR decreasing from 150 kg/h to 133 kg/h and the filling ratio increasing from 0.23 to 0.33, and the area fraction of primary carbide in ESR ingot decreased to 1.05%. Compared with the original ESR process, the area fraction of primary carbide decreased by 23%.

Fig. 2.15 Effect of fill ratio on the volume fraction of primary carbides

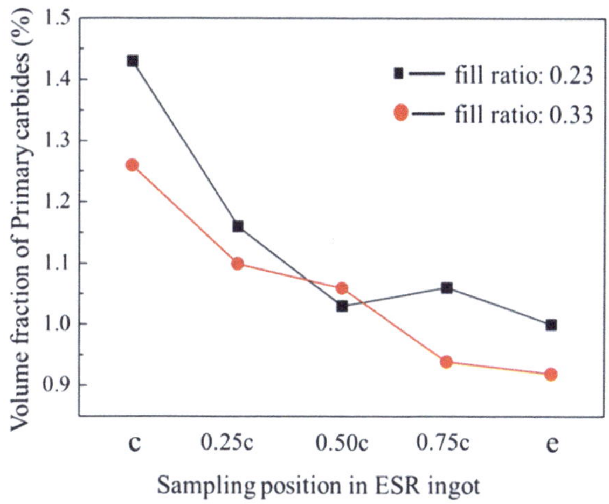

2.1.3 *Effect of Cooling Intensity of ESR on the Primary Carbides*

The cooling intensity has an important influence on the primary carbide precipitation during the solidification process of electroslag remelting steel. Reasonable cooling intensty can effectively avoid or reduce the crack, shrinkage cavity and element segregation of ESR ingot.

2.1.3.1 8Cr13MoV High Carbon Martensitic Stainless Steel

The effects of three cooling intensities of 600, 800 and 1000 L/h on carbide distribution in ESR ingot were studied. The distribution of carbide in ESR ingot with different cooling intensities is shown in Fig. 2.16.

It can be seen from Fig. 2.16 that the distribution of primary carbides in ESR ingots is uneven under low cooling intensity. The primary carbides are seriously aggregated

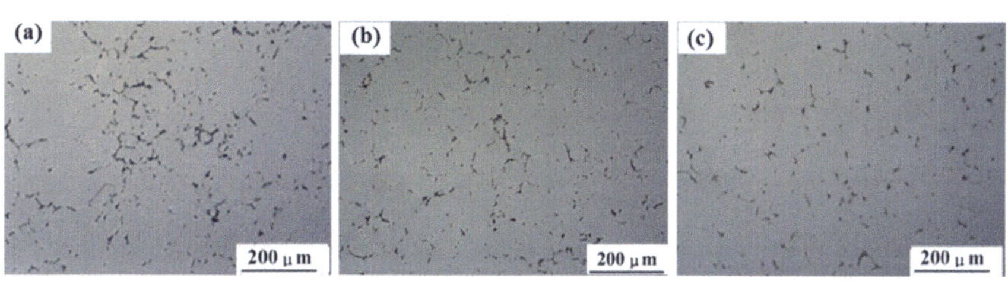

Fig. 2.16 Distribution of carbide in ESR ingot with different cooling intensities

Table 2.4 Basic and characteristic parameters of carbide

Relational expression	Unit	Parameter meaning
$V_v = A_A$	%	Volume fraction of carbides
$N_v = N_A/\overline{D}$		Amount of carbides per unit volume

in some places, and obvious network is formed between primary carbides. With the increase of cooling intensity, the distribution of carbides tends to be uniform.

The basic parameters of carbide include: volume fraction of carbide V_V, number of carbides in unit volume N_V, average diameter of carbide. The relationship between them is shown in Table 2.4. The basic parameters and structure characteristic parameters of quantitative metallography of carbides are shown in Table 2.5. Where N is the number of carbides in the field of vision, A is the field area, L is the field length, W is the field width, and t_0 is the carbide spacing.

As shown in Table 2.5, the metallographic parameters and microstructural characteristic parameters were calculated. Under lower cooling intensity, the carbides volume fraction V_V is larger. The average diameter $\bar{D}$ was larger, and the spacing of carbides was smaller when the number of carbides per unit volume was equal. With decreasing the cooling intensity, not only the amount but also the size of carbides increased obviously, and the crowding intensity of carbides increased.

The three-dimensional morphology of carbides in different ESR ingots is shown in Fig. 2.17.

Table 2.5 Basic parameters and Characteristic parameters of carbides

Sample No.	Basic parameters of quantitative metallography					Characteristic parameters of carbides		
	N	A, μm^2	$\bar{D}$, μm	L, μm	W, μm	V_V, %	N_V	t_0, μm
600 L/h	814	14786.10	18.17	1103.21	815.09	1.64	4.98 e^{-5}	27.18
800 L/h	567	9945.95	14.95	1103.21	815.09	1.11	4.22 e^{-5}	28.72
1000 L/h	665	8153.58	14.38	1103.21	815.09	0.91	5.06e^{-5}	27.02

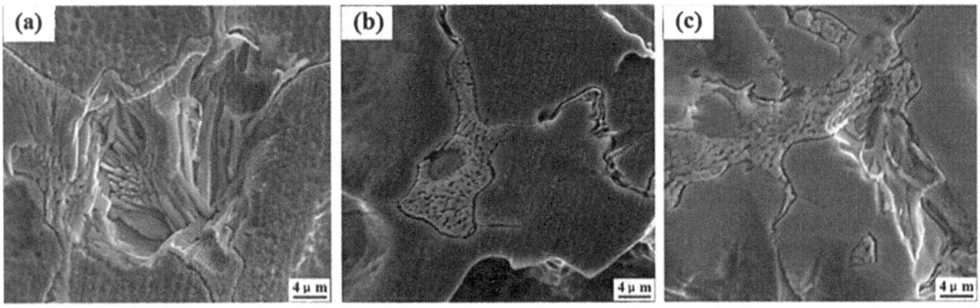

Fig. 2.17 Morphology of carbides under different cooling intensity: **a** 600 L/h; **b** 800 L/h; **c** 1000 L/h

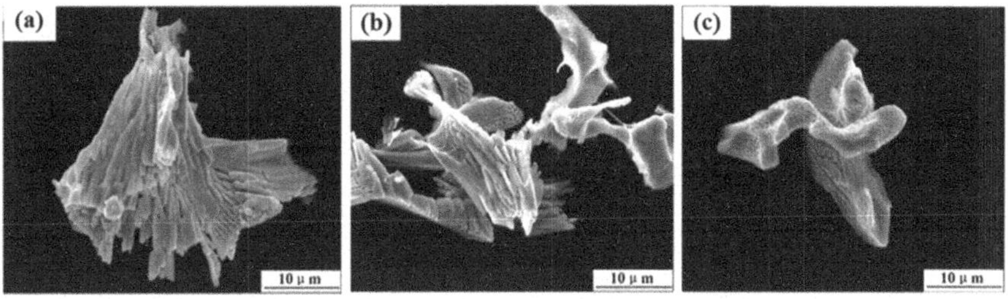

Fig. 2.18 Three-dimensional morphology of carbides under different cooling intensity: **a** 600 L/h; **b** 800 L/h; **c** 1000 L/h

It can be seen from Fig. 2.17 that the morphology of primary carbides is obviously different under different cooling intensities. At low cooling intensity, the primary carbides are coarse. The results show that the volume of carbide in ESR ingots with the cooling intensity of 600 L/h is obviously larger than that of carbide in ESR ingots with cooling intensity of 800 and 1000 L/h. Furthormore, the internal structure of carbides can be seen clearly after being eroded. The carbides in the three samples are eutectic structure. The carbides in ESR ingots with the cooling intensity of 600 L/h are in the form of skeleton, which are formed by many long carbides connected with each other. The skeleton carbides in ESR ingots with the cooling intensity of 800 and 1000 L/h are composed of many short rods or grains.

The three-dimensional morphology of carbides observed by SEM is shown in Fig. 2.18.

As can be seen from Fig. 2.18, most of the carbide in the ESR ingots with low cooling intensity are angular, and some of them have smooth contours. The skeleton-like carbides in the ESR ingots with cooling intensity of 600 L/h are composed of many long strips of small carbides connected to each other. The size of carbide in the ESR ingot with cooling intensity of 800 L/h is relatively small. A few carbide structures are similar to the former, but most of them are the same as those in the ESR ingot with cooling intensity of 1000 L/h. Most of the carbides in ESR ingots with cooling intensity of 1000 L/h are smooth in outline without obvious edges and corners. The carbides are smaller in size and have a granular and short rod-like structure inside. The change of carbide morphology was caused by the increase of cooling intensity, grain refinement and segregation reduction. 8Cr13MoV belongs to hypereutectoid steel. Primary austenite precipitates from liquid phase during solidification. Due to the composition segregation, when the carbon content at the solidification front reaches about 1% in the liquid phase, liquid phase composition has conforms to the eutectic alloy. The austenite precipitates continuously in the liquid phase. When the liquid composition reaches the eutectic point, eutectic reaction ($L \rightarrow M_7C_3 + \gamma$) takes place and carbide precipitates.

The schematic view for the growth of primary carbides is shown in Fig. 2.19. The shadow area around dendritic crystal represents solute enrichment area. The grain size of ESR ingot with cooling intensity of 600 L/h is large and the den-drite spacing is wide. When the composition of residual liquid steel reaches eutectic point,

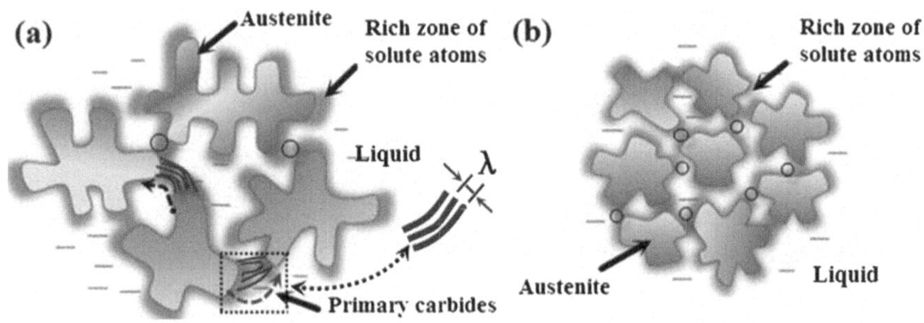

Fig. 2.19 Schematic view for the growth of primary carbides: **a** 600 L/h; **b** 800 L/h (λ: the distance between eutectic structure of carbides)

primary carbides will precipitate along pre-existing austenite. The concentration of solute elements in the whole liquid phase can not be completely consistent. Due to the concentration gradient, the eutectic growth direction of the first nucleation will grow from the edge of one grain to the edge of one or more adjacent grains, as shown by the dotted arrow in Fig. 2.19a, which is the reason for the directional growth of carbides in ESR ingot with cooling intensity of 600 L/h. However, in some regions, the concentration gradient of solute is small, so the eutectic growth has no fixed direction, as shown in the circle position in Fig. 2.19b. The results show that the cooling intensity of 1000 L/h is high, the grain size of ESR ingot is small and the spacing is short. The concentration gradient of solute is small, and the eutectic growth will lose its fixed direction due to more grain boundaries.

In addition to the overall appearance of carbides, the com-pact degree of eutectic structure was different in ESR ingots produced under different cooling intensity. When the eutectic reaction proceeded, solute atoms (C, Cr and Mo) were dis-charged from liquid steel and formed another phase, i.e., M_7C_3-type carbides. Therefore, the radial diffusion along sol-id-fluid interface perpendicular to the primary carbides will become dominant and effectively decrease the solute concen-tration on the solidification front. This radial diffusion caused the decrease in the distance (λ) between eutectic structures of carbides. According to the result $\lambda = 7.94 \times R^{-0.19}$, λ decreased with increasing cooling intensity obviously. With increasing cooling intensity, the segregation of solute atoms was decreased, consequently the solute concentration in remaining liquid steel decreased. This effect was similar with radial diffusion of solute atoms. Therefore, with increasing cooling intensity, λ decreases and the eutectic structure of carbides becomes more compact.

2.1.3.2 H13 Hot Working Die Steel

Because that the content of alloying elements in H13 steel can reach about 8%, and it belongs to hypereutectoid steel, the carbon and some alloy elements, such as V, Mo, Cr, in the grain boundary or dendrite region, lead to the enrichment and

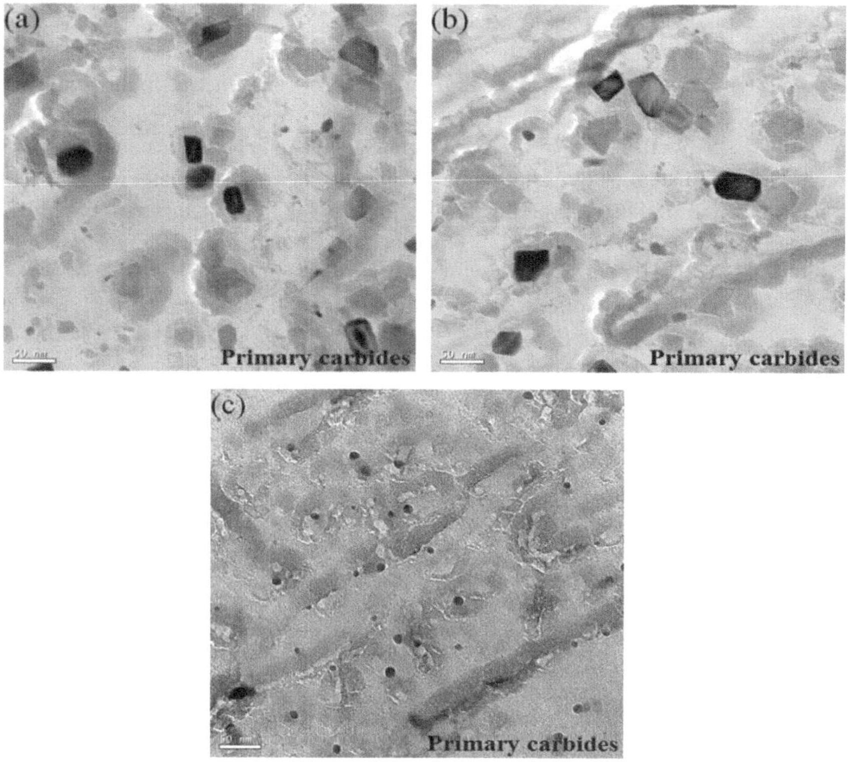

Fig. 2.20 TEM photographs of carbides in ESR ingots produced with different cooling intensities: **a** 600 L/h; **b** 800 L/h; **c** 1000 L/h

precipitation of carbides. The TEM images of carbide in ESR ingot under different cooling intensities of 400, 800 and 1200 L/h are shown in Fig. 2.20. It can be seen from Fig. 2.20a that when the cooling intensity is low, a large amount of carbides precipitate in the steel. The carbides are mostly plate-shaped and strip-shaped, and the carbides are mostly aggregated and precipitated. This shows that the segregation of carbon and alloy elements is serious and the carbides formed are coarse in low cooling system. With the increase of cooling intensity, the segregation of carbides is obviously improved and the distribution is more uniform, and the size of carbides is also reduced, as shown in Fig. 2.20c. Therefore, increasing the cooling intensuty can reduce the carbon segregation and promote the uniform distribution of carbides in the steel. This is because that when the cooling intensity increases, the supercooling degree of molten steel will also increase. During the solidification process, the carbides are distributed regularly along the cooling direction and have directionality.

It is found that V_8C_7 and Fe_3Mo_3C are the main precipitated carbides in the ESR ingot of H13 steel. The carbides precipitated after annealing are square and spherical VC, Mo_6C and $Cr_{23}C_6$. That is to say, a large amount of $Cr_{23}C_6$ secondary carbides will be precipitated after annealing, and most of the primary carbides will be enriched and separated out in ESR ingot. The carbides in the network segregation zone of the metallographic sample of ESR ingot were observed by SEM. It was found that most of the carbides precipitated along the grain boundary, most of which were square and strip-shaped with rich V in the size about 5 μm. Some carbides containing Mo precipitated with smaller size and quantity. The carbides rich in V and Mo can also precipitate with the size of about 8 μm. As shown in the EDS spectrum in Fig. 2.21, each carbide contains three alloying elements: V, Cr and Mo. Different types of carbides have different mass fractions of alloying elements.

The SEM-EDS results of carbide in ESR ingots with different cooling systems after electrolysis are shown in Fig. 2.22.

The carbides in Fig. 2.22a are strip-shaped carbides rich in V and Mo and Cr. The carbides in Fig. 2.22b are spherical Cr-rich carbides. The carbides in Fig. 2.22c are granular Mo rich carbides. The size of V-rich carbides is larger than that of Cr and Mo rich carbides.

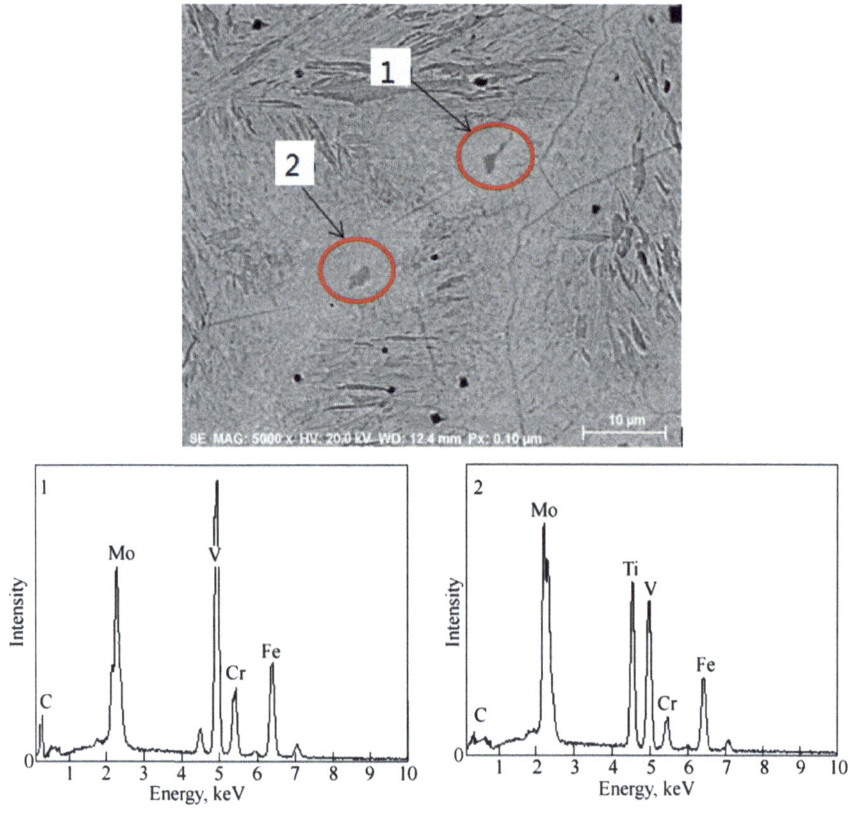

Fig. 2.21 SEM-EDS results of carbide in ESR ingot

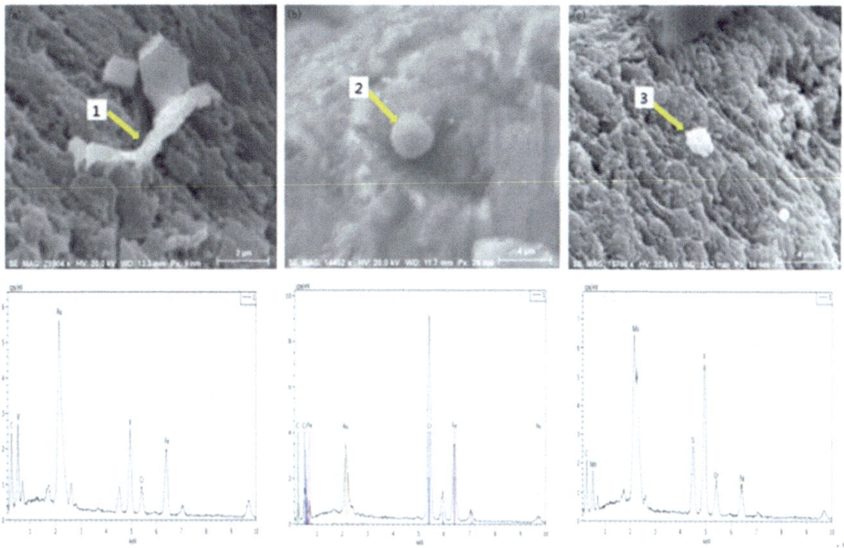

Fig. 2.22 SEM-EDS results of different types of carbides

The distribution of carbides with different cooling intensity is shown in Fig. 2.23. It can be seen from Fig. 2.23 that the carbide size is mostly distributed between 1–2 μm. Most of the carbide size in the center is above 0.5 μm, and a small amount of carbide size at the edge is below 0.5 μm. It is due to the small temperature gradient in the center of electroslag ingot, which affects the precipitation of carbide. With the increase of cooling intensity, the amount of carbides in the center increases about

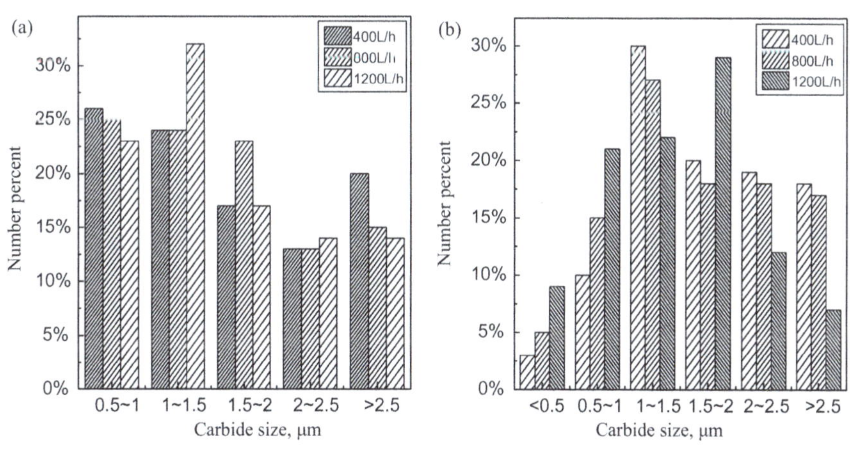

Fig. 2.23 Carbide size distribution in H13 steel: **a** Carbide size statistics at the center of ESR ingot; **b** Carbide size statistics at the edge of ESR ingot

1 μm, while the number of carbides at the edge of 0.5–1 μm increases obviously, and the size of carbides decreases.

The results show that increasing the cooling intensity can shorten the local solidification time, prevent the growth of secondary dendrites, refine the grains and reduce the segregation of elements. It is helpful to reduce the quantity and size of primary carbides in steel and make their distribution more uniform.

2.2 Effect of Continunous Directional Solidification of Electroslag Remelting on the Carbide Segregation

The goal of ESR ingot structure is controlled to obtain fine columnar crystal structure. However, due to the limitation of solidification conditions in the electroslag remelting process, there is inevitably a certain amount of equiaxed crystal in the center, which leads to serious segregation at the center axis of ESR ingot. In order to improve the behavior of central segregation, it is necessary to apply directional solidification electroslag remelting process. The primary dendrite spacing, secondary dendrite spacing, columnar crystal growth direction and dendrite growth morphology are important factors affecting carbon segregation in ESR ingots. The morphology of dendrites in ESR ingots, the distance between dendrites and the growth direction of dendrites are controlled by the temperature gradient, concentration gradient, cooling rate and growth rate of dendrites at the front of solid-liquid interface during solidification of ESR ingots [19, 20].

2.2.1 Effect of Directional Solidification of Electroslag Remelting on the Dendrite Arm Spacing

The comparison of ESR ingot between directional solidification electroslag remelting process and traditional electroslag process is shown in Fig. 2.24.

Figure 2.24a shows the current single loop fixed mold used in the traditional electroslag remelting process, while Fig. 2.24b shows the continuous directional solidification electroslag remelting process adopts the double loop ingot extraction conductive mold, and the ingot is forced to cool by secondary water spray after being extracted from the lower end of the mold. The principle of continuous directional solidification electroslag remelting process is to control the shape of liquid metal pool, increase the axial temperature gradient and reduce the degree of solute segregation in the process of solidification by changing the solidification conditions of traditional electroslag remelting.

The microstructure morphology of cross section and longitudinal section of the ingot remelted through ESR and ESR-CDS process are presented in Figs. 2.25 and 2.26, respectively.

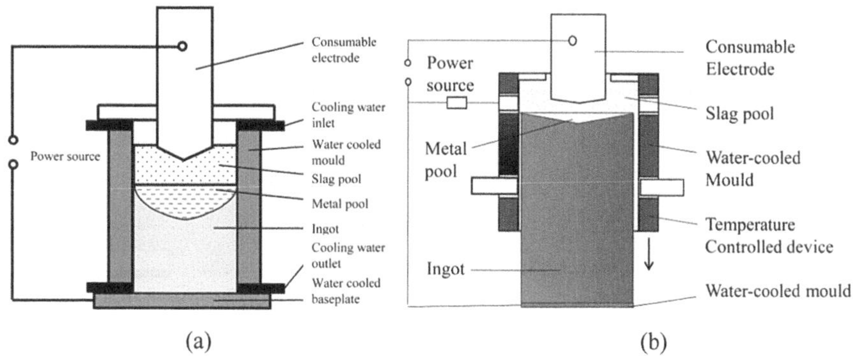

Fig. 2.24 Schematic diagrams of two types of electroslag remelting: **a** ESR; **b** ESR-CDS

Figures 2.25a and 2.26a shows that the morphology of dendrite in ingot remelted through traditional ESR process, while Figs. 2.25b and 2.26b exhibit fine microstructure what influenced by directional solidification remelted through ESR-CDS process. Based on the comparison of cross-section dendrite structure between the Fig. 2.25a, b, it shows that the dendrite in the center of ESR ingot produced by traditional electroslag process completely grows into coarse dendrite, and there are a large number of equiaxed and columnar crystals alternately distributed; while the dendrite in the ESR ingot produced by directional solidification electroslag process distributes uniformly and finely from the edge to the center, and the morphology and size of dendrite in the 1/2 radius and the center are similar. Based on the comparison the longitudinal section dendrite structure between Fig. 2.26a, b, it shows that the dendrite structure in the ESR ingot produced by the traditional electroslag process presents fine columnar crystal from the edge of the ingot to 1/2 radius, and gradually transforms into coarse columnar crystal and part of equiaxed crystal, until the final center is chaotic equiaxed crystal structure, as shown in (a-1) (a-2) (a-3) in Fig. 2.26a. However, the dendrite structure in the ESR ingot produced by continuous directional solidification process is columnar structure with uniform distribution from the edge to the center of the ingot, and the morphology of dendrite structure at 1/2 radius is similar to that at the center, as shown in (b-1) (b-2) (b-3) in Fig. 2.26b. Therefore, there are columnar grains and equiaxed grains in the center of ESR ingot produced by traditional electroslag remelting process. It is easy to form element segregation in the axial center of ESR ingot, resulting in the precipitation of coarse primary carbide and eutectic carbide, as well as the accumulation of inclusions. Based on the above analysis, the dendrite morphology of ESR ingots produced by continuous directional solidification is columnar crystal which grows approximately parallel to the axial direction of the ingot, and the dendrite morphology from 1/2 radius to the center is similar, and the dendrite spacing is similar. However, the dendrite morphology of ESR ingot produced by traditional electroslag process gradually changes from columnar crystal to mixed crystal morphology of columnar crystal and equiaxed crystal from edge to center, and the dendrite growth direction is chaotic.

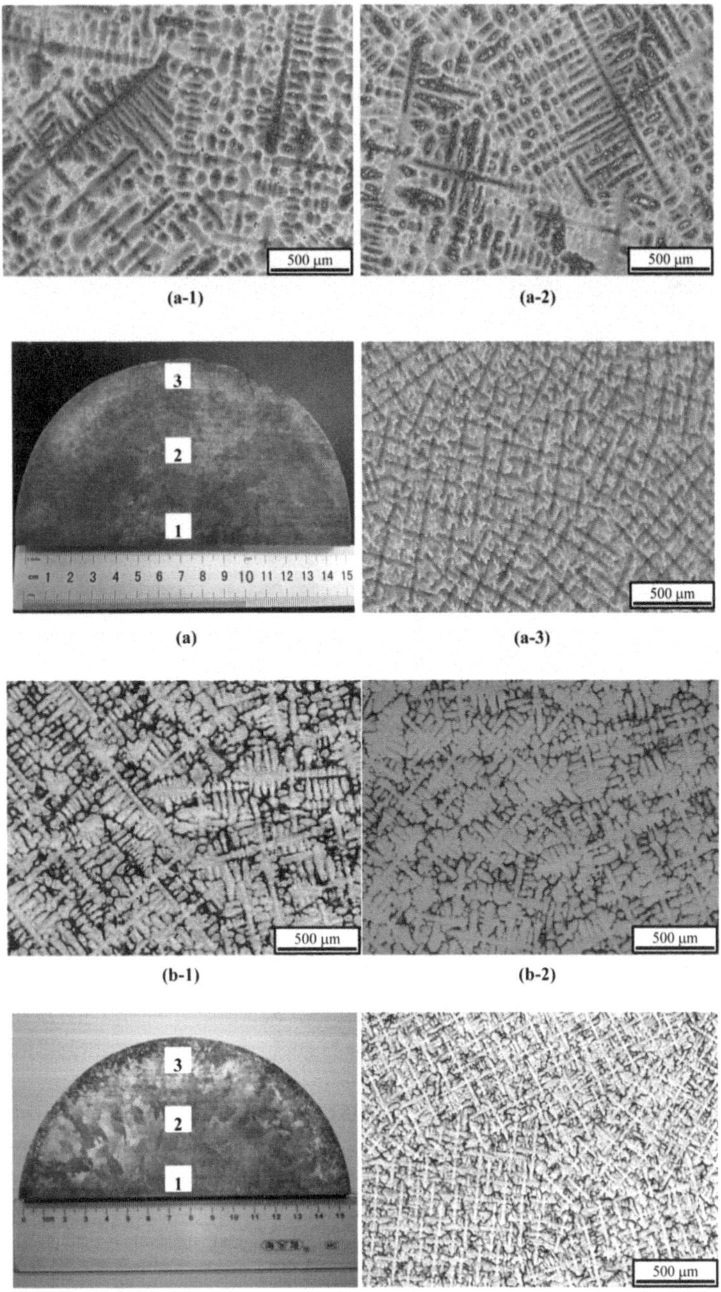

Fig. 2.25 Microstructure in the transversal section of ingots remelted through different processes: **a** ESR; **b** ESR-CDS

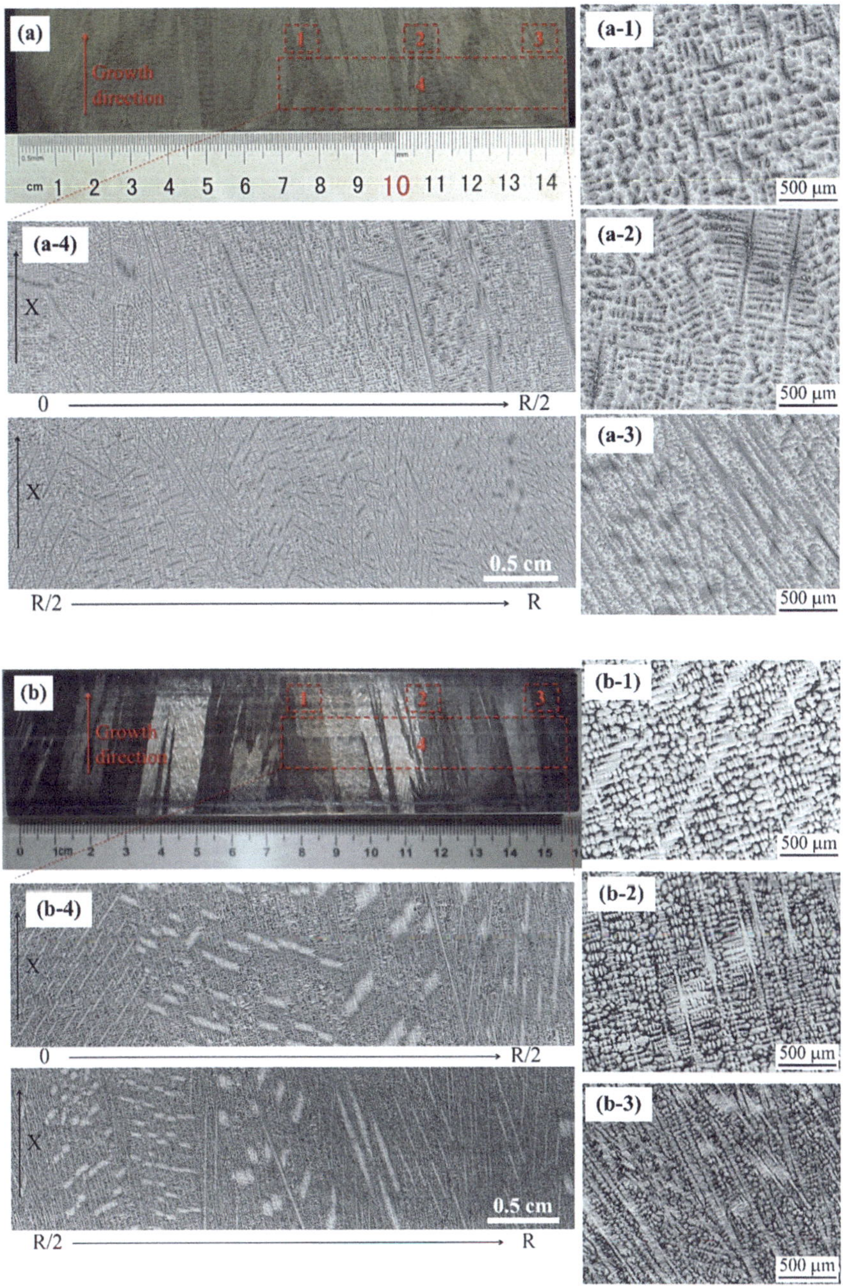

Fig. 2.26 Microstructure in the longitudinal section of ingots remelted through different processes: **a** ESR; **b** ESR-CDS

The growth behavior of dendrite in ESR ingot is mainly affected by the shape of liquid metal pool, the depth of liquid metal pool, ESR melting rate, local solidification time and temperature gradient distribution. Based on the above analysis, the two electroslag processes are controlled by the same electroslag remelting rate, so the effect of remelting rate on dendrite growth can be ignored. Previous studies [21, 22] have shown that the shape of shallow liquid metal pool can be obtained by mold rotation or consumable electrode rotation, which can improve the quality of ESR ingot. The condition of dendrite growth along the axial direction of ingot can be achieved by controlling the shallow liquid metal pool in the electroslag process [23].

The solidification process of ESR is simulated. The parameters of ESR process and the physical parameters of metal and slag required for simulation are shown in Tables 2.6 and 2.7. In the remelting process, the remelting rates of ESR and ESR-CDS were controlled to similar by adjusting the current intensity.

The primary dendrite spacing and secondary dendrite spacing in traditional ESR and continuous directional solidification ESR ingots are shown in Figs. 2.27 and 2.28. It can be seen from Figs. 2.27 and 2.28 that compared with the dendrite size in traditional ESR ingot, the primary dendrite spacing and secondary dendrite spacing in continuous directional solidification ESR ingot are smaller.

The primary dendrite arm spacing, secondary dendrite arm spacing and growth direction of columnar grain were important parameters to characterize the quality of

Table 2.6 Process parameters in the process of remelting hot-work die steels with ESR and ESR-CDS

Process parameters	ESR	ESR-CDS
Diameter of mold, mm	160	160
Diameter of electrode, mm	120	120
Voltage, V	38	38
Current, A	2400	2100
Weight of slag, kg	6	5
Melting rate, kg/min	1.65	1.6

Table 2.7 Thermo-physical properties for the metal and slag

Physical properties	Metal	Slag
Liquid density, kg/m^3	7500	2626
Solid density, kg/m^3	8146	2790
Liquidus temperature, K	1653	1723
Solidus temperature, K	1503	1618
Liquid vol. thermal exp. Coeff., K^{-1}	1.5×10^{-4}	9.0×10^{-5}
Liquid thermal conductivity, W/(m·K)	30.52	0.5
Solid thermal conductivity, W/(m·K)	16.72 (773 K)	7.8 (773 K)
Electrical conductivity, $\Omega^{-1} \cdot m^{-1}$	7.6×10^5	239

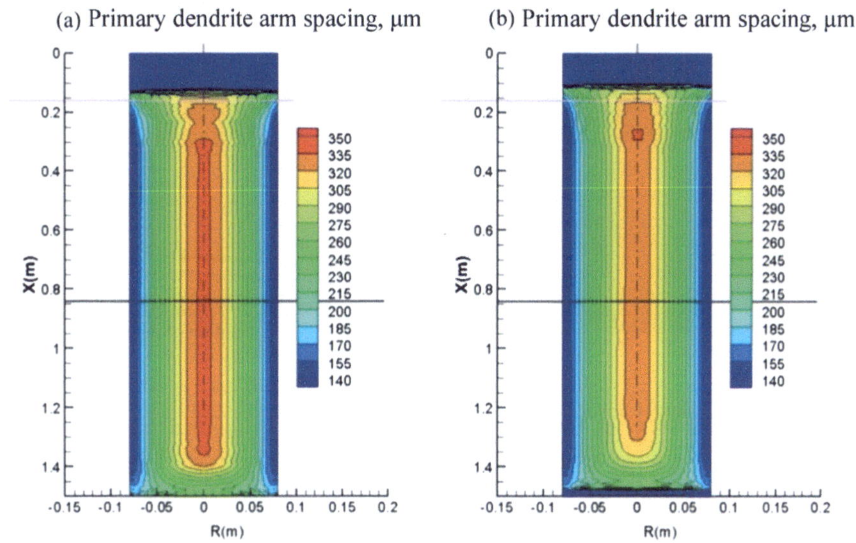

Fig. 2.27 Primary dendrite arm spacing simulated by MeltFlow-ESR: **a** ESR; **b** ESR-CDS

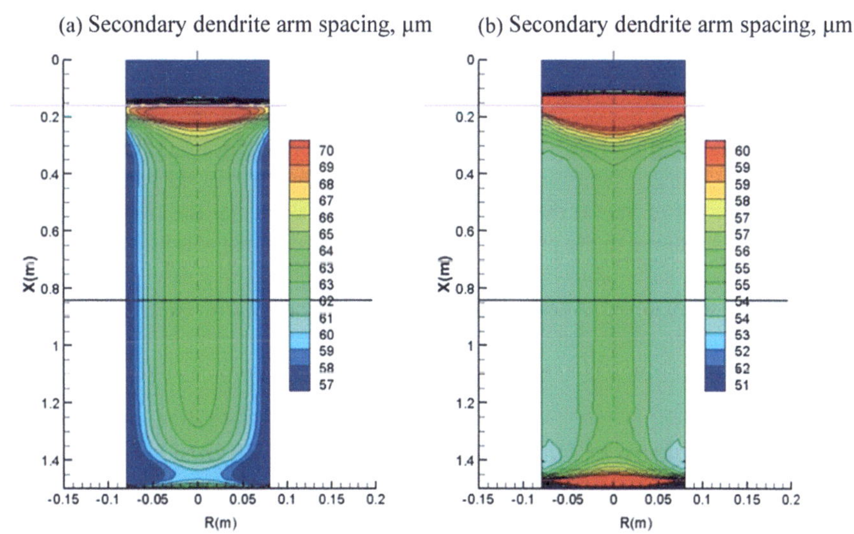

Fig. 2.28 Secondary dendrite arm spacing simulated by MeltFlow-ESR: **a** ESR; **b** ESR-CDS

electroslag remelting process. Moreover, dendrite spacing could be used to characterize micro-segregation in the ingot. Thus, the primary dendrite arm spacings (d_I) and secondary dendrite arm spacing (d_{II}) in longitudinal section of ingot were used to evaluate the microstructure in this study and conducted in a statistical averaging

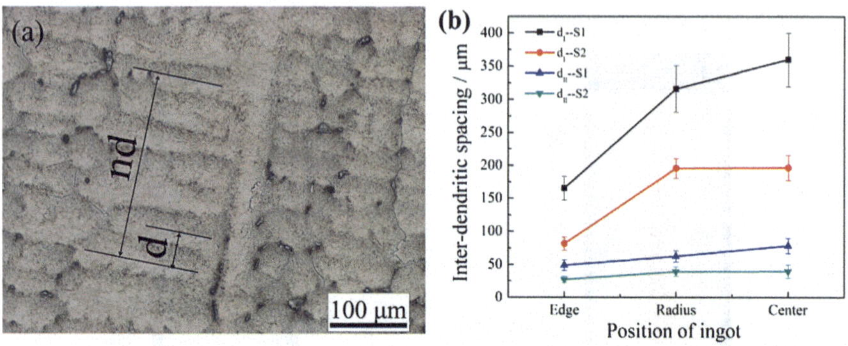

Fig. 2.29 **a** Schematic view of dendritc arm spacing measuring method and **b** dendrite arm spacing at different positions of S1 and S2

way to prevent randomness. The schematic view of dendrite arm spacing measuring method and correspondingly statistical results are shown in Fig. 2.29.

The statistical results of dendrite spacing in Fig. 2.29b show that continuous directional ESR process can reduce the primary dendrite spacing from 360 μm to 197 μm and the secondary dendrite spacing from 78 μm to 40 μm in the core of traditional ESR ingot S1. At the same time, the primary dendrite spacing and secondary dendrite spacing of traditional ESR ingot S1 increase from the edge of ingot to the core of ingot, while the dendrite spacing of continuous directional solidification ESR ingot S2 is consistent from 1/2 radius of ingot to the core of ingot, which indicates that continuous directional solidification ESR Process can obtain denser and more uniform dendrite structure ingot. At the same time, the dendrite structure at the 1/2 radius of ESR ingot S2 is similar to that at the core, which also indicates that the shape of liquid metal pool is shallower in the process of continuous directional solidification. There is a quantitative relationship between dendrite spacing and macrosegregation and interdendritic segregation. The smaller the dendrite spacing, the more uniform the structure and composition distribution.

2.2.2 Morphology of Dendrite Growth in ESR Ingots

It can be seen from Figs. 2.25 and 2.26, there are quite differences in dendrite morphology ingots smelted by different ESR processes. The growth morphology of dendrite is mainly affected by the growth rate of dendrite, the temperature gradient distribution at the front of liquid-solid interface and the concentration of solute atoms in the liquid phase at the front of liquid-solid interface. The distribution of supercooling at the front of liquid-solid interface during solidification is shown in Fig. 2.30.

It can be seen from Fig. 2.30 that during the solidification process, as the liquid-solid interface continues to move forward, solute atoms continue to accumulate in the

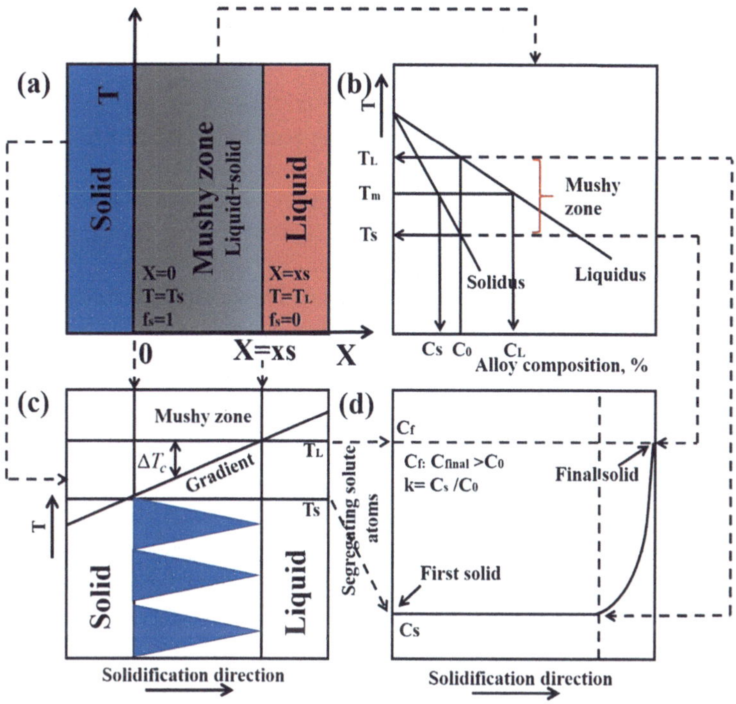

Fig. 2.30 Schematic diagram of solidification conditions distribution at the front of liquid-solid interface during solidification: **a** Boundary conditions of solidification; **b** Element segregation behavior in two-phase equilibrium phase diagram; **c** Temperature gradient and component supercooling at the front edge of liquid-solid interface; **d** Concentration distribution of segregation elements after solidification

front of the liquid-solid interface, resulting in the increase of solute concentration at the end of solidification, meanwhile, resulting in the continuous increase of component supercooling. The undercooling not only can be used as the driving force of equiaxed crystal nucleation, but also the driving forece of columnar crystal growth. When the proportion of equiaxed grains in the front of liquid-solid interface is enough to block the growth of columnar crystal tip, the nucleation and growth of equiaxed grains are dominant. Due to the limitation of heat transfer direction in the center of traditional ESR ingot, the heat transfer in the center is slow and non-directional, which inhibits the further advance of columnar crystal tip and creates conditions for the nucleation and growth of equiaxed crystal. The temperature gradient and the transfer direction in the front of the liquid-solid interface of the columnar crystal growth can be controlled by the directional solidification, and these make the ESR ingot obtain columnar crystal structure growing along the axial direction.

2.2.3 *Effect of Directional Solidification of Electroslag Remelting on the Carbon Segregation*

Continuous directional solidification electroslag remelting technology [24–27] can control the temperature gradient at the front of solid-liquid interface, local solidification time and cooling rate during solidification, which can obtained a shallow and flat liquid metal pool, columnar crystal structure with approximate axial growth. Furthermore, the central solidification paste zone can be eliminated, and the element segregation degree in ESR ingot can be improved. The amount, size and distribution of precipitates between dendrites can be controlled, and the hot working performance of alloy ingot can be improved.

Within the process, the liquid metal pool is solidified directionally that results in uniform and relatively fine dendritic structure. The dendritic structure is mostly influenced by the cooling rate in the process, temperature gradient ahead of the crystallization front, and the intensity of interdendritic flow. In order to investigate the solidified conditions in ESR and ESR-CDS process, the numerical simulation calculated by Melt-Flow software was applied to analyze the temperature fields and velocity fields. The pool profile and distribution of temperature field in ESR and ESR-CDS process are shown in Figs. 2.31 and 2.32.

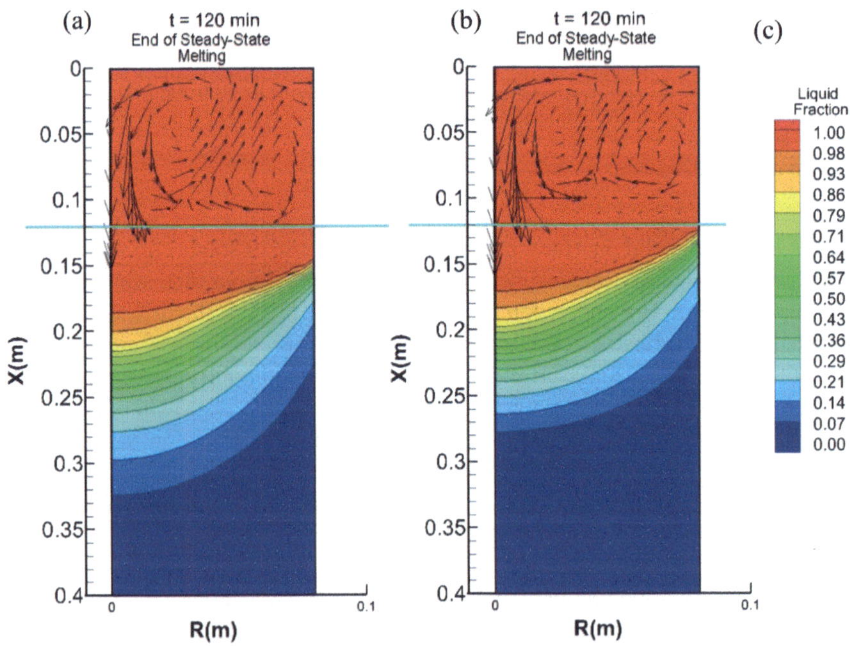

Fig. 2.31 Pool profile in different remelting process: **a** ESR; **b** ESR-CDS; **c** Liquid phase ratio scale

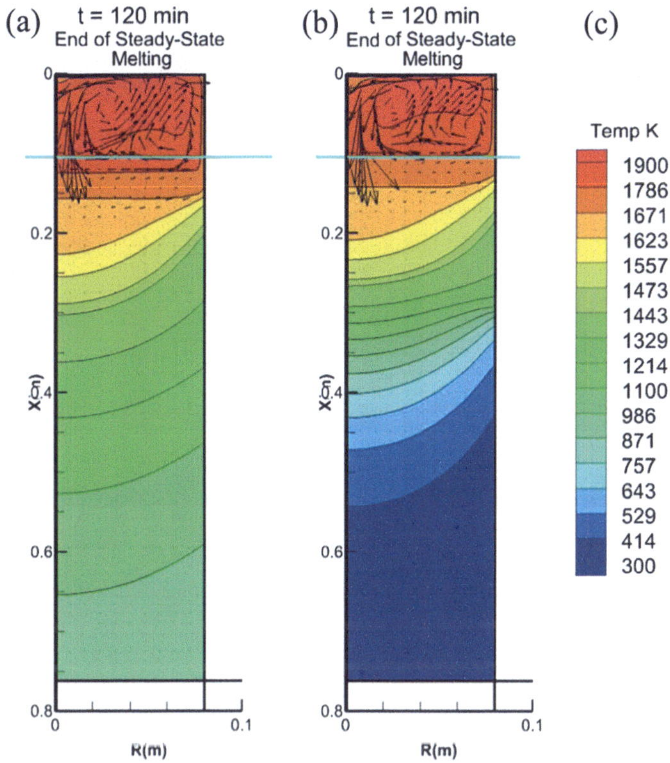

Fig. 2.32 Temperature field in different remelting process: **a** ESR; **b** ESR-CDS; **c** Temperature ratio scale

The pool profile in different remelting process are shown in Fig. 2.31. The mushy zone and solidified ingots could be clearly recognized. The liquidus and solidus maximum depth are 67 and 250 mm individually in ESR process. However, the maximum depth of solidified pool in ESR-CDS process are 50 and 180 mm, which decreases by 25.4 and 28% compared with conditions in ESR process, respectively. The two-phase zone crystallized from liquid to solid becomes narrower in ESR-CDS process compared to ESR process. It is obviously that the shape of liquid metal pool profile would be shallow and flat remelting through ESR-CDS process. In conclusion, shallow and flat liquid metal pool could be obtained during the ESR-CDS process by controlling the direction of thermal flow, the temperature gradient of solidification front and the solidification rate.

The distribution of temperature field in ESR and ESR-CDS process is displayed in Fig. 2.32 The temperature gradient of ingot in ESR-CDS process is evidently larger than that in ESR process. This difference indicated that spraying secondary aerosol cooling water in ESR-CDS process could improve intensity and change direction of heat transformation, thus controlling the growth direction of columnar grains to parallel to the axis of ingot and refining the microstructure. Moreover, precipitated

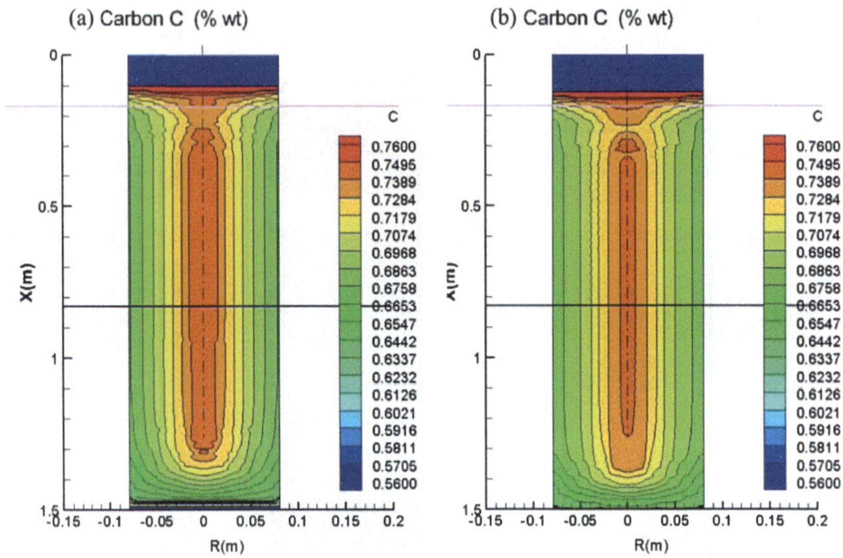

Fig. 2.33 Carbon segregation degree: **a** Traditional ESR; **b** Continuous directional solidification ESR

carbides along grain boundaries could be always reduced by smaller segregation degree of solute atoms in inter-dendrites. In conclusion, microstructure in terms of columnar grains and precipitated carbides could be refined remelting through ESR-CDS process.

The comparison of carbon segregation degree between traditional ESR and continuous directional solidification ESR ingots is shown in Fig. 2.33. It is obvious from Fig. 2.33 that the continuous directional solidification electroslag remelting process reduces the degree of carbon segregation in the ingot. The main reason is that the growth morphology of dendrite in continuous directional solidification ESR ingot affects the distribution of carbon in ESR ingot and reduces the segregation degree of solute element carbon.

The element segregation in ESR ingot is mainly divided into macro segregation and micro segregation in dendrite gap. In the solidification process, due to the different growth direction of dendrite, solute atoms continuously gather in the liquid-solid interface. With the solidification, the solute atoms are constantly pushed to the final solidification area, thus forming the macro segregation phenomenon of ESR ingot. At the same time, the element segregation also exists between the dendrite stem and the dendrite gap, that is, the solute atoms are poor in the pre solidified dendrite stem and enriched in the dendrite gap at the end of solidification, thus forming the micro segregation phenomenon in the dendrite gap of ESR ingot. The effect of directional solidification electroslag remelting process on the segregation behavior of elements was analyzed by in original position analysis (OPA) and electron probe microanalysis (EPMA).

(1) Macro segregation

The 2D contour maps of carbides formed elements distribution scanned by OPA in the same transverse section of ingots were shown from Figs. 2.34, 2.35, 2.36 and 2.37.

The degree of the carbon contents in ingots was represented by different colors, which increases with the transition of colors change from cold to warm. It can be seen by comparing Fig. 2.34a, b that more uniform distribution of carbon and less segregation were achieved in the case of the ingot E-2 subjected to the ESR-CDS process. It is well known that the elements Cr, Mo and V are strong former of carbides, and the segregation of these elements are beneficial to the formation of carbides during solidification. Thus, to control the characteristic (i.e. amount, size, morphology and distribution) of carbides precipitated from residual liquid during solidification, the segregation of carbon and carbides formed elements should be alleviated.

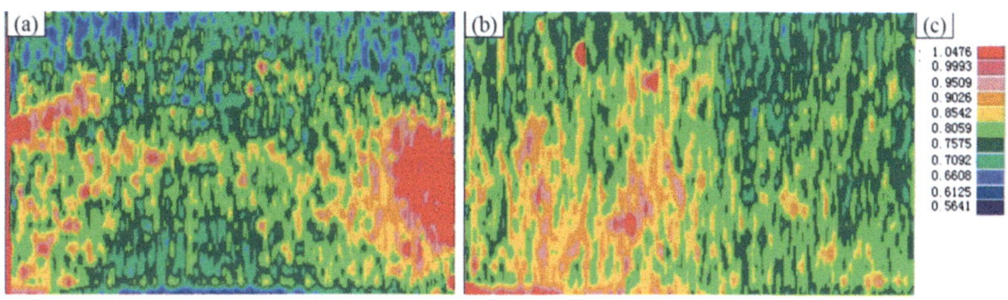

Fig. 2.34 Distribution maps of carbon in the transverse section of ingots: **a** S-1; **b** S-2; **c** Scale

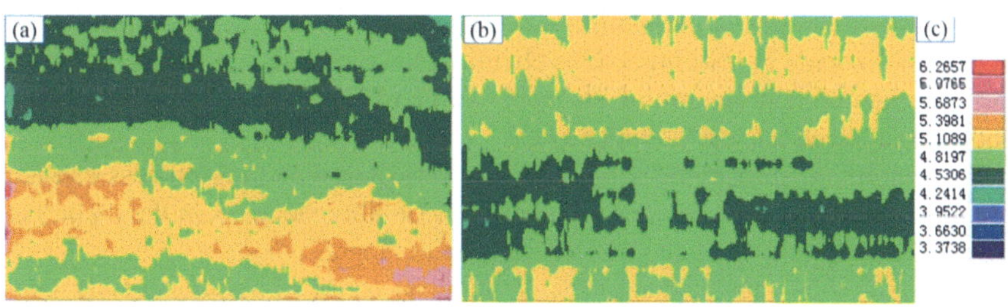

Fig. 2.35 Distribution maps of chromium in the transverse section of ingots: **a** S-1; **b** S-2; **c** Scale

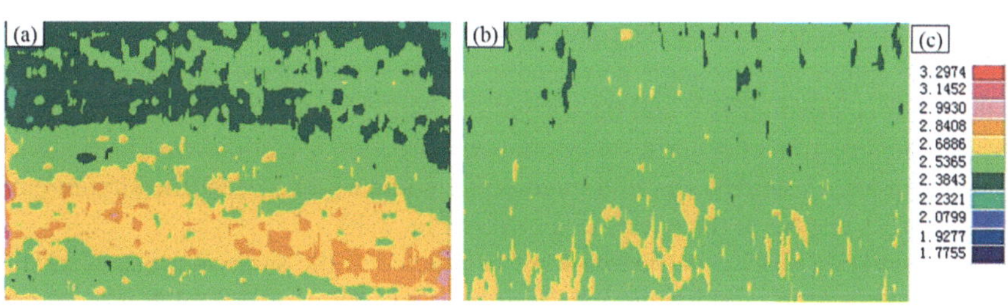

Fig. 2.36 Distribution maps of molybdenum in the transverse section of ingots: **a** S-1; **b** S-2; **c** Scale

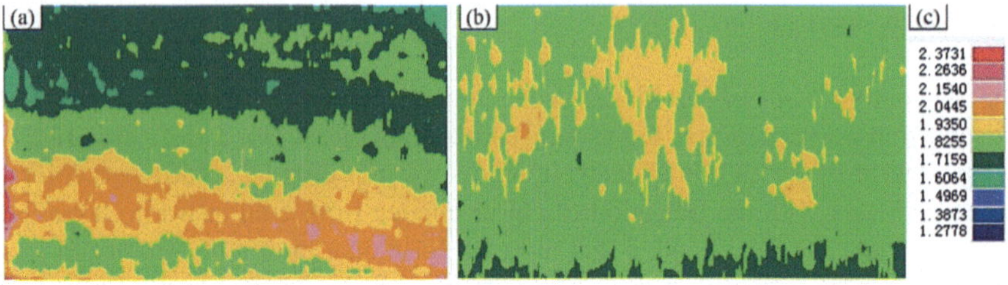

Fig. 2.37 Distribution maps of vanadium in the transverse section of ingots: **a** S-1; **b** S-2; **c** Scale

(2) Dendrite segregation

The segregation of elements in residual liquid during solidification in ESR and ESR-CDS process were detected through EPMA, and the results were shown in Fig. 2.35.

It can be seen from Fig. 2.36c, d that fluctuation degree of scanning line of elements in ingot E-1 is much more intense than that in ingot E-2, especially for elements V and Mn. Fluctuation degree of scanning line means the difference of the elements mass percent distributed between dendrite arm and interdendritic area. It can be concluded that the dendritic segregation of elements in ingot E-2 was alleviated through ESR-CDS process compared to that in ingot E-1 produced through ESR process.

During the solidification process, the solute atoms are generally enriched in the dendrite gap at the end of solidification, while the solute atoms are poor in the first solidified dendrite stem area. Therefore, the size of dendrite spacing can be characterized by the oscillation times of EPMA line scanning curve. The comparison of the oscillation times of carbon linear scanning curve in Fig. 2.38c, d shows that the dendrite spacing in ESR ingot S2 is obviously smaller than that in ESR ingot S1, that is, the dendrite structure in ESR ingot can be obviously refined by directional solidification electroslag remelting process.

The continuous directional solidification electroslag remelting process can not only effectively reduce the segregation behavior of alloy elements in the ingot, but also significantly refine the dendrite structure in the ingot.

The results show that the dendrite structure in S1 center of traditional ESR ingot is chaotic, the growth direction of dendrites is disordered. There are obvious cross section of equiaxial and columnar crystals, and the secondary dendrite spacing reaches 80 μm. The inner part of S2 ingot of directional solidification electroslag ingot is columnar crystal which is approximately parallel to the axial growth from the edge to the center, and the angle between the growth direction and the axial direction of the ingot is less than 13°, and the secondary dendrite spacing is less than 40 μm. The growth direction of dendrites of continuous directional solidification ESR ingots is shown in Fig. 2.39. The electron microscope comparison of dendrite structure of ESR ingot S1 and S2 in the center is shown in Fig. 2.40.

ESR and ESR-CDS experiments were conducted at the same melting rate. The schematic view of the solidification behavior during ESR and ESR-CDS process are

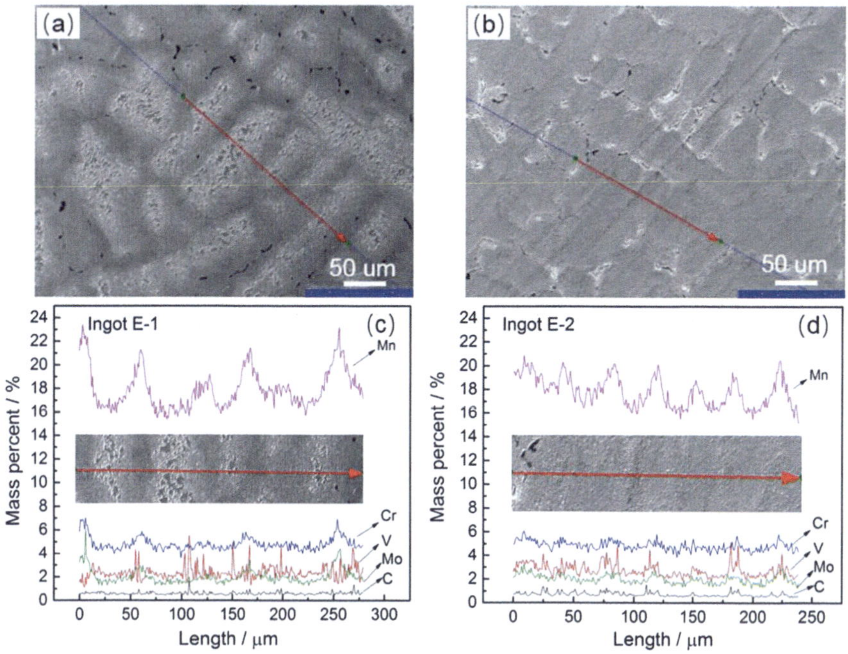

Fig. 2.38 SEM images and segregation of elements of ingots E-1 and E-2: **a**, **c** Ingot E-1; **b**, **d** Ingot E-2

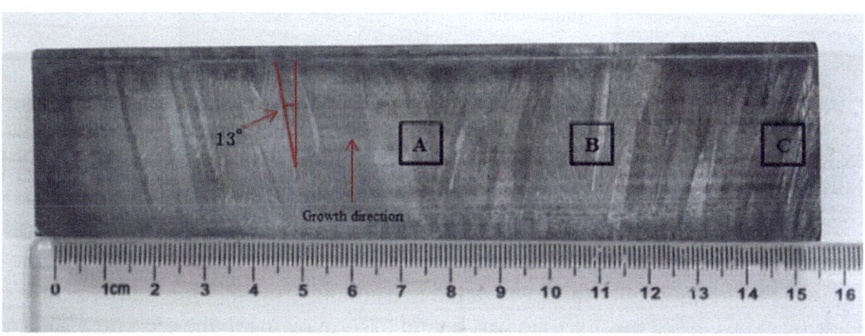

Fig. 2.39 Dendrite growth direction of continuous directional solidification electroslag ingot

shown in Fig. 2.41a, b, respectively. The difference in the microstructure of ESR-1 and ESR-2 ingots could be attributed to the following aspects: (1) different local solidification time caused by the shape of the liquid metal pool; (2) different cooling rate controlled by external conditions; (3) concentration gradient at the solid/liquid interface. A deep V-shape liquid metal pool formed during the ESR process influenced the growing direction of dendrite, removal of nonmetallic inclusion and morphological

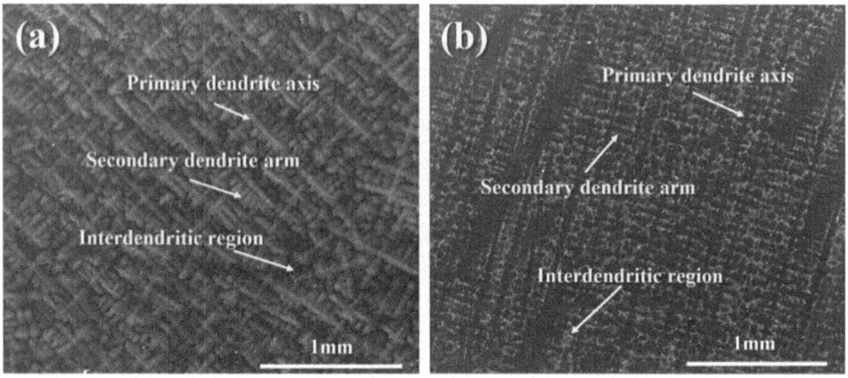

Fig. 2.40 Comparison diagram of dendrite structure in ESR ingot core (SEM): **a** Traditional ESR ingot S1; **b** Directional solidification ESR ingot S2

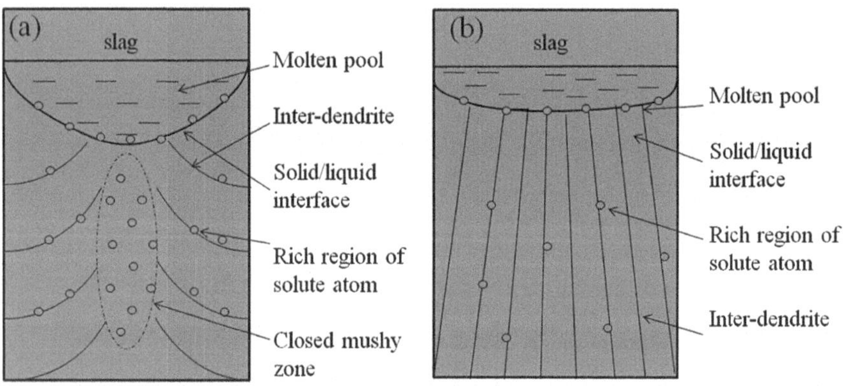

Fig. 2.41 Schematic view for the solidification behavior during ESR and ESR-CDS process: **a** ESR; **b** ESR-CDS

characteristics of precipitates. The direction of temperature gradient was perpendicular to the solid/liquid interface, which led to a certain growing direction of primary dendritic axis. Closed mushy zone was bound to form owing to the crossing primary dendritic arms. Shallow liquid metal pool was obtained during the ESR-CDS process by controlling the direction of thermal flow, the temperature gradient of solidification front and the solidification rate. The growing direction of primary dendritic axis was nearly parallel to the axis of solidified ingot, which could avoid the formation of closed mushy zone.

2.3 Carbide Control in Steel Based on Continuous Directional Solidification Electroslag Remelting

2.3.1 Effect of Continuous Directional Solidification Electroslag Remelting on the Size and Amount of Carbide

It was commonly known that the secondary dendritic spacing has a great influence on the size, amount and distribution of the carbides in the interdendritic region of ingots. Flemings [28] revealed that the local solidification time and dendritical axial spacing have the following relation:

$$\log d = k_1 + k_2 \log T \tag{2.1}$$

where d represents the dendritic axial spacing, μm; k_1 and k_2 are constant, determined by the alloy element content; T is the local solidification time, min.

According to Eq. (2.1), a smaller dendrite spacing indicates a shorter local solidification time. Compared with the traditional electroslag remelting process, the unidirectional solidification electroslag remelting can shorten the local solidification time and reduce the segregation of elements. The formation of primary carbide or eutectic carbide during solidification process is mainly dependent on the segregation of elements.

To further study the effect of ESR and ESR-CDS technique on primary carbides in ESR ingots, the microstructure of specimens taken from ESR-1 and ESR-2 ingots and carbides extracted from ESR ingots were analyzed by SEM-EDS. The microstructure of ESR ingots and three-dimensional structure of carbides were shown in Fig. 2.42. Figure 2.42a, b presented the distributions of primary carbides in Fig. 2.40a, b, respectively.

Figure 2.42a showed that the carbides precipitated along the equiaxed grain boundaries and distributed densely. Figure 2.42b showed that the carbides precipitated along the columnar grain boundaries and distributed dispersedly, and their size was smaller than that in Fig. 2.42a.

The basic parameters and characteristic parameters of primary carbides in 20 photos are counted by image analysis software(IPP), and the values are listed in Table 2.8 and Fig. 2.43.

Comparing the characteristic parameters of carbides, it can be found that the amount and size of carbides in the steel decreased when using ESR-CDS process. This illustrated that finer primary carbides and more diffuse carbides can be obtained by ESR-CDS process.

Figure 2.42c, d showed the three-dimensional structure of carbides extracted from ingots ESR-1 and ESR-2. The composition of the carbides analyzed by EDS was showed in

It could be seen from Table 2.9 that two types of carbides were generated in the

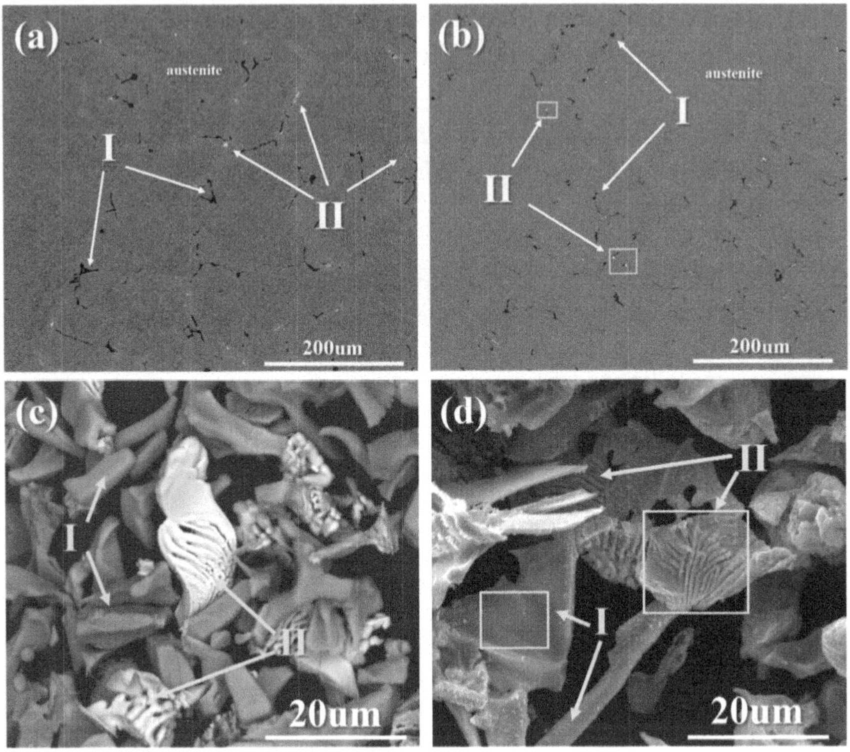

Fig. 2.42 Microstructure of ESR ingots and three-dimensional carbides structure in ESR-1 and ESR-2 ingots: **a**, **c** ESR-1; **b**, **d** ESR-2

Table 2.8 Basic parameters and characteristic parameters of carbides

Sample	Basic parameters of quantitative metallography							Characteristic parameters of carbides	
	Statistic of carbides parameters					Size of photo			
	N	*A*, μm	$\overline{D}$, μm	*l*, μm	*w*, μm	*L*, μm	*W*, μm	V_V, %	N_V
S-1	527	12226.43	5.85	9.35	3.65	595.92	507.29	4.04	0.035
S-2	521	5517.55	3.94	5.65	2.44	595.92	507.29	1.83	0.016

ingot S-1 and S-2. The type I was primary MC-type carbides which were vanadium-rich containing a certain amount of Cr and Mo elements. The type II was M_2C-type eutectic carbides which were molybdenum-rich containing a certain amount of Mn, Cr and V elements.

XRD was used to further confirm the types of carbides, and the results were shown in Fig. 2.44.

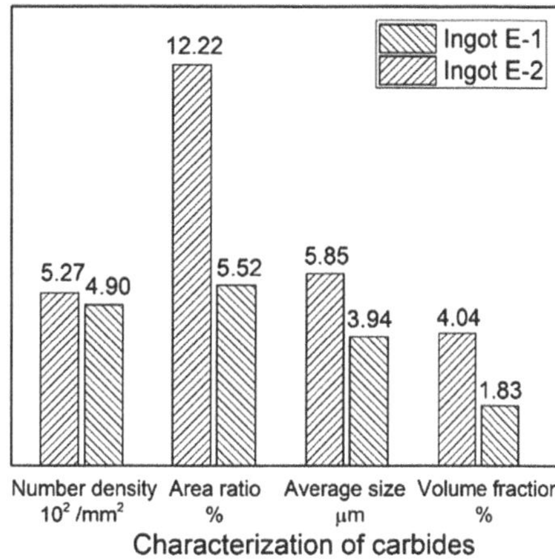

Fig. 2.43 Comparison of characterizations of carbides in ingot S-1 and S-2

Table 2.9 Compositions of the carbides in ingots ESR-1 and ESR-2 (wt%)

Element	Precipitated phase	C	Fe	Mo	Mn	Cr	V
S-1	Point I	15.27	4.39	10.72	1.95	2.36	65.31
	Point II	13.68	25.24	30.33	10.93	10.02	9.80
S-2	Point I	13.61	3.80	19.16	1.67	6.47	55.28
	Point II	14.82	12.27	33.90	10.78	13.57	12.19

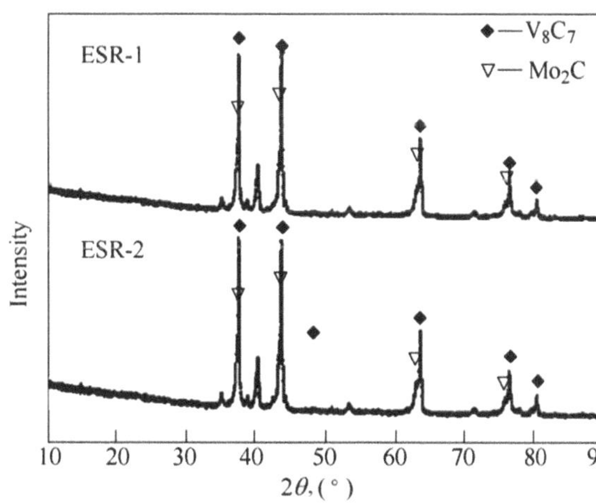

Fig. 2.44 XRD patterns of the carbides powder

The XRD results shown in Fig. 2.44 indicated that the MC-type and M_2C-type carbides were V_8C_7 (Type I) and Mo_2C (Type II), respectively.

Figure 2.4c showed that the morphology of type I carbides in ESR-1 and ESR-2 was short rod-shaped and disc-shaped with multi-angles, respectively. The

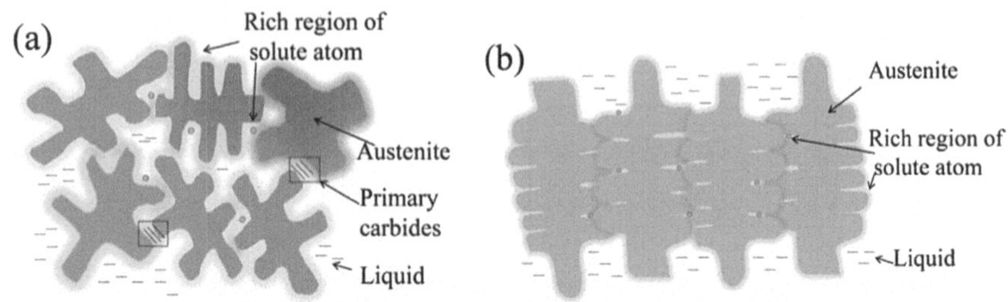

Fig. 2.45 Schematic of dendritic growth during solidification: **a** ESR; **b** ESR-CDS

morphology of type II carbides in ESR-1 was lamellar fish-skeleton-like, while that of type II in ESR-2 was thin-leaf skeleton-like and it was similar to that of type II in ESR-1, which were typical eutectic carbides. The difference of the morphology of carbides can be explained by the modes of dendritic growth during solidification The schematic of dendritic growth during solidification in ESR and ESR-CDS process are schematically presented in Fig. 2.45.

The growing direction of dendrites in conventional ESR process were chaotic even in the small areas as shown in Fig. 2.45a, whereas dendrites in ESR-CDS process was parallel to each other throughout the whole ingot as shown in Fig. 2.45b. Compared to the microstructure in ESR process, the dendritic arm was finer and the dendritic spacing was narrower in the ESR-CDS process, as shown in Fig. 2.45. The shadow area around dendritic crystal represented solute enrichment area. When the composition of residual liquid steel reached eutectic point during solidification of liquid steel, primary carbides precipitated along pre-existed austenite. Due to the difference in the concentration gradient of carbon content, the premiere carbides might grow up from one grain boundary to other grain boundaries. The driving force for carbides growing located in the grain boundaries. Therefore, the driving force of carbides growth in conventional ESR process was confused. On the contrary, it was single direction nearly parallel to crystal < 001 > in ESR-CDS process. This was the reason for the difference in morphologies of eutectic carbides obtained in ESR and ESR-CDS process, as shown in Fig. 2.42c, d.

In conclusion, compared with the traditional ESR process, the continuous directional solidification ESR process can not only reduce the size of primary and eutectic carbides, reduce the number of carbides formed, improve the distribution of carbides in the ESR ingot, but also change the morphology of primary carbides from short rod shape to disc shape, but can not change the morphology of primary carbide The type of secondary carbides.

2.3.2 *Effect of Continuous Directional Solidification Electroslag Remelting on the Three-Dimensional Structure of Carbides*

To understand the effect of continuous directional solidification electroslag remelting process on the formation and growth behavior of carbides, the two-dimensional and three-dimensional structure of carbides were extracted to detect the feature in morphology of carbides. The morphologies of precipitated carbides were shown in Figs. 2.46 and 2.47.

The three-dimensional morphology of MC-type carbides shown in Fig. 2.48c was short rod-shaped, while that in Fig. 2.48d was disc-shaped with multi-angles. Meanwhile, from the in situ observation of carbides morphology in deep-etched samples shown in Fig. 2.48a, b, it is obviously known that the MC-type carbides precipitated along preexisted austenitic grain boundaries. The morphology of MC-type carbides was influenced by the solidification rate and local concentration of solute atoms, as well as pre-formed austenite dendritic interface [29]. During solidification process, the growth orientation of austenitic dendrite in ESR process was irregular, while that in ESR-CDS process was nearly parallel to the axis of the ingot <001>. Thus, the concentration gradient of solute atoms in residual liquid in ESR process was irregular, while that in ESR-CDS process was nearly parallel to crystal <001>, due

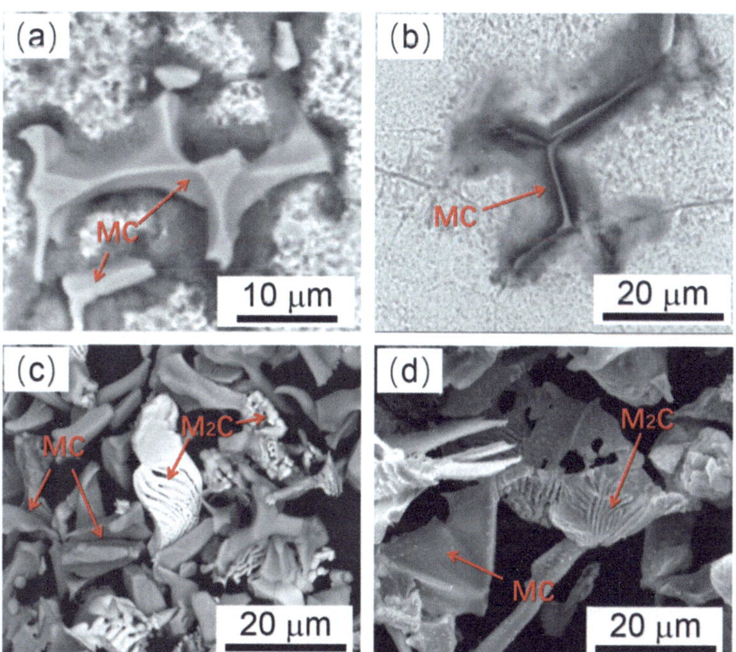

Fig. 2.46 Morphology of MC-type carbides (SEM): **a**, **c** Ingot S1; **b**, **d** Ingot S2

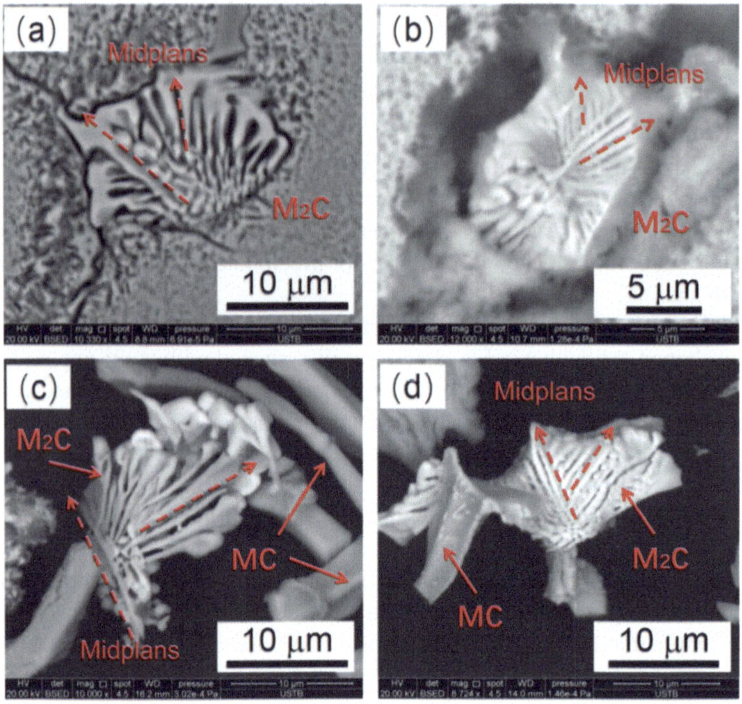

Fig. 2.47 Morphology of M_2C-type carbides (SEM): **a**, **c** Ingot S1; **b**, **d** Ingot S2

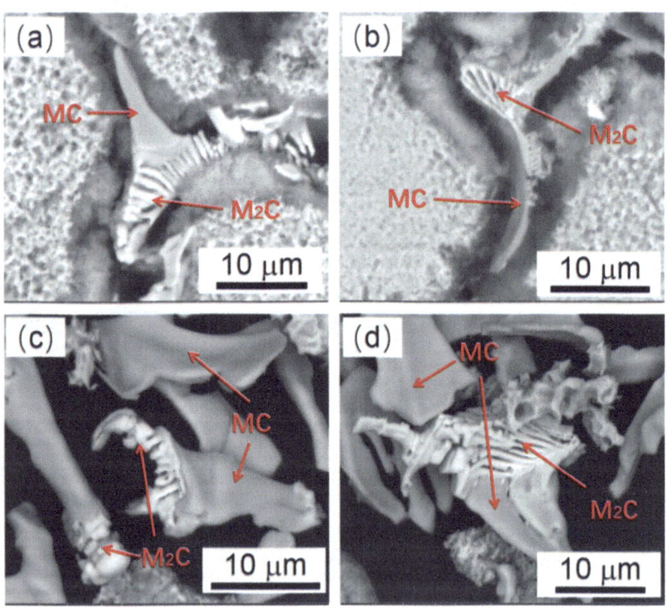

Fig. 2.48 Mix Morphology of MC and M_2C-type carbides (SEM): **a**, **c** Ingot S1; **b**, **d** Ingot S2

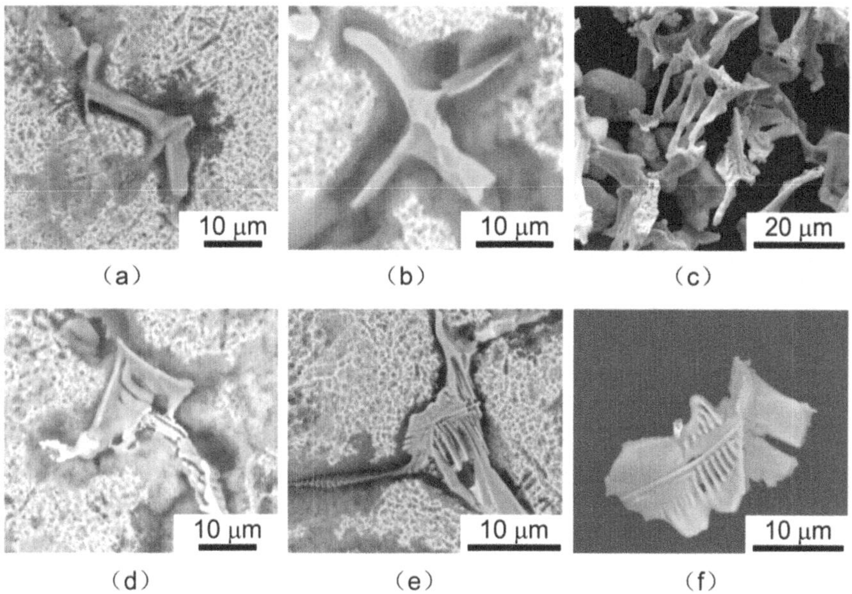

Fig. 2.49 Morphologies of MC-type carbides in ingots S1 and S2: **a**, **b**, **c** ingot S1; **d**, **e**, **f** ingot S2

to solute atoms were generally rejected into interdendritic liquid area during solidification process. Therefore, the directions of driving force for precipitated carbides growth in ESR process were irregular. On the contrary, those were single direction parallel to <001> in ESR-CDS process. This was one reason for the difference in morphologies of MC-type carbides produced in ESR and ESR-CDS process. In addition, the solidification rate in ESR-CDS process is greatly larger than that in ESR process may be another factor to control the morphology of MC-type carbides.

The two-dimensional morphology of M_2C-type carbides in ingot S-1 and S-2 shown in Fig. 2.49a, b were similar to each other, they were both fibrous-shapes. While from the three-dimensional morphologies shown in Fig. 2.49c, d, they were lamellar and long strip shape arranged in parallel. It can be known that the M_2C-type carbides exhibits the dendritic structure. This dendritic morphology of M_2C-type carbides may be attributed to the temperature difference between the surface and center of the ingot when M_2C-type carbides began to precipitate during solidification. Figure 2.49 shows that a midplane of M_2C-type carbides forms firstly along the direction of the heat flow, then a series of branches grow from it all around. When the branches of M_2C-type carbides meet each other or come across to austenite matrix and the branches of MC-type carbides (shown in Fig. 2.49), M_2C-type carbides stops growing.

The mix morphology of MC and M_2C-type carbides were shown in Fig. 2.49. It indicated that the precipitation of MC-type carbides occurred accompanying with the formation of M_2C-type carbides. The reason for the morphology of MC and M_2C-type carbides could be explained as follows. When the eutectic reaction occurring during the solidification process of electroslag remelting, the solute atoms from each phase in the eutectic mixture were exhausted to the liquid and depart from the front of solid-liquid interface by diffusion, then deposit on the other phase finally. MC-type carbides need plenty of alloy element V atoms to ensure its growing, and alloy element V atoms in sounding region spreads to where MC-type carbides exist and the alloy elements Mo and Cr atoms are left around MC-type carbides. Meanwhile, the eutectic M_2C-type carbides needs plenty of alloy elements Mo and Cr during its growth and alloy element V is left around M_2C-type carbides. The alloy elements Mo and Cr atoms diffused from the region of MC-type carbides are employed for the growth of M_2C-type carbides, and the alloy element V transfers to the place where MC-type carbides grow up. Therefore, the different eutectic carbides mix together and grow corporately can be observed in Fig. 2.48.

The morphology of MC-type eutectic carbides is a usually outstanding concern, but the growth mechanism of carbides is rarely reported. Therefore, the growth behavior of carbide during solidification process was analyzed and the mechanism of carbide growth was discussed.

It can be seen from equilibrium phase diagram that the precipitated temperature of carbides (i.e., MC, M_2C, M_7C_3 and $M_{23}C_6$) in austenitic hot-work die steel in sequence during equilibrium solidification condition were 1380 °C, 1210 °C, 920 °C, 835 °C and 660 °C, respectively. However, during actual solidification process, the segregation of elements and super cooling phenomenon are inevitable to occur. Thus, the temperatures of precipitated carbides have a certainly distinction. The non-equilibrium phase diagram shows that MC and M_2C carbides precipitated from the residual liquid with mass fraction of solid from 0 to 1. The correspondingly eutectic reactions as follows: L → γ+MC occurs at 1210 °C ($f_s = 0.86$) and L → γ+MC + M_2C occurs at 1160 °C ($f_s = 0.95$). However, M_7C_3 and $M_{23}C_6$-type carbides cannot be found in microstructure, which may be attributed to rapid solidification resulting to precipitation of carbides from austenite phase scarcely. Thus, the microstructure consists of austenite, eutectic MC and M_2C-type carbides.

The precipitation temperatures of MC and M_2C-type carbides indirect verify the explanation about the morphology of carbides affected through ESR-CDS process. The precipitation temperature of MC-type carbides is higher than that of M_2C-type carbides, which results to the influence by local concentration of solute atoms and local solidification rate on morphology of MC-type carbides is greatly larger than that on morphology of M_2C-type carbides. Therefore, under the influence of directional solidification in ESR-CDS process, the morphology of MC-type carbides was changed, while that of M_2C-type carbides was similar in comparison with ESR process. In addition, the eutectic reactions of precipitated carbides were in agreement with the illustration about formation behavior of carbides in previous section [30].

It was known that MC-type carbides take an FCC structure with the most stable plane of {111}. The octahedron consisting of {111} faces is therefore the equilibrium shape. The growth patterns of MC-type carbides present dendritic morphology, the correspondingly morphology were shown in Fig. 2.49. The difference in growth patterns of MC-type carbides between ESR and ESR-CDS process is dendritic morphology with rod shaped (shown in Fig. 2.49a–c) and dendritic morphology

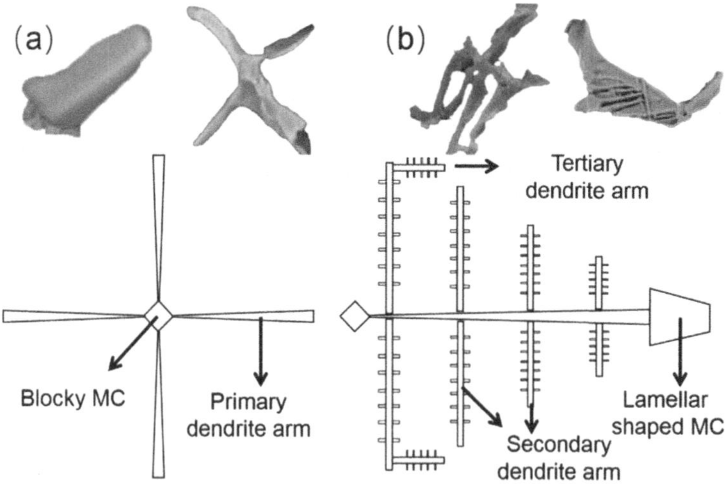

Fig. 2.50 Schematic view of growth patterns of MC-type carbides. **a** Growth of carbides with secondary dendritic arms; **b** Growth of carbides with tertiary dendritic arms

with lamellar shaped (shown in Fig. 2.49d–f). The reason for dendritic carbides is the carbides grow along the <001> directions more quickly than other directions for the least atoms in the {001} faces of the FCC crystal. The secondary and tertiary dendritic arms were observed in Fig. 2.49. The Schematic view of growth patterns of MC-type carbides was shown in Fig. 2.50. The growth steps were illustrated at the tips of the carbides, the development and connection of these secondary and tertiary dendritic arms form the flake-shaped MC-type carbides in the end.

M_2C-type carbides is the products of the eutectic reaction at the end of solidification process. Their three-dimensional morphology are maze like or skeleton like, and the Schematic view of growth patterns of M_2C-type carbides is as shown in Fig. 2.51.

During the solidification process of molten steel, solute atoms continuously accumulate in the liquid phase at the front edge of the liquid-solid interface with the decrease of temperature and the growth of columnar crystals. The eutectic reaction will begin on the condition that the concentration of solute reaches the eutectic component point. When the theoretical solid fraction(f_s) is 0.95, the binary eutectic reaction $L \rightarrow \gamma + M_2C$ or ternary eutectic reaction $L \rightarrow \gamma + MC + M_2C$ will begin and M_2C-type carbide will precipitate. M_2C-type carbides are distributed in the clearance of dendrite. The dendritic backbones in the three-dimensional structure are formed along the direction of heat flow. Then the carbides grows from the backbone to the surrounding, forming a maze or skeleton three-dimensional structure.

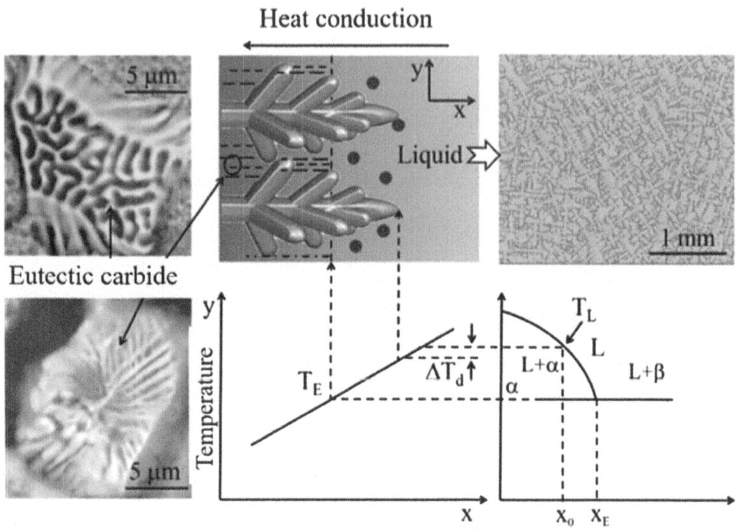

Fig. 2.51 Schematic view of growth patterns of M_2C-type carbides

2.3.3 Performance Analysis of Directional Solidified ESR Ingots

Due to the low hardness of the matrix, austenitic hot work die steel needs solution strengthening with alloy elements and precipitation strengthening with secondary carbides characterized by the fine size and dispersed distribution. The composition of austenitic hot work die steel contains a lot of carbide forming elements Cr, Mo and V, which lead to the formation of a large number of primary carbides or eutectic carbides during solidification. Traditional ESR ingot S1 and continuous ESR ingot S2 undergo the same forging and annealing treatment, followed by solution heat treatment and aging heat treatment. The heat treatment process is shown in Fig. 2.52, and the samples after heat treatment are defined as H1 and H2 respectively.

The SEM microstructure photographs of ingots E-1 and E-2 under aging treatment state were shown in Fig. 2.53.

After heat treatment, the microstructure of austenitic hot work die steel is composed of austenitic matrix, retained primary carbide and precipitated secondary carbide. Among them, there are a large number of primary carbides in H1, which are coarse in size, while the primary carbides in H2 decompose completely, and the residual primary carbides are small in size. At the same time, the secondary carbides precipitated in ingot H2 characterized with smaller size, more quantity, and distribute more dispersly compared with those in ingot H1. M_2C carbides cannot be found after heat treatment of samples H1 and H2, mainly due to the complete decomposition of M_2C carbides during heat processing and heat treatment.

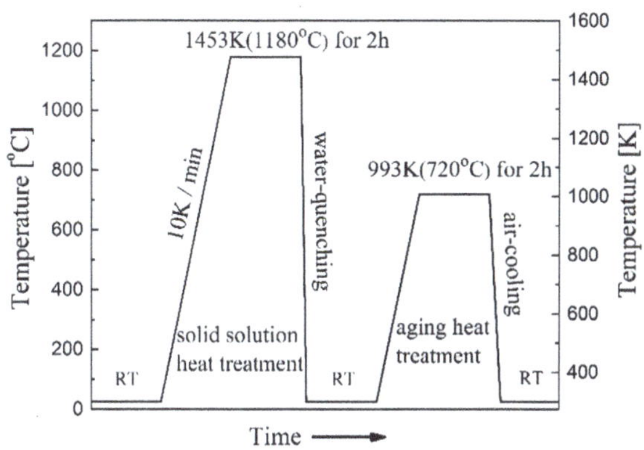

Fig. 2.52 Schematic view of heat treatment process

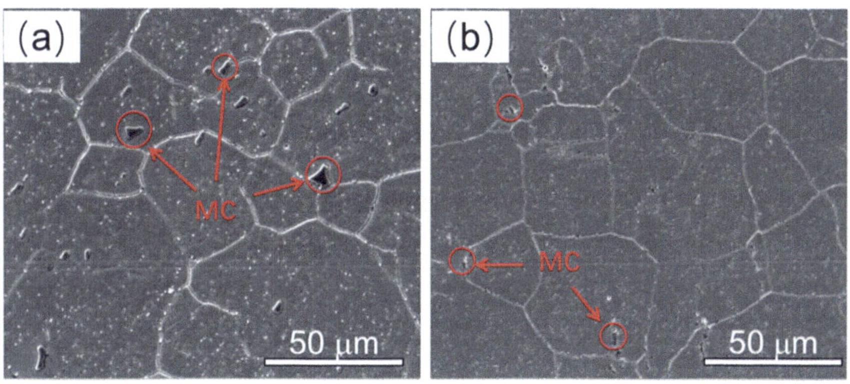

Fig. 2.53 SEM microstructure photographs of ingots E-1 and E-2 under aging treatment state: **a** Ingot E-1; **b** Ingot E-2

The differences of heat treatment microstructure between sample H1 and sample H2 are mainly due to the different morphology of MC carbide in ESR ingot, which leads to the different decomposition degree during heat treatment. The disc or lamellar MC carbide in ESR ingot S2 is easier to be decomposed or eliminated than the short rod MC carbide in ESR ingot S1, which leads to the non-decomposition of primary carbon with smaller size and less quantity in sample H2 after heat treatment. Meanwhile, the carbon and alloy elements are more uniformly dissolved in austenite matrix, the distribution of secondary carbides is more dispersed. Figure 2.54 shows the comparison of hardness and impact energy values of sample H1 and sample H2 after the same solution aging heat treatment.

It can be seen from Fig. 2.54 that the hardness and impact energy of sample H2 after the same heat treatment are higher than those of sample H1. The hardness and impact energy of ESR ingot S2 after the same heat treatment are higher than those

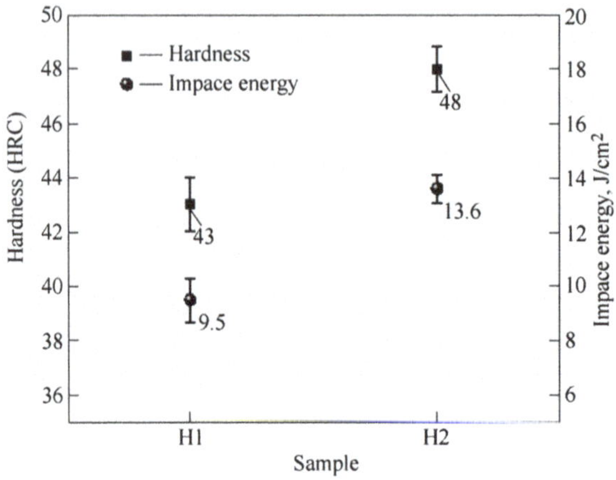

Fig. 2.54 Hardness and impact energy values of samples after the same solution aging heat treatment

of ESR ingot S1. According to the hardness value in Fig. 2.54 and the microstructure analysis in Fig. 2.53, the hardness of sample H2 is higher than that of sample H1, which mainly attributes to the primary carbide decomposed more completely in sample H2. The solid solution of carbon and alloy elements in austenite matrix provide a reinforcement effect with solid solution. Meanwhile, after aging treatment, the secondary carbides with smaller size, more dispersed distribution and more quantity precipitated in the sample H2 provide a reinforcement effect with precipitation strengthening. The impact fracture morphology of sample H1 and sample H2 are shown in Fig. 2.55.

The SEM analysis on fracture surface indicates that, the fracture characteristic of ingot E-1 is quasi-cleavage in Fig. 2.55a after aging heat treatment, while for ingot

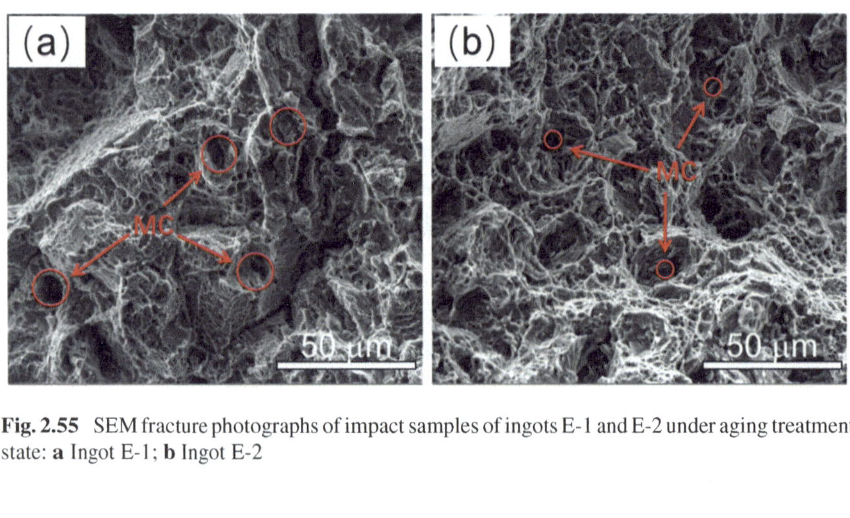

Fig. 2.55 SEM fracture photographs of impact samples of ingots E-1 and E-2 under aging treatment state: **a** Ingot E-1; **b** Ingot E-2

E-2, it is quasi-cleavage fracture with mass small dimples in Fig. 2.55b. The basic characteristics of quasi-cleavage fracture are small cracked grain and tear ridges. The cracks generally derive from area of stress concentration such as inclusions and carbides. The fracture surface presented much more amounts of carbides and larger size of carbides in Fig. 2.55a than that in Fig. 2.55b. Thus, it is suggested that superior toughness is associated with minimization of large primary carbides.

In conclusion, compared with the traditional ESR process, the continuous directional solidification ESR process has more compact structure, finer dendrite gap and more uniform composition. At the same time, it can also reduce the size and quantity of carbides formed in the solidification process, improve the morphology of primary MC carbides. These carbides characterize make them easier to be decomposed in the later heat treatment process. The above aspects are the main reasons for the high hardness and impact toughness of austenite hot work die steel produced by continuous directional solidification electroslag remelting process compared with traditional electroslag remelting process.

References

1. Zhu QT, Li J, Shi CB et al (2017) Precipitation behavior of carbides in high-carbon martensitic stainless steel. Int J Mater Res 108:20–28
2. Suh SH, Choi J (1986) Effect of melting rate on the carbide cell size in an electroslag remelting high speed steel ingot. ISIJ Int 26(4):305–309
3. Chen X, Jiang ZH, Liu FB et al (2017) Effect of melt rate on surface quality and solidification structure of Mn18Cr18N hollow ingot during electroslag remelting process. Steel Res Int 88(2):1600186
4. Wang Q, Li BK (2015) Numerical investigation on the effect of fill ratio on macrosegregation in electroslag remelting ingot. Appl Therm Eng 91:116–125
5. Kharicha A, Ludwig A, Wu M (2005) Shape and stability of the slag/melt interface in a small DC ESR process. Mater Sci Eng, A 413–414:129–134
6. Wang Q, Cai H, Pan LP et al (2016) Numerical investigation of influence of electrode immerse depth on heat transfer and fluid flow in electroslag remelting process. JOM 68(12):1343–1349
7. Mitchell A, Hernandez-morales B (1990) Electromagnetic stirring with alternating current during electroslag remelting. Metall Trans B 21(4):723–731
8. Fezi K, Yanke J, Krane MJM (2015) Macrosegregation during electroslag remelting of alloy 625. Metall Mater Trans B 46(2):766–779
9. Ridder SD, Reyes FC, Chakravorty S et al (1978) Steady state segregation and heat flow in ESR. Metall Trans B 9(3):415–425
10. Kelkar KM, Patankar SV, Srivatsa SK et al (2013) Computational modeling of electroslag remelting(ESR) process used for the production of high-performance alloys. Paper presented at proceedings of the 2013 international symposium on liquid metal processing & casting, USA, pp 3–12
11. Liang Q, Chen XC, Fu R et al (2011) Introduction of theoretical-basis and application of numerical simulation software Meltflow-ESR in electroslag remelting process. J Mater Metall 10(S1):106–111
12. Corey JOC, John JD, David GE et al (2017) Industrial-scale validation of a transient computational model for electro-slag remelting. Paper presented at Liquid Metal Processing and Casting, Pennsylvania, pp 93–102

13. Mitchell A, Joshi S (1972) The densities of melts in the systems $CaF_2 + CaO$ and $CaF_2 + Al_2O_3$. Metall Trans B 3(8):2306–2307
14. Dong YW, Jiang ZH, Liu FB et al (2012) Simulation of multi-electrode ESR process for manufacturing large ingot. ISIJ Int 52(12):2226–2234
15. Mills KC, Keene BJ (1981) Physicochemical properties of molten CaF_2-based slags. Int Metal Rev 26(1):21–26
16. Karimi-sibaki E, Kharicha A, Bohacek J et al (2015) A dynamic mesh-based approach to model melting and shape of an ESR electrode. Metall Mater Trans B 46(5):2049–2061
17. Jiang ZH, Dong YW, Gneg X et al (2015) Electroslag Metallurgy. Science Press, Beijing
18. Weber V, Jardy A, Dussoubs B et al (2009) A comprehensive model of the electroslag remelting process: description and validation. Metall Mater Trans B 40(3):271–280
19. Qi YF, Li J, Shi CB et al (2018) Effect of directional solidification in electroslag remelting on the microstructure and cleanliness of an austenitic hot-work die steel. ISIJ Int 58(7):1275–1284
20. Qi YF, Li J, Shi CB et al (2017) Effect of directional solidification of electroslag remelting on the microstructure and primary carbides in an austenitic hot-work die steel. J Mater Process Tech 249:32–38
21. Shi XF, Chang LZ, Wang JJ (2015) Effect of mold rotation on the bifilar electroslag remelting process. Int J Min Met Mater 22(10):1033–1042
22. Chumanov VI (2010) Chumanov IV (2010) Increasing the efficiency of the electroslag process and improving the metal quality by rotating a consumable electrode: Part 1. Russ Metall 6:499–504
23. Fu R, Li FB, Yin FJ et al (2015) Microstructure evolution and deformation mechanisms of the electroslag refined-continuous directionally solidified (ESR-CDS®) superalloy Rene88DT during isothermal compression. Mater Sci Eng, A 638:152–164
24. Dong YW, Jiang ZH, Li ZB (2009) Segregation of niobium during electroslag remelting process. J Iron Steel Res Int 16(1):7–11
25. Chen XC, Fu R, Ren H et al (2011) Study on nonmetallic inclusions of ESR-CDS FGH96 alloy. China New Tech Prod 10:1–2
26. Zhan LC, Chi HX, Ma DS et al (2013) The as-cast microstructure of ESR-CDS M2 high speed steel. J Mater Eng 7:29–34
27. Fu R, Chen XC, Ren H et al (2011) Microstructure and hot deformation behavior of René88DT alloy processed by ESR-CDS. J Aeronaut Mater 31(6):8–13
28. Flemings MC (1981) Solidification processing. Metallurgical Industry Press, Beijing
29. Li XW, Wang L, Dong JS et al (2014) Effect of solidification condition and carbon content on the morphology of MC carbide in directionally solidified nickel-base superalloys. J Mater Sci Technol 30(12):1296–1300
30. Qi YF, Li J, Shi CB (2018) Characterization on microstructure and carbides in an austenitic hot-work die steel during ESR solidification process. ISIJ Int 58(11):2079–2087

Chapter 3
Carbide Control in Rolling Process

Abstract It is difficult to completely avoid the formation of primary carbides in the special steels produced by conventional solidification process. It is therefore necessary to control the primary carbides in subsequent rolling and heat treatment processes. The primary carbides will be broken and redistributed during blooming process. High temperature diffusion annealing process could not only effectively reduce the amount of primary carbides, but also promote the uniform distribution of carbon and alloying elements and increase the content of martensite in the steel. Reducing the rolling finishing temperature, increasing the strain rate and the cooling rate of finish rolling could refine the grains, which is beneficial to the subsequent steel processing. The rolling temperature has a great influence on the microstructure of steel. The reduction of the rolling temperature will significantly increase the deformation stress, which is favorable to broken primary carbides in the steel. With the decrease in the thickness of the cold rolled sheet, the size and amount of carbides decrease, and the distribution of carbides in the steel becomes more uniform.

Keywords Blooming · High temperature diffusion annealing · Dissolution · Hot rolling · Cold rolling

Employing unidirectional ESR process, decreasing the melting rate and increasing the fill ratio can effectively reduce the amount of primary carbides in high-carbon tool and die steel [1, 2]. However, it is very difficult to avoid the formation of primary carbides during ESR process [3–8]. The primary carbides need to be further controlled in the subsequent process. Usually, forging and hot rolling process can broke and spread the primary carbides around the primary austenite. The effect of forging and hot rolling on the microstructure of ESR ingot is shown in Fig. 3.1.

The microstructure of ESR ingot after forging is shown in Fig. 3.1a. Primary carbides are broken by the forging pressure, and distribute along a certain direction. There is a certain amount of secondary carbides precipitated near the primary carbide. Because that a small amount of primary carbide will dissolve into the surrounding matrix during heating process, the concentration of alloying elements around primary carbide will increase, and these alloying elements will precipitate in the form of secondary carbide during cooling process. The microstructure of the ingot after hot rolling is shown in Fig. 3.1b. The grains are elongated along the rolling direction.

J. Li and C. Shi, *Carbide in Special Steel*, Engineering Materials,
https://doi.org/10.1007/978-981-16-1456-9_3

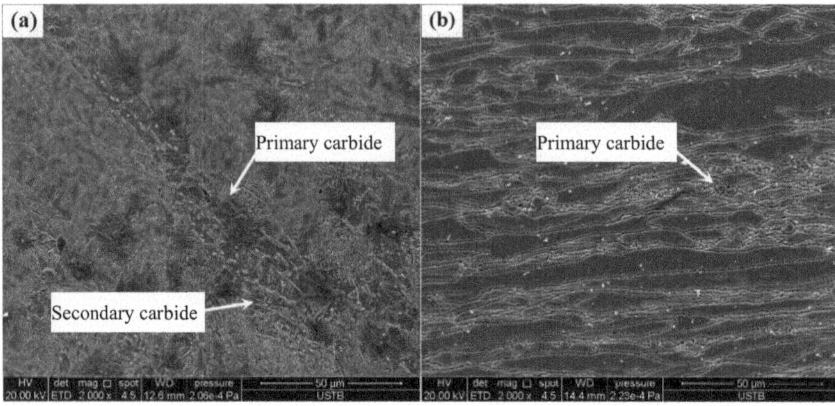

Fig. 3.1 Microstructure of ESR ingot after forging and hot rolling: **a** After forging; **b** After hot rolling

The primary carbides are further broken and their size reduced. Considering that the cooling rate of rolling slab is fast, the secondary carbides in the microstructure is not obvious. High temperature diffusion annealing after forging and cogging down can promote the dissolution of primary carbide. High temperature diffusion annealing plays an important role in decreasing element segregation and promoting the dissolution of primary carbides [9, 10].

Taking the cold rolling slab of 8Cr13MoV high-carbon martensite steel as an example, the hot rolling process is as follows: (1) cogging down process. ESR ingots are heated to 1200 °C in a stepping furnace and helld for 2 h before rolling. The cogging down temperature is about 900 °C. After 7 passes of rolling, the square ESR ingots with sides of 210 mm are rolled into primary rolling slab with the thickness of 30 mm. (2) Finishing rolling process. Primary rolling slab with the thickness of 30 mm are heated to 1180 °C for 30 min and then rolled. The starting temperature is about 900 °C. After 7 passes of rolling, the fine rolling slab with a thickness of 3.5 mm are formed. The process of cold rolling is as follows: the incoming and outlet thickness of material is 3.0 mm and 2.0 mm, respectively. The annealing temperature of the rolling plate is 860–880 °C. During the second rolling, the incoming material thickness is 2.0 mm and the outlet thickness is 1.5 mm. No. 2 cold rolling mill is 4 high irreversible cold rolling mill. The thickness of the incoming material is 1.5 mm, and it is rolled at 0.2 mm drop volume each time. The annealing temperature is 860–880 °C, and the thickness of the cold rolled finished products is 2.5 mm, 2 mm, 1.5 mm, 0.9 mm and 0.7 mm.

The heating and holding stage of ESR ingots before hot rolling and before finish rolling have obvious effect on the dissolution of primary carbide. The primary carbide is further broken by the hot rolling process. The study on the diffusion annealing process of ESR ingots and hot rolling slab, as well as the effect of hot rolling process on primary carbide is helpful to reduce the content of primary carbide in steel and reduce the harm of primary carbides on the processing and service performance of steel. In the process of cold rolling, the material structure and cardides (amount, size, distribution and shape) change constantly in the process of multi-pass rolling and repeated annealing. There are few reports on the factors and mechanism of its changes.

Therefore, on the basis of the research and control of the microstructure and carbide in the hot rolling slab, the further research on the evolution of the material microstructure and carbide in the multi-pass cold rolling process and their influence on the properties can provide guidance for the control of the microstructure and carbide in the final cold rolling slab.

3.1 Effect of Cogging Down and High Temperature Diffusion Annealing Process on the Carbide

3.1.1 Effect of Cogging Down on the Carbide

The morphology and distribution of primary carbide on 8Cr13MoV primary rolling slab are shown in Fig. 3.2. The primary carbide is dark gray and the steel matrix is light gray under the backscattering diffraction condition of SEM. The primary carbides in the ESR ingots are mostly agglomerated rod-like. After hot rolling process, the primary carbides are obviously broken and dispersed, and they are arranged in a discontinuous line along the rolling direction.

Ten fields of view of 1 mm^2 were randomly selected from the sample of the priming-plate, and the area fraction of primary carbide in the sample was counted by image-Pro Plus Image analysis software (IPP). The results are shown in Table 3.1.

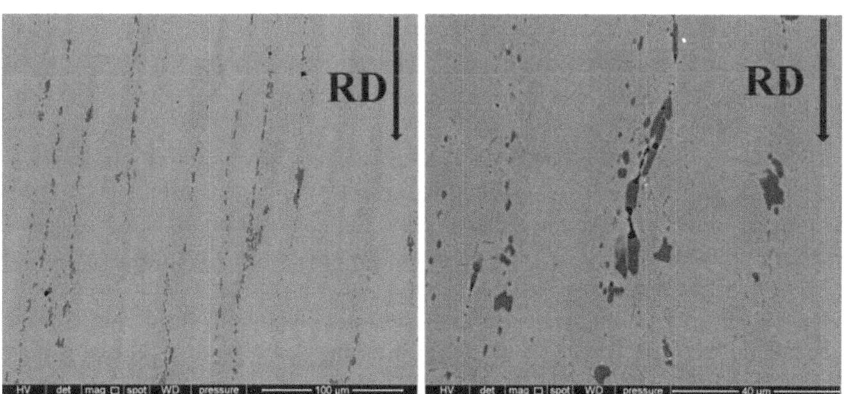

(a) (b)

Fig. 3.2 Primary carbide after hot rolling: **a** 1000 times; **b** 3000 times (RD stands for rolling direction)

Table 3.1 Statistics of area fraction of primary carbide in primary rolling slab

Number	1#	2#	3#	4#	5#	6#	7#	8#	9#	10#
Area fraction, %	2.52	2.15	2.19	2.94	2.70	2.24	2.61	2.10	2.23	2.01

The primary carbide is formed in the solidification process of molten steel. It can be seen from Table 3.1 that the average area fraction of primary carbide in the primary rolling slab and ESR ingots is 2.37% and 1.37%, respectively. The area fraction of primary carbide increases after hot rolling process. This is because that the primary carbides in ESR ingots are mostly aggregated. They are broken, extended and dispersed during the primary rolling process.

The morphology and distribution of primary carbides in the finish rolling slab are shown in Fig. 3.3, in which (a) and (c) are 500 times and 1000 times photos under the conditions of backscattering diffraction by SEM, respectively. The dark gray region is primary carbides. The primary carbides in (a) and (c) identified and reversed displayed by image analysis software are shown in (b) and (d), respectively, and the primary carbides are marked as bright white.

It can be seen from Fig. 3.3 that (a) and (b), (c) and (d) have a high matching degree of primary carbide recognition. From Fig. 3.3b and d, the amount, distribution, morphology and other characteristic parameters of primary carbide can be clearly identified. Compared with the primary rolling slab, the distribution of primary carbides in the steel after finish rolling process are more discrete, and the distribution along the rolling direction has almost disappeared.

3.1.2 Effect of Diffusion Annealing of ESR Ingot on Primary Carbide

Samples were taken at the center, 1/2 radius and edge of 8Cr13MoV ESR ingot (melting rate: 150 kg/h, electrode diameter: 110 mm, filling ratio: 0.23). The schematic view is shown in Fig. 3.4.

The samples were heated to 1180 °C at the heating rate of 10 °C/min and held for 2 h, and then cooled in the furnace. The samples were then ground and eroded. The morphology of primary carbides at 1/2 radius of the ESR ingots before and after high-temperature diffusion annealing were observed, and the results were shown in Fig. 3.5. Many secondary carbides precipitated on the matrix of the ESR ingots after they were taken out and buried in sand for annealing. Considering that the annealing temperature was relatively low, the primary carbides in ESR ingots have basically preserved its original morphology, which is in the shape of large bar or block, and has strong continuity. After high temperature diffusion annealing, the continuity between primary carbides is greatly weakened, and a large number of primary carbides with rod structure are dissolved and broken, forming spherical or granular primary carbides, while the primary carbides with block structure remains

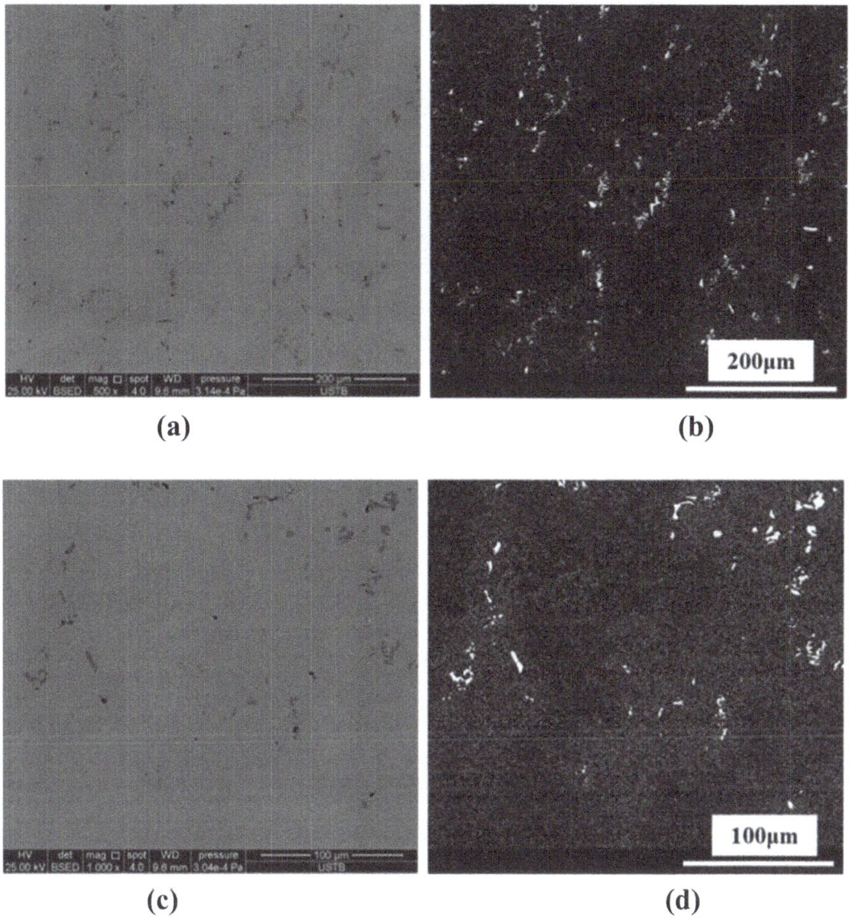

Fig. 3.3 Primary carbides in finish hot rolling process: **a** Primary carbides in 500 times field of view; **b** Primary carbide identified and reversely displayed by image processing software; **c** Primary carbide at 1000 times field of view; **d** Fig. **c** primary carbide identified and reversely displayed using image processing software

Fig. 3.4 Schematic view of ESR ingot in high temperature diffusion annealing process

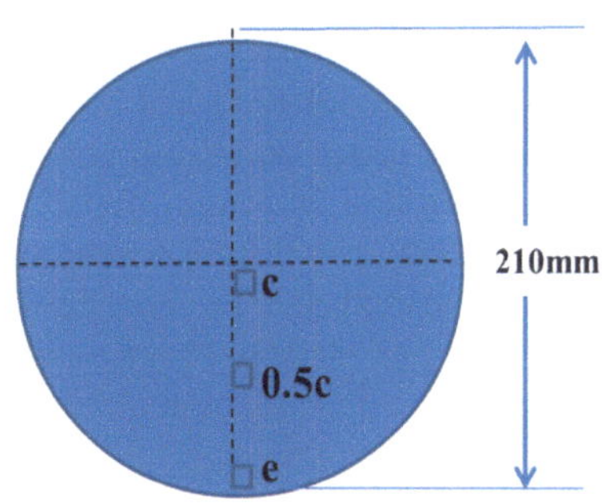

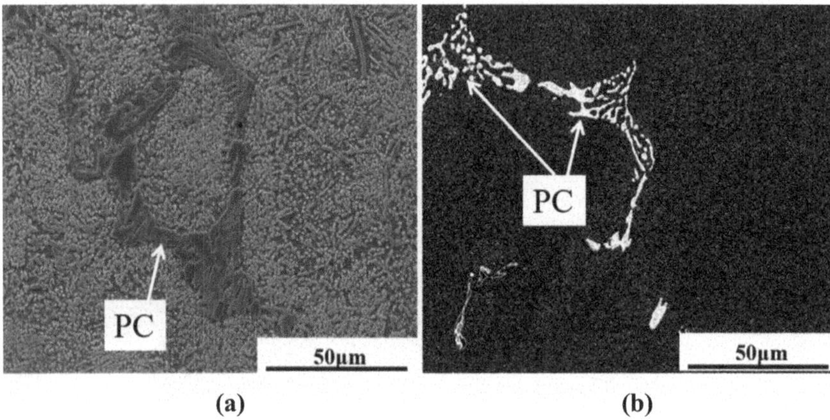

Fig. 3.5 Influence of high temperature diffusion annealing on the morphology of primary carbide at 1/2 radius of ESR ingot (PC represents primary carbide): **a** Before annealing; **b** After annealing

basically unchanged. According to the formation and growth mechanism of primary carbides, the primary carbides with bar structure are mainly eutectic carbides, which precipitated at the eutectic composition of molten steel, while the primary carbides with block structure are the pre-eutectic carbide precipitated when the molten steel reaches hypereutectic composition. Compared with common eutectic carbides, these bulk carbides have a higher dissolution temperature.

Scanning electron microscope was used to observe the dissolution of primary carbide at different positions of ESR ingot by high-temperature diffusion annealing. IPP image analysis software was used to calculate the content and dissolution rate of primary carbides at different positions of ESR ingot. The observation and statistical results are shown in Figs. 3.6 and 3.7, respectively.

According to the comprehensive analysis of Figs. 3.6 and 3.7, diffusion annealing at 1180 °C for 2 h can effectively dissolve the primary carbide in the ESR ingots. The rod-shaped eutectic carbides are dissolved and broken, and the dissolution ratio of primary carbide at each positions are above 50%. Because that the sizes of the original primary carbides at the center and 1/2 radius of the ESR ingot are larger, the dissolution efficiency of the high temperature diffusion annealing is lower. The original size of primary carbides at the edge of ESR ingots are fine, and there are with more rod structures and less block structures.

After diffusion annealing process at high temperature, the average content of primary carbide decreased from 1.37 to 0.66%. The fine primary carbide can be dissolved better in diffusion annealing process at high temperature. Therefore, it is an effective method to promote the dissolution of primary carbide that decreasing the depth of liquid metal pool and the secondary dendrite spacing, as well as shortening the local solidification time in the ESR process.

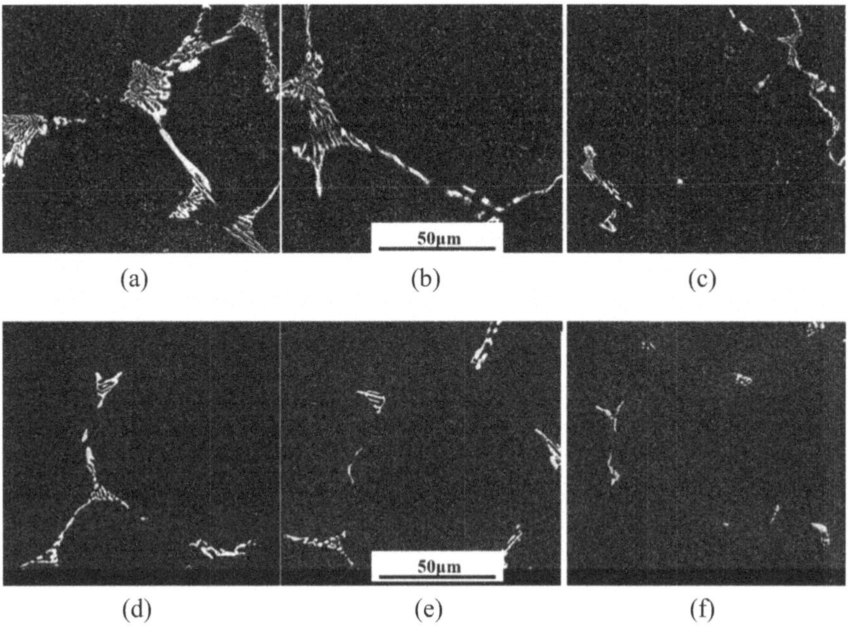

Fig. 3.6 Effect of elevated temperature diffusion annealing on the dissolution of primary carbides at different positions of ESR ingot: **a** and **d** are the center of ESR ingot before and after high temperature diffusion annealing respectively; **b** and **e** are 1/2 radius of the ESR ingot before and after high-temperature diffusion annealing respectively; **c** and **f** are the edges of ESR ingots before and after high-temperature diffusion annealing

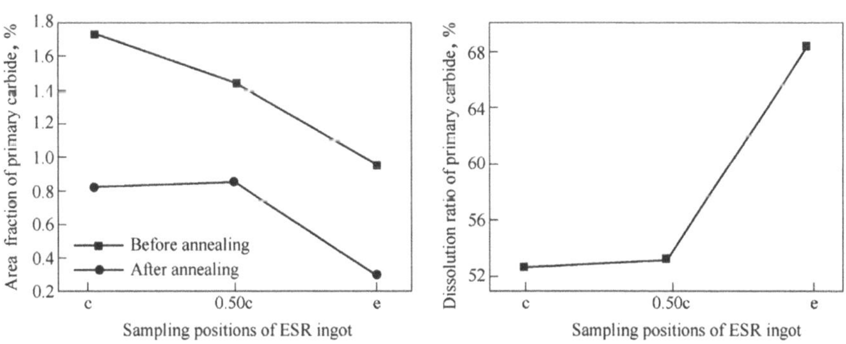

Fig. 3.7 Effects of high temperature diffusion annealing on the content and dissolution ratio of primary carbide at different positions of ESR ingots

3.1.3 *Effect of Diffusion Annealing of Hot Rolling Slab on Primary Carbide*

High temperature diffusion annealing is carried out on hot rolling slab 30 mm thick after hot rolling casting. Four samples in the size of 15 mm × 15 mm × 30 mm were randomly selected from the hot rolling slab and heated to 1180 °C in a heating furnace for 30 min, 60 min, 90 min and 120 min respectively before air cooling. The metallographic structure and XRD pattern of 8Cr13MoV hot rolling slab are shown in Fig. 3.8 after air cooling at 1180 °C for 30 min. The black and white aciculate tissue in the figure is martensite, and the flat gray and yellow tissue is residual austenite. There are many broken primary carbides distributed in the grain boundary.

The microstructure and XRD patterns show that the residual austenite content in the hot rolling slab is very high after 30 min of diffusion annealing. According to the determination method of residual austenite content in GB 8362—87, the residual austenite content is 57.5%. According to the calculation in Sect. 1.2.2 of this book on the change of the element content in the newly-formed austenite with temperature during the solidification of 8Cr13MoV steel, the content of carbon and alloying elements in the austenite increases gradually from the grain center to the edge. The increase in the content of carbon and alloying elements can improve the stability of supercooled austenite [11, 12] and cause the C curve to move to the right. Therefore, the residual austenite is mainly caused by the enrichment of carbon and alloying elements near the grain boundary, which leads to the high stability of austenite in this region and the absence of martensite transformation during the cooling process. In the process of air cooling after diffusion annealing at high temperature, the regions with low carbon content are transformed into martensite, and the regions with high carbon content form residual austenite which is retained to room temperature.

The metallographic structure of 8Cr13MoV hot rolling slab after heating to 1180 °C for different time in diffusion annealing is shown in Fig. 3.9.

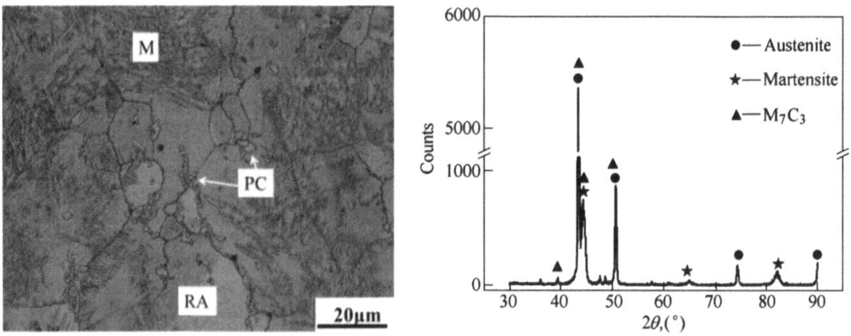

Fig. 3.8 Metallographic structure and XRD patterns of 8Cr13MoV hot rolling slab after 30 min diffusion at high temperature (M, RA and PC respectively represent martensite, residual austenite and M_7C_3 primary carbide)

Fig. 3.9 Metallographic structure of 8Cr13MoV hot rolled plate after high temperature diffusion annealing: **a** 30 min; **b** 60 min; **c** 90 min; **d** 120 min

As can be seen from Fig. 3.9, when the high-temperature diffusion annealing was conducted for 30 and 60 min, the metallographic structure did not change much, and the residual austenite content in the microstructure was relatively high. When the heat preservation time reached 90 min, the residual austenite content in the microstructure significantly decreased. When the heat preservation time is 120 min, the residual austenite content has been reduced to less than 5%. The decrease of residual austenite content indicates that the enrichment area of carbon and alloying elements in steel decreases and the distribution of carbon and alloying elements tends to be homogenized. Uniform distribution of carbon and alloying elements can also avoid the phenomenon of carbide aggregation caused by carbon segregation, which is beneficial to obtain uniform distribution of secondary carbides in subsequent spheroidizing annealing process. In the diffusion annealing process of 8Cr13MoV hot rolling slab at 1180 °C, the dissolution process of primary carbide is shown in Fig. 3.10.

It can be seen from Fig. 3.10 that the primary carbides generated in the electroslag remelting process are mostly distributed at the grain boundary, and the primary carbides with coiled and rod-like structure at the grain boundary are broken and dispersed around the grain boundary in the hot-rolling process. As the diffusion

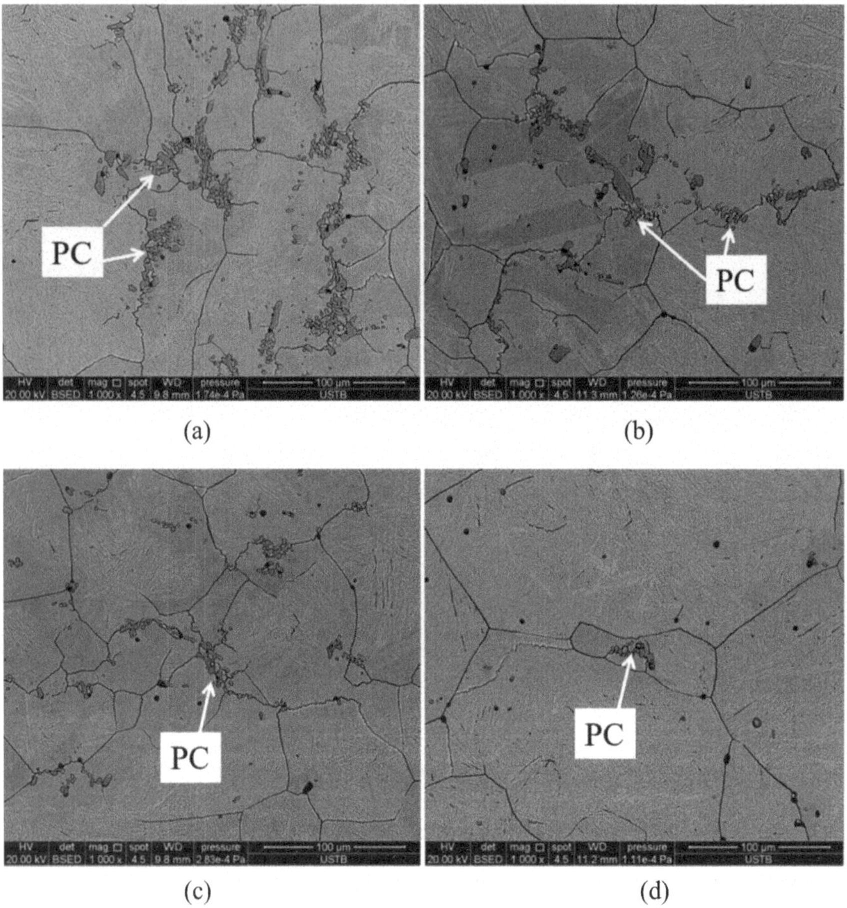

Fig. 3.10 Primary carbide dissolution process (PC represents primary carbide) in the diffusion annealing process of 8Cr13MoV hot rolled plate at 1180 °C: **a** Heat preservation for 30 min; **b** Heat preservation for 60 min; **c** heat preservation for 90 min; **d** Heat preservation for 120 min

annealing time was 30 min, the content of primary carbide at the grain boundary was still high, and the primary carbide was agglomerated around the grain boundary in the form of blocks or small particles. At 60 min of heat preservation, a large number of primary carbides far away from the grain boundary dissolved, and the primary carbides at the grain boundary did not change significantly. At the temperature for 90 min, the primary carbide at the grain boundary dissolved, and the size and quantity decrease. At 120 min of heat preservation, most primary carbides were dissolved, and only a small amount of primary carbides remained at the grain boundary.

In the process of diffusion annealing at high temperature, the primary carbide which is far away from the grain boundary first dissolves. During the solidification of liquid steel, the austenite nuclei grow gradually, and the grain boundary is the final solidification region. The content of alloying elements is the lowest in the austenite formed at the center of grain, and the content of alloying elements such as carbon and chromium is the highest in the austenite finally formed. Therefore, the concentration gradient of carbon, chromium and other elements near the primary carbide far away from the grain boundary is larger. In the process of high-temperature diffusion annealing, carbon, chromium and other elements in the primary carbide disperse faster to the matrix. Therefore, in the process of high temperature diffusion annealing, primary carbides far away from grain boundary are preferred to be dissolved. With the extension of holding time, primary carbides at grain boundary are gradually dissolved.

Image analysis software was used to calculate the influence of heat diffusion annealing holding time on the content of primary carbide in steel. The results are shown in Fig. 3.11.

It can be seen from Fig. 3.11 that the area fraction of primary carbide in 8Cr13MoV hot rolling slab shows an obvious decreasing trend with the extension of the heat diffusion annealing holding time. The original primary carbide area fraction in hot rolling slab is 2.37%. After 120 min of high temperature diffusion annealing, the primary carbide area fraction decreases to 0.17% and the primary carbide reduction efficiency reaches 92.8%. It can be seen that the high temperature diffusion annealing process of hot rolling slab plays an important role in the dissolution of primary carbide.

In Sect. 3.1.1, high temperature diffusion annealing can reduce the area fraction of primary carbide by 52%, while in this section, the same process of high temperature diffusion annealing on hot rolling slab reduces the area fraction of primary carbide by 92.8%. Therefore, the quantity of primary carbide decreases more when the hot rolling slab is subjected to high temperature diffusion annealing. The main reasons are as follows: the content of carbon and alloying elements is higher in the original position of primary carbide in ESR ingots, the element concentration gradient is smaller in the high temperature diffusion annealing, and the dissolution of primary

Fig. 3.11 Effect of 8Cr13MoV hot rolling slab holding time on primary carbide area fraction by high temperature diffusion annealing

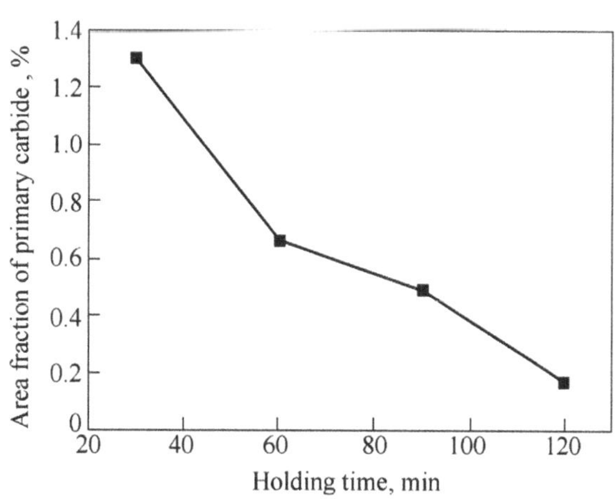

carbide is slower. After hot rolling, the primary carbide is broken and distributed around the austenite with low content of carbon and alloy elements. At this time, the concentration gradient between the primary carbide and the surrounding matrix is larger, which is more conducive to the diffusion of elements in the heating and heat preservation process.

3.1.4 Effect of High Temperature Diffusion Annealing on Network Carbides in High Carbon Steel

Network carbides in high carbon steel are generated during the cooling after rolling. It is because that the solubility of carbon in austenite decreases as the temperature decreases, which causes the precipitation of carbon along the austenite grain boundary during cooling. The formation of network carbides has been studied by many researchers. A common consensus has been given by these researchers. It is that network carbides mainly arise from central segregation during solidification of billet. The central segregation of continuous casting billet cannot be completely eliminated. In order to avoid the generation of martensite at the core of high carbon steel, slow cooling is often used in the late phase of phase transition. However, according to the continuous cooling curve, secondary cementite is easily precipitated in the slow cooling process. At the same time, in the process of hot rolling or annealing, due to the high heating temperature and long holding time, the austenite grain size is coarse. The carbide precipitates along the grain boundary in the subsequent slow cooling process, and then forming the network carbides. In addition, if the finish rolling temperature is too high, in the following slow cooling process will be easy to form the network carbide. The generation of the network carbides is closely related to the chemical composition of steel and the degree of original carbides segregation. The greater the degree of the original carbide segregation in the casting billet, the more likely the network carbides will be formed in the carbides dense area. The subsequent improper heat treatment processing will also aggravate the degree of the network carbide segregation. Studies have shown that the network carbides in steel tend to be coarsened and continuous due to the small shape variable, the high finish rolling temperature and slow cooling rate after rolling.

Network carbide increases the non-uniformity of the chemical composition of the steel, which is easy to cause excessive structural stress in the quenching during the heat treatment, leading to the deformation and cracking of the parts. The presence of the network carbide greatly weakens the connection between the matrix grains and then adversely affects the mechanical properties of the steel. The impact performance is the most important one, which decreases with the increase of the network carbide level. The network carbide also has a significant effect on the limit of bending strength and tensile strength. For example, it causes the cracking of the tool steel and shortens the service life of rolled pieces. Besides, with the increase of the network carbide level, the contact fatigue strength of steel decreases. When there keeps coarse

Table 3.2 Chemical composition of 82B steel (wt%)

C	Si	Mn	Cr	P	S
0.81–0.83	0.21–0.25	0.73–0.77	0.25–0.27	≤0.017	≤0.012

Table 3.3 Central carbon segregation degree of different samples

Sample no.	1#	2#	3#	4#	5#	6#	7#	8#
Central carbon segregation degree	1.10	1.08	1.07	1.06	1.05	1.03	0.98	0.95

network carbide, the contact fatigue strength of the longitudinal specimens decreases by about 30%. Furthermore, the service life of the part is reduced by about 1/3 with each increase of the network carbide level. The severe network carbide cannot be eliminated in the spheroidizing annealing process, and only by normalizing process can the network carbide be eliminated or improved. The network carbide with less bad structural properties, even if parts of the network carbide can break or even be spheroidized in spheroidizing annealing, still causes the larger carbide and the non-uniformity of carbide.

Taking 82B steel as an example, the effect of central carbon segregation on network carbides of billets is ascertained. The chemical composition of 82B steel is shown in Table 3.2 Central carbon segregation of the steel is shown in Table 3.3 Metallographic structure of different central carbon segregation billet is shown in Fig. 3.12.

As can be seen from Fig. 3.12, there are several coarse and closed carbides connected into a piece on the casting billet sample with the central carbon segregation above 1.07, forming an obvious network carbide. When the central carbon segregation decreases to 1.06 and 1.05, a small amount of carbides which do not form a closed net can still be seen on the casting billet. but when the central carbon segregation decreases to 1.03, carbides are basically invisible. When the center of the casting billet is negative segregation, the amount of the network carbide also tends to increase with the aggravation of the negative segregation.

In practical production, the carbide segregation is generally controlled by high temperature diffusion annealing in the heating furnace, and then the formation of network carbides is controlled. The high temperature diffusion annealing process is as follows: the preheating time of the preheating section is about 33 min, and the temperature is between 910 and 940 °C. The heating time in the heating section is 42 min and the temperature is between 1130 and 1150 °C. The soaking time of the soaking section is 25 min, and the temperature is between 1250 and 1280 °C. The initial rolling temperatures are about 1020, 1060 and 1080 °C. The carbon content of the rolled billets is analyzed. The sampling position is shown in Fig. 3.13. The results are listed in Table 3.4.

As shown in Table 3.4, the severe central segregation in billets is improved to some extent after heating of the billets. The improvement effect is relatively close at these three heating cycles. The carbon content is decreased from 0.97% to about

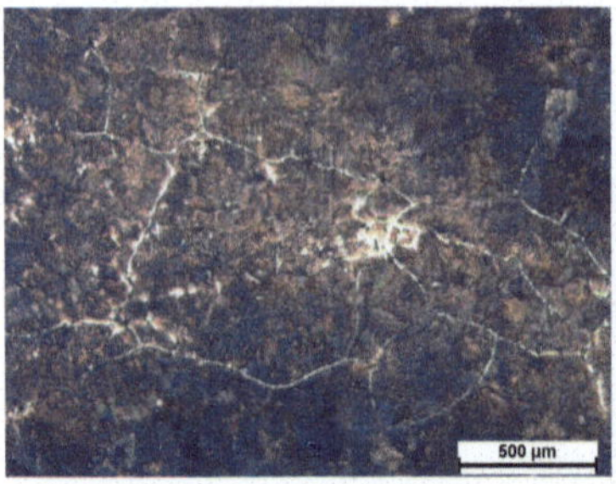

1#(Central carbon segregation degree is 1.10)

2#(Central carbon segregation degree is 1.08)

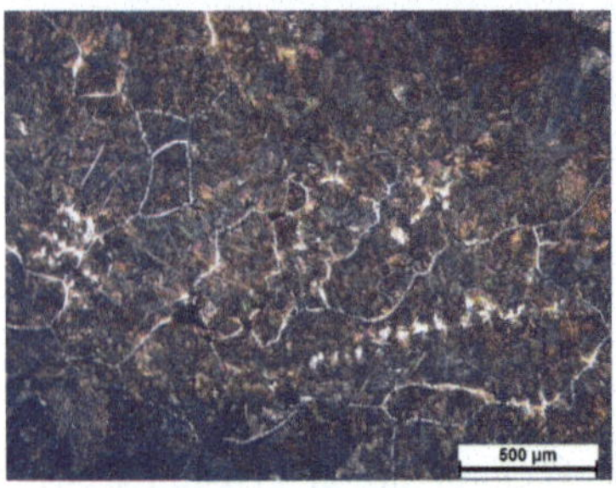

3#(Central carbon segregation degree is 1.07)

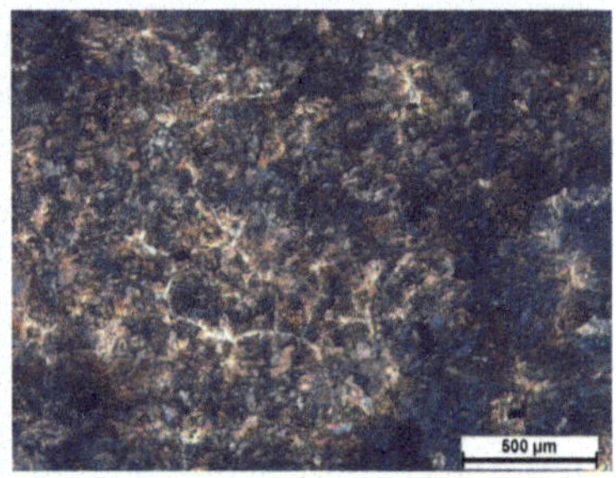

4#(Central carbon segregation degree is 1.06)

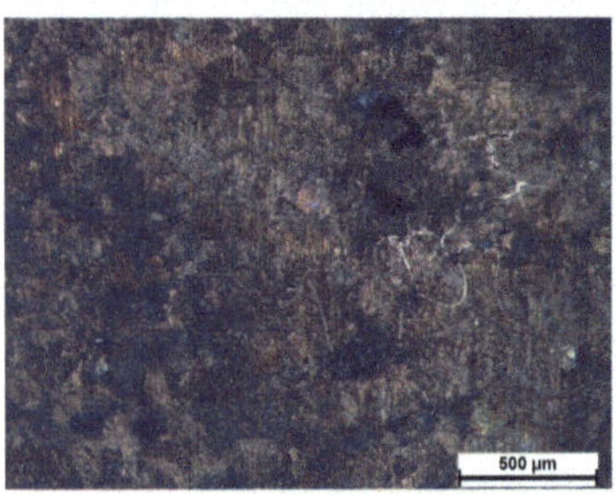

5#(Central carbon segregation degree is 1.05)

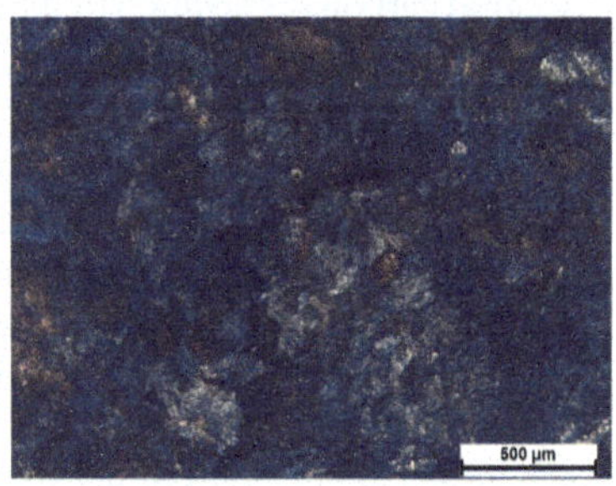

6#(Central carbon segregation degree is 1.03)

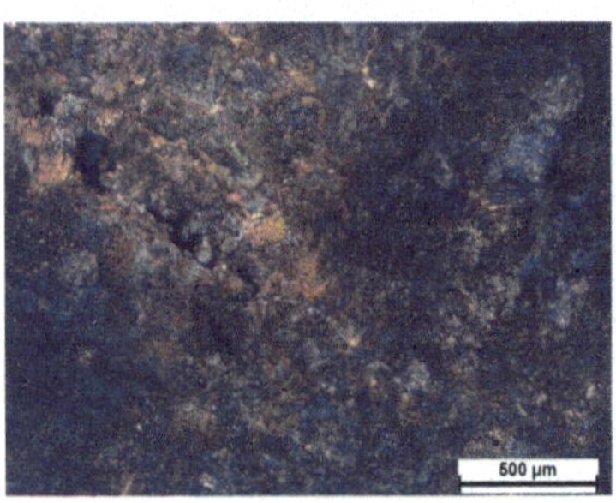

7#(Central carbon segregation degree is 0.98)

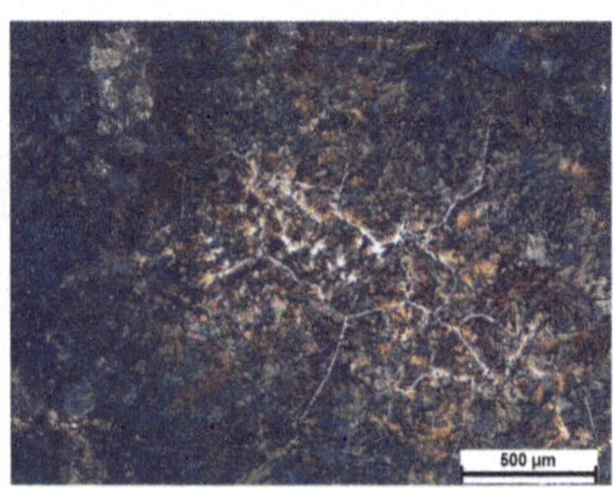

8#(Central carbon segregation degree is 0.95)

Fig. 3.12 Microstructure of the billets with different central carbon segregation

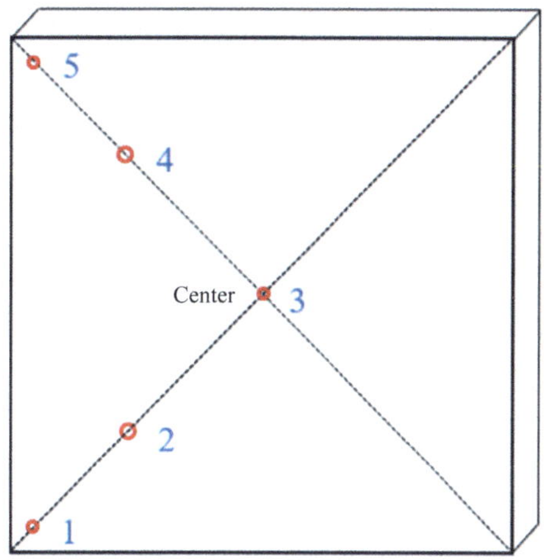

Fig. 3.13 Sampling position at the cross section of billet

Table 3.4 Carbon segregation index of the billets after heating

Number	Processing parameter	Carbon content at different position, wt%					Segregation index
		1	2	3	4	5	
3307934	Without heating	0.84	0.88	0.97	0.89	0.83	1.100
1080#	Initial rolling temperature 1080 °C	0.81	0.85	0.89	0.86	0.82	1.052
1060#	Initial rolling temperature 1060 °C	0.80	0.85	0.88	0.85	0.81	1.050
1020#	Initial rolling temperature 1020 °C	0.80	0.83	0.88	0.85	0.79	1.060

0.89%. The central segregation index is decreased from 1.10 to 1.06. The segregation of the rolled billets at initial rolling temperatures 1080 and 1060 °C is lower than that at 1020 °C.

The microstructure of the rolled billets at inital rolling temperatures 1020, 1060 and 1080 °C is shown in Figs. 3.14, 3.15 and 3.16.

As known in Figs. 3.13, 3.14 and 3.15, the metallographic structure of the casting billet with initial rolling temperature about 1080 °C shows no network carbide, the same is true of the initial rolling temperature about 1060 °C. However, when the initial rolling temperature drops to 1020 °C, there keeps the obvious network carbide.

In order to meet the high initial rolling temperature, it is necessary to increase the heating temperature when the cooling conditions of the casting billet are not different from each other. The carbon atoms at the center diffuse more fully toward the edges and thus reduce the formation of the central network carbide for the reason that the increase of heating temperature promotes the diffusion of carbon atoms.

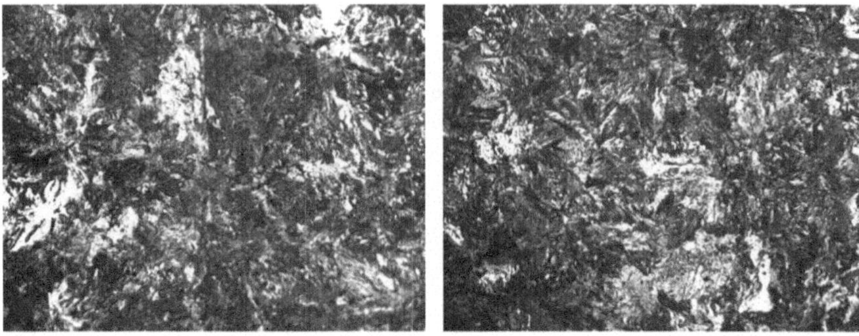

Fig. 3.14 Metallographic structure of the casting billet at the inital rolling temperature 1080 °C

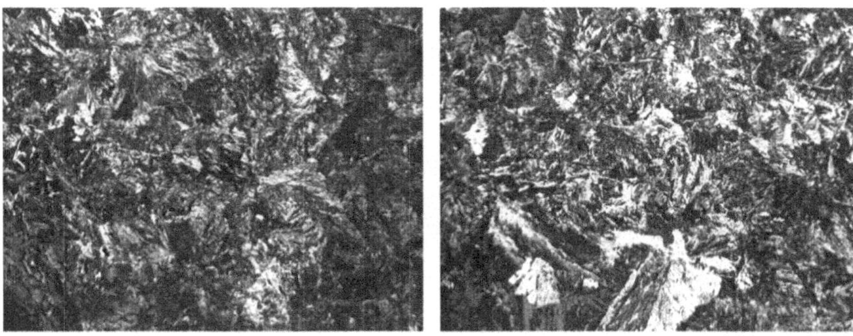

Fig. 3.15 Metallographic structure of the casting billet at the inital rolling temperature 1020 °C

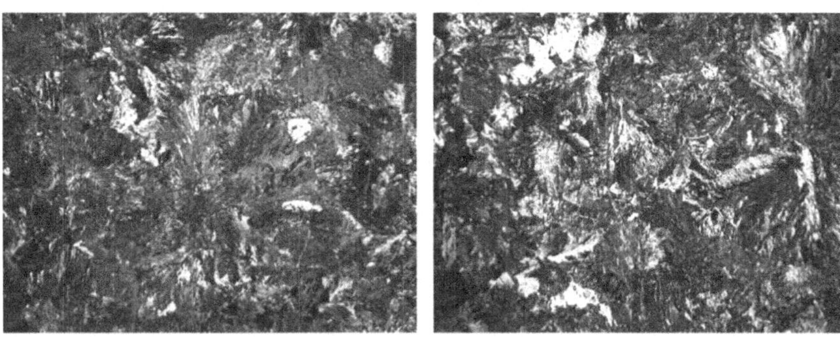

Fig. 3.16 Metallographic structure of the casting billet at the inital rolling temperature 1060 °C

3.2 Effect of Hot Rolling Process on Carbide

Hot rolling process is an important link in the production of high-carbon martensite stainless steel. In this process, not only the primary carbide is further broken, but also the grain size is affected by the hot rolling process [13, 14].

3.2.1 *Effect of the Deformation in Hot Rolling on Carbide*

(1) Effect of hot rolling deformation on carbides of high carbon martensite stainless steel

8Cr13MoV inital rolling slab with initial thickness of 30 mm is adopted. The slab is processed into 100 mm × 100 mm × 30 mm for standby use. Requirements for hot rolling heating process: heating rate shall not exceed 100 °C/h before 600 °C, heating to 1200 °C for 2 h. The different deformation amounts of hot rolling are grouped, and the specific parameters are shown in Table 3.5.

Figure 3.17 shows the microstructure of specimens with different deformation during hot rolling.

It can be seen from Fig. 3.17 that under the condition of the same inital rolling temperature and small deformation, the finish rolling temperature is higher, enough time is allowed for recrystallization, and the grains are evenly distributed equiaxial crystals. With the increase of deformation amount, the grain cannot recover after deformation and elongation under the action of external force [15], and the grain will be further refined due to the hindrance of carbide in the deformation process. In unit area, sample No. 3 has far more grains than sample No. 1. It can be seen from Fig. 3.17d and f that the size of carbide in sample No. 1 with the minimum deformation is significantly larger than that of the other two samples.

The optical microstructure of the carbide of the sample is shown in Fig. 3.18. The average area of a single carbide from sample No. 1 to No. 3 was 14.47 μm^2, 9.89 μm^2 and 7.87 μm^2, respectively.

The above results show that the carbide size decreases with the increase of hot rolling deformation. A larger amount of pressure decreases the degree of primary carbide extrusion, and the deformation temperature decreases with the increase of

Table 3.5 Parameters of different hot rolling deformation

Sample no.	Inital rolling temperature, °C	Pass	Thickness, mm	Reduction rate, %
No. 1	1200	1	20	33.3
No. 2	1200	2	12.5	58.3
No. 3	1200	4	5	83.3

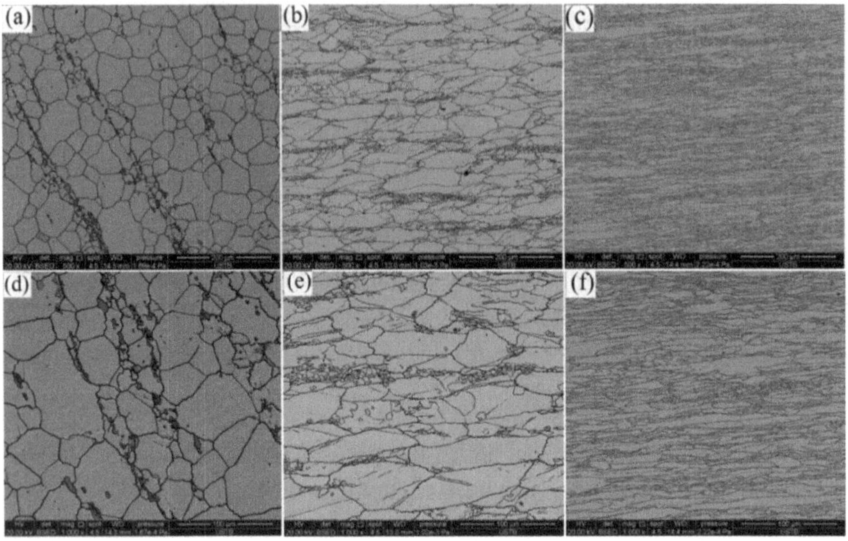

Fig. 3.17 Scanning electron microscope (SEM) photos of microstructure of hot rolling samples with different deformation amounts: **a** and **d** No. 1; **b** and **e** No. 2; **c** and **f** No. 3

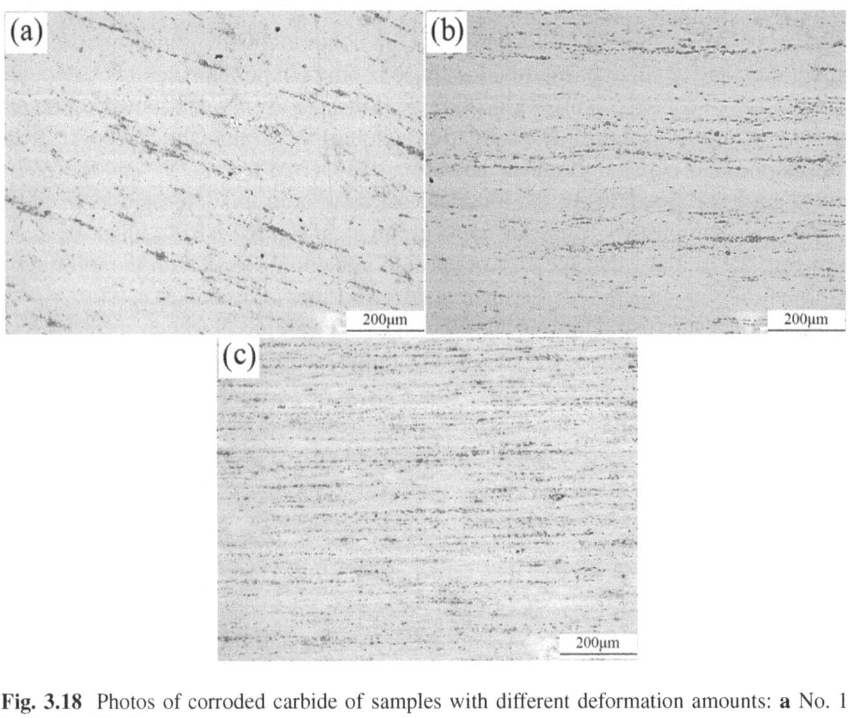

Fig. 3.18 Photos of corroded carbide of samples with different deformation amounts: **a** No. 1; **b** No. 2; **c** No. 3

pressure. The extrusion force of the matrix on the primary carbide increases accordingly. The deformation of hot rolling in industrial production should be determined according to the requirements of cold rolling products. Increase the deformation of hot rolling as much as possible so as to reduce the rejection rate of cold rolling.

3.2.2 Effect of Hot Rolling Temperature on Carbide

3.2.2.1 Effect of Hot Rolling Temperature on Carbides in High Carbon Martensite Stainless Steel

(1) Effect of hot rolling start temperature on the microstructure and carbide

Different initial rolling temperatures were grouped, results are shown in Table 3.6.

SEM-BESD pictures of samples at different inital rolling temperatures are shown in Fig. 3.19.

It can be seen from Fig. 3.19 that the elongation of grain becomes more obvious with the decrease of the inital rolling temperature. The finish rolling temperatures of Samples No. 4 and No. 6 are basically the same, respectively 890 and 869 °C. The higher the finish rolling temperature, the higher the dynamic recrystallization degree of the material structure is. The comparison of carbides in the microstructure showed that the higher the initial rolling temperature was, the larger the size of carbides was. The specimen is corroded by a special method of etching carbides, and the light mirror image shown in Fig. 3.20 is obtained.

According to the simple comparison of photos in Fig. 3.20, with the decrease of the inital rolling temperature, the amount of carbide in the sample also seems to decrease. IPP image software was used to calculate the volume fraction of carbide. The results showed that the carbide volume fractions of No. 4–No. 6 samples were 2.36%, 1.50% and 1.13%, respectively. In the range of rolling temperature, the dissolution rate of carbides into the matrix should be very low, and only the carbides small enough can be completely dissolved. Particularly fine carbides are also ignored by the software. However, this is sufficient to indicate that the higher the rolling temperature, the more adverse3.20 to carbide crushing. On the contrary, the reduction of rolling temperature is beneficial to carbide crushing and dissolution of carbide into matrix.

Table 3.6 Different inital rolling temperature parameters

Sample no.	Inital rolling temperature, °C	Pass	Thickness, mm	Reduction rate, %
No. 4	1200	4	5	83.3
No. 5	1100	4	5	83.3
No. 6	900	4	5	83.3

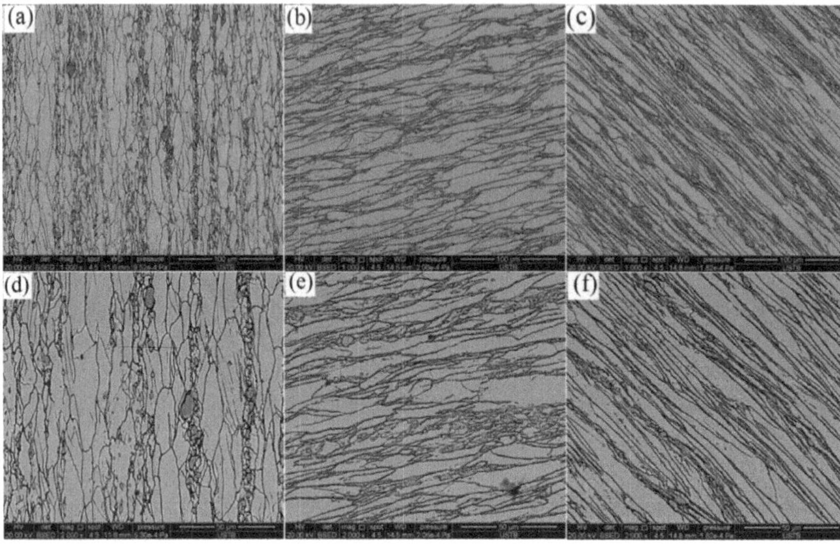

Fig. 3.19 SEM-BSED images of hot rolling at different initial rolling temperatures: **a** and **d** No. 4; **b** and **e** No. 5; **c** and **f** No. 6

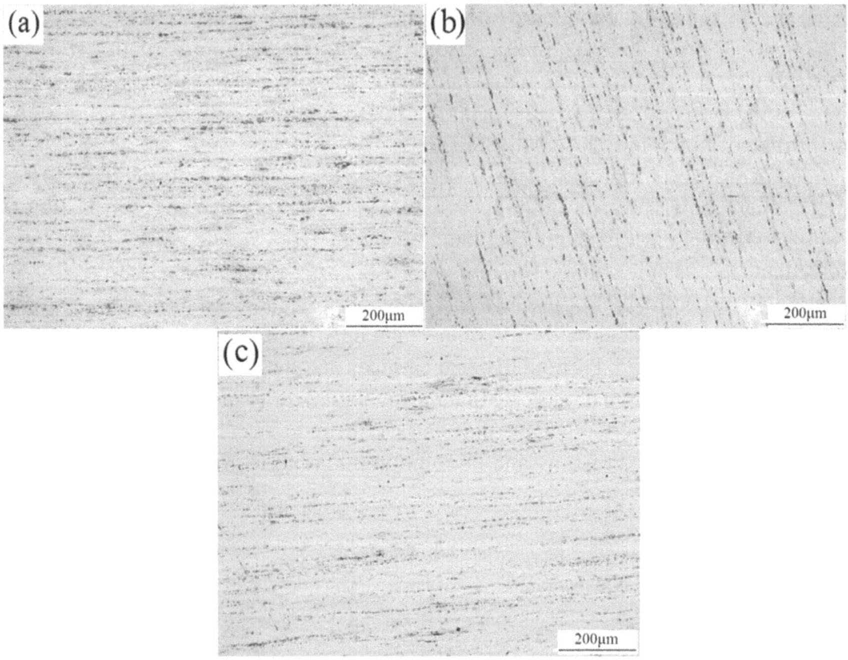

Fig. 3.20 Photos of corroded carbide of samples at different inital rolling temperatures: **a** No. 4; **b** No. 5; **c** No. 6

Table 3.7 Parameters of different finish rolling temperature

Sample no.	Finish rolling temperature, °C	Pass	Thickness, mm	Reduction rate, %
No. 7	900	4	5	83.3
No. 8	800	4	5	83.3
No. 9	700	4	5	83.3

(2) Effect of finish rolling temperature on microstructure and primary carbide

Different finish rolling temperatures of hot rolling are shown in Table 3.7.

The effect of finish rolling temperatures on the microstructure is shown in Fig. 3.21.

As can be seen from Fig. 3.21, with the decrease of the finish rolling temperature, the grains in the structure were elongated more obviously and the grain size was more uneven. It is found that the lower the finish rolling temperature is, the smaller and more irregular the grains around the carbide are, as shown in Fig. 3.21d–f. From the magnified grain boundary, the corrosion of grain boundary of sample No. 9 with the lowest finish rolling temperature is the most serious, especially around the carbide, which indicates that there are more defects caused by deformation around the carbide. These defects were not sufficiently recovered during cooling due to the low finish rolling temperature. In contrast, with high finish rolling temperature, these defects can be eliminated during static recrystallization and recovery. The fine grains around the carbide grow by recrystallization. As shown in Fig. 3.21e, some grains grow through recrystallization to enclose a carbide. The carbide corroded microstructure is shown in Fig. 3.22.

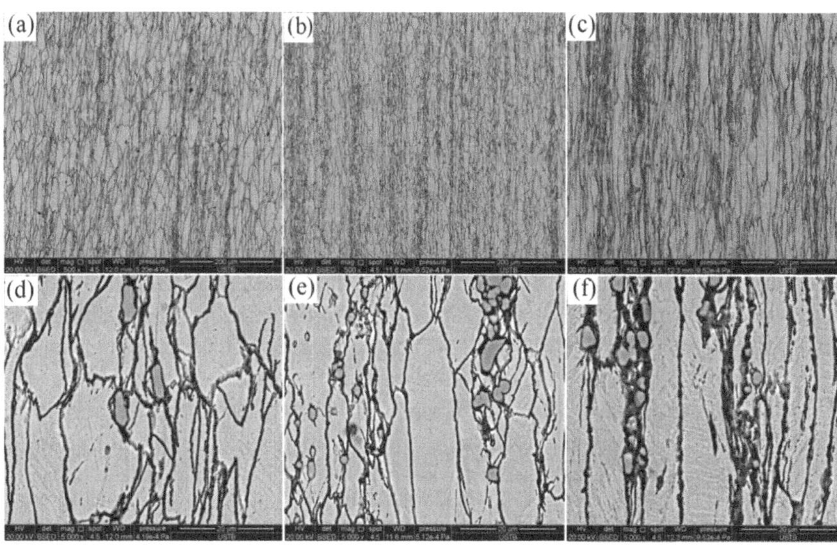

Fig. 3.21 SEM-BSED images of hot rolling samples at different finish rolling temperatures: **a** and **d** No. 7; **b** and **e** No. 8; **c** and **f** No. 9

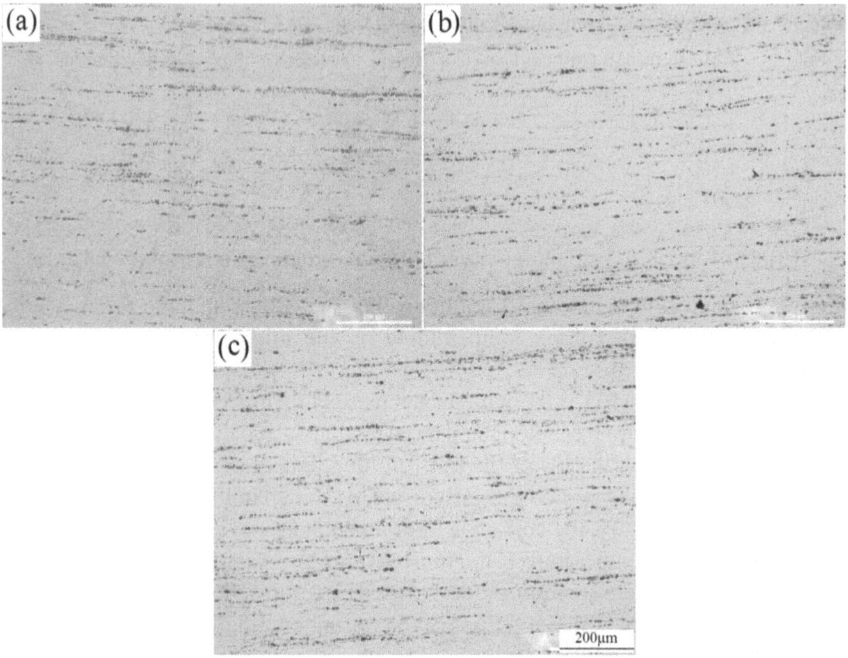

Fig. 3.22 Photographs of corroded carbides of samples at different finish rolling temperatures: **a** No. 7; **b** No. 8; **c** No. 9

As can be seen from Fig. 3.22, with the decrease of the finish rolling temperature, the particle size of carbide after crushing becomes smaller, which is also reflected in Fig. 3.21. According to the statistics of carbide size by IPP software, the average area of single carbide from No. 7 to No. 9 is 12.67 μm^2, 9.14 μm^2 and 9.11 μm^2, respectively. Under the condition of high finish rolling temperature, the size of carbide in the microstructure is large.

Although low finish rolling temperature is helpful for carbide crushing to some extent, the lower the finish rolling temperature is, the more defects will be generated in the microstructure, especially around the carbides, dislocation movement is hindered to gather, in addition, the carbides are mostly distributed in the grain boundary position, and the bonding with the matrix is weak. Therefore, in actual production, if the finish rolling temperature is too low, it may lead to cracks at the edge of hot rolling slab or internal microcracks, which will affect the product yield. Moreover, after the temperature reduction, the required rolling force will also be increased, which also has certain requirements on equipment capacity. Therefore, consideration should be given to the yield in the production process, and the finish rolling temperature should be appropriately reduced without decreasing the yield if the equipment capacity allows [16].

3.2.3 *Effect of Rolling Temperature on Reticulated Carbides in GCr15 Bearing Steel*

The effect of different rolling temperature on the microstructure of bearing steel GCr15 is shown in Figs. 3.23 and 3.24.

It can be seen from Figs. 3.23 and 3.24, the grain size of bearing steel increases gradually with the increasing of finishing temperature. The effect of finishing temperature on network carbide in GCr15 is shown in Fig. 3.25.

Figure 3.25 shows that when the finishing rolling temperature is 900 °C, the lamellar pearlite structure is coarse and the carbide network degree is serious. When the finishing rolling temperature reduces to 800 °C, the lamellar pearlite structure

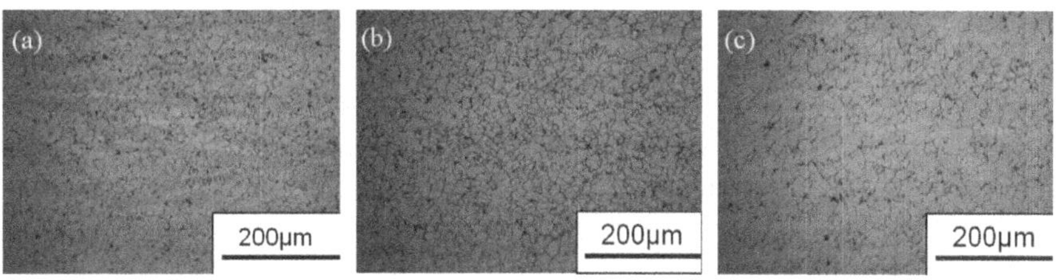

Fig. 3.23 Microstructure of GCr15 bearing steel at different finish rolling temperatures: **a** 700 °C; **b** 800 °C; **c** 900 °C

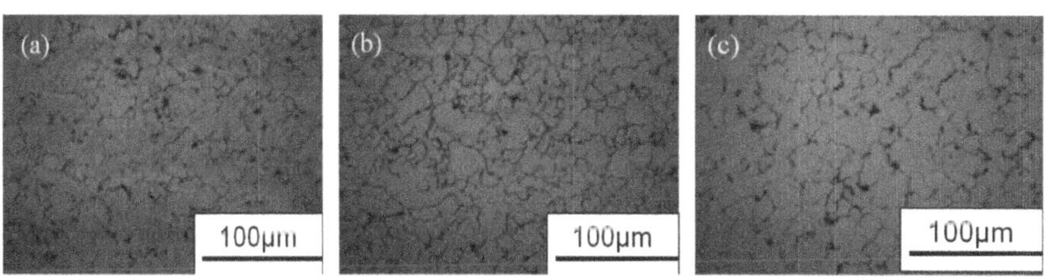

Fig. 3.24 Microstructure of GCr15 bearing steel at different finish rolling temperatures: **a** 700 °C; **b** 800 °C; **c** 900 °C

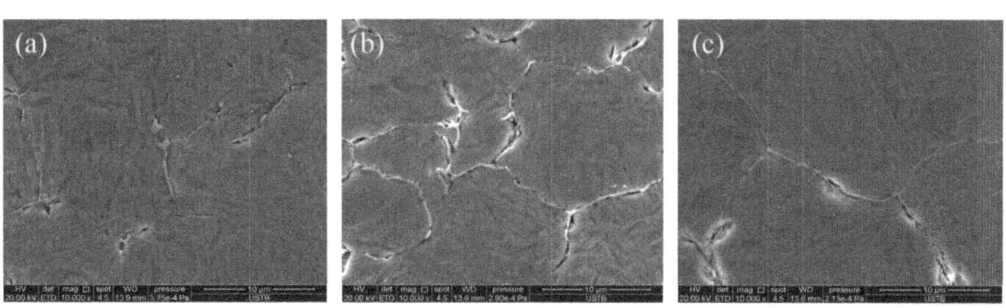

Fig. 3.25 SEM morphology of GCr15 bearing steel at different finish rolling temperatures: **a** 700 °C; **b** 800 °C; **c** 900 °C

is gradually refined and the carbide network degree is reduced. When the finishing rolling temperature reduces 700 °C, the pearlite structure is very refined and the carbide network degree is further reduced, and the carbide is in strip distribution Some of them have been transformed into spherical or near spherical carbides. This is mainly due to the high content of carbon and chromium in bearing steel GCr15. During the continuous cooling process after rolling, the solubility of carbon in austenite gradually decreases with the decreasing of temperature. Therefore, excess carbon will precipitate from austenite to form chromium rich proeutectoid secondary carbides. The formation of proeutectoid secondary carbides is affected by the diffusion rate of elements. Due to the many defects at the grain boundary, the diffusion speed of carbon and chromium in the grain boundary is much faster than that in the grain. The proeutectoid secondary carbides nucleate and grow preferentially at the grain boundary, and then connect with each other to form a network structure.

When the finishing temperature is 900 °C, the austenite matrix located in the complete recrystallization zone, which leads to the refined grain after recrystallization, and increases the position of carbide precipitation correspondingly. However, a large number of carbides will precipitate along austenite grain boundary during cooling after high rolling temperature, which leads to the serious carbide network in bearing steel GCr15. When the finishing rolling temperature is 800 °C, the austenite matrix located in the intermediate region between the high temperature recrystallization zone and non-recrystallization zone, the bearing steel GCr15 is in the non recrystallization zone at the end of rolling, which leads to increase the deformation bands and the carbide nucleation sites in the deformed austenite grains. Therefore, the precipitation of carbides is more dispersed, and the refined pearlite or degenerated pearlite is formed, and the microstructure is gradually refined. When the finishing rolling temperature is 700 °C, the bearing steel GCr15 located in the two-phase temperature range of carbide and austenite, the deformation of carbide and austenite without recrystallization occurs at the same time during the rolling process, which leads to increase the distortion degree and dislocation density in austenite grains. Therefore, the precipitation of carbides is more dispersed and the pearlite pellets are more refined. In addition, the first precipitated carbides also deform. Therefore, the network carbides gradually dissolve and fracture, and finally form dispersed strip carbides.

With the increasing of finishing rolling temperature, the degree of grain boundary corrosion of bearing steel gradually decreases, especially in the area near carbide. This indicates that defects are easy to form and gather in the area near carbides. With the increasing of finish rolling temperature, the atomic energy of defects increases, the movement becomes more active and the dislocation migration is promoted during the reddening process of bearing steel GCr15. Therefore, the defects near carbides can be effectively recovered and removed during the recrystallization process. Conversely, with the decreasing of finishing temperature, the defects near carbides can not be removed due to the inhibition of dislocation migration. Meanwhile, due to the decrease of finishing temperature, it is difficult to recrystallize the fine grains, especially the fine grains near carbides, so the distribution of grain size is not uniform.

3.3 Effect of Hot Rolling Process on Carbide

3.3.1 Carbide and Microstructure Analysis of Cold Rolling Slab

Figure 3.26 shows the microstructure of the 7Cr17MoV strips obtained via hot rolling, cold rolling, and annealing.

Figure 3.26 indicates that the microstructure of 7Cr17MoV after rolling and annealing consisted of pearlite and globular carbides. After cold rolling and annealing, the number of large carbide particles decreased and the distribution of the particles became more uniform, i.e., the microstructure became uniform and fine. The microstructure and the carbide particles of the 0.7 mm cold-rolled strip were finer than those of the 1.5-mm strip.

The carbides in cold rolling slab mainly came from hot-rolled strips. Cold rolling distorted and broke the grains and carbides. Moreover, the rolled strip grains were in a high energy state and easily recrystallized due to the deformation energy [17]. Annealing resulted in static recovery and static recrystallization, the nucleation and growth of new grains, and the dissolution and separation of carbides. Thus, the microstructure became denser and finer, and the carbides were refined and distributed more uniformly.

The SEM morphologies of carbides particles in 7Cr17MoV strip formed by rolling is shown in Fig. 3.27.

It can be seen from Fig. 3.27 that the carbides in the cold rolling slab are nearly spherical, and the size of the carbides tends to be uniform. The large-size carbides are obviously reduced, with the average size of about 1 μm. There are certain nano sized carbides. The specific size statistics and quantitative analysis of element composition are shown in Table 3.8.

It can be seen from Table 3.8 that the average number of carbides in the cold rolling slab is 1759 under 3000 times field of view. Compared with the hot rolling slab, the

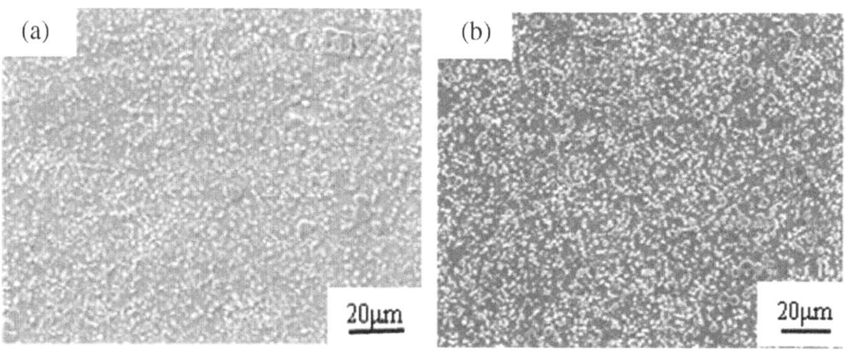

Fig. 3.26 Microstructure of 7Cr17MoV strips obtained by rolling: **a** Cold rolled (1.5 mm); **b** Cold rolled (0.7 mm)

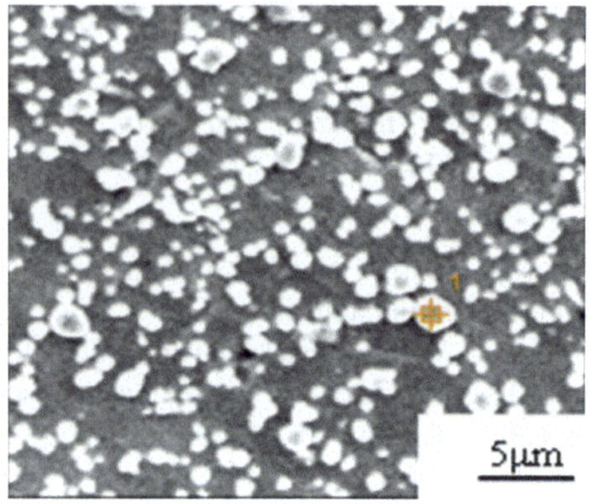

Fig. 3.27 SEM morphologies of carbides particles in 7Cr17MoV strip formed by cold rolled (0.7 mm)

Table 3.8 Quantitative analysis results of carbide in 7Cr17MoV steel cold rolling slab

Quantitative analysis of morphology	Area, μm^2	Width, μm	Height, μm	Total amount	
	3.561	1.017	0.939	1759	
Element composition, %	C	V	Cr	Fe	Mo
	12.65		49.49	37.86	

amount of carbides increases significantly, and the particle size becomes small and uniform. This is mainly due to the comprehensive effect of plastic processing leading to carbide crushing and refining and carbide recrystallization [18].

If the size of carbide in the cold rolling slab is too large, it is necessary to control the large carbide particles in ESR ingot from the source, and proper rolling and annealing process should be adopted. From the quantitative analysis results of elements in Table 3.8, it can be concluded that the carbides contained in the 7Cr17MoV steel strip after cold rolling are still mainly (Fe, Cr)$_{23}$C$_6$, and it is difficult to find the carbides formed by molybdenum and vanadium. This may be due to the dissolution of relatively fine molybdenum and vanadium carbides after cold rolling and annealing.

3.3.2 *Effect of Carbon Content on Carbides in Cold Rolling Slab*

The effect of carbon content of cold rolling slab on carbides in high carbon stainless steel is shown in Fig. 3.28, and the size of carbides is shown in Table 3.9.

It can be seen from Fig. 3.28 that the distribution of carbides in the cold rolling slab is more uniform and finer than that in the hot rolling slab. The aggregation basically disappears, and the large particle carbides are significantly reduced. The size of carbides in hot rolling slab affects the size of carbides after cold rolling. It can be

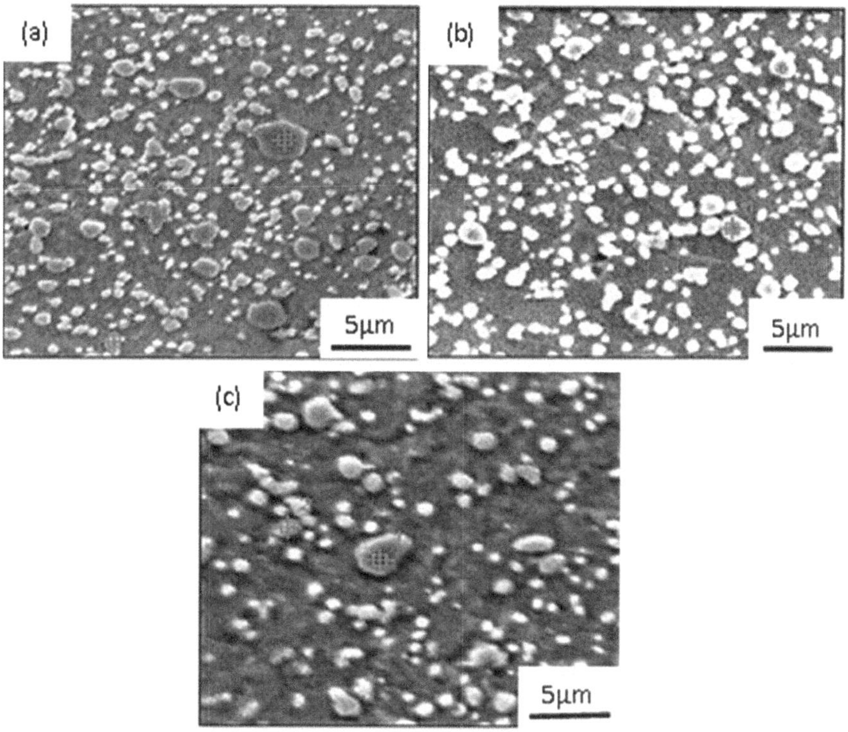

Fig. 3.28 Carbide morphology of tool shear materials with different carbon content after cold rolling: **a** 6Cr13; **b** 7Cr17MoV; **c** 8Cr13MoV

Table 3.9 Quantitative analysis of carbides in high carbon stainless steel knife shear materials with different carbon contents after cold rolling

Parameter	6Cr13	7Cr17MoV	8Cr13MoV
Length, μm	0.847	1.017	1.257
Height, μm	0.727	0.939	1.097
Length/Height	1.17	1.08	1.14

seen from Table 3.9 that the carbide size in 6Cr13, 7Cr17MoV and 8Cr13MoV cold rolling slab is about 0.8 μm, 1.0 μm, and 1.1 μm, respectively. With the increase of carbon content, the average size of carbide particles in cold rolling slab also increases. The average size difference is smaller than that of hot rolling slab. This is due to the fact that large carbides are more likely to deform or even break during cold rolling, and their sizes are more average after annealing [19]. Therefore, when the cold rolling slab reaches a certain thickness, the influence of carbide particle size on the reduction is reduced.

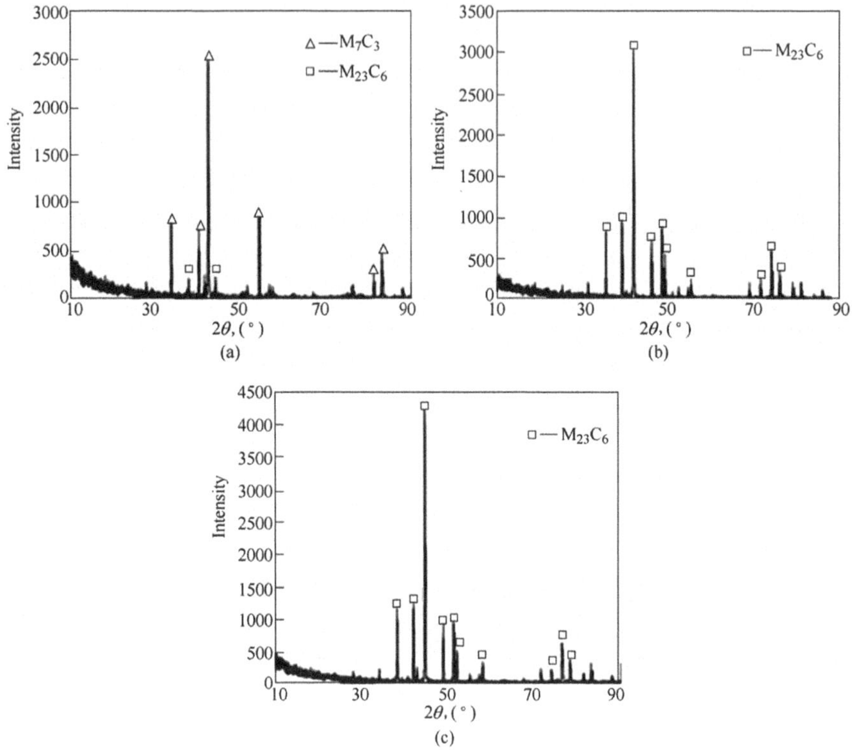

Fig. 3.29 XRD patterns of carbide in tool shear materials with different carbon content after cold rolling: **a** 6Cr13; **b** 7Cr17MoV; **c** 8Cr13MoV

According to the XRD pattern in Fig. 3.29, the carbides in 6Cr13, 7Cr17MoV and 8Cr13MoV cold rolling slab are still mainly M_7C_3 and $M_{23}C_6$-type, respectively. Compared with that in hot rolling slab, the intensity of XRD diffraction peak after cold rolling is higher, which may be due to the more carbides after cold rolling.

3.3.3 Effect of Thickness of Cold Rolling Slab on Carbide

Figure 3.30 shows the typical SEM morphology of carbides in 7Cr17MoV strips of different thicknesses obtained by cold rolling and annealing.

It can be seen from Fig. 3.30 that the carbides in the rolled strip were nearly spherical. Decreasing the strip thickness reduced the size of the carbide particles and generated many fine, nano-sized carbide particles. As reduced to a determinate thickness of the cold-rolled strip, the trend in the decrease of carbides size faded. The sizes of carbides declined invisibly after the specimens were rolled to a thickness of 1.5 mm. Furthermore, the number of carbide particles markedly increased as the

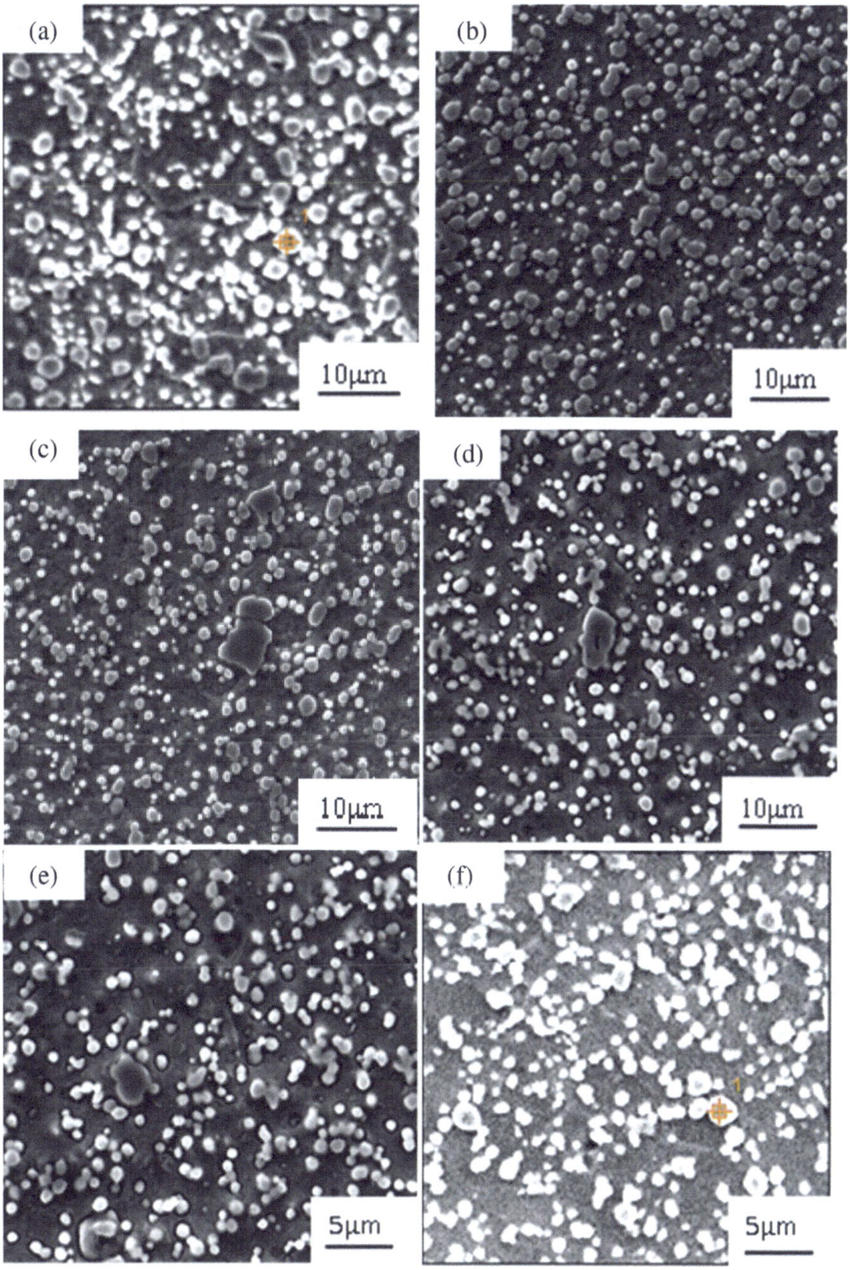

Fig. 3.30 SEM morphologies of carbides particles in 7Cr17MoV strip formed by rolling: **a** Hot rolling (3 mm), **b** Cold rolling (2.5 mm), **c** Cold rolling (2.0 mm), **d** Cold rolling (1.5 mm), **e** Cold rolling (0.9 mm), and **f** Cold rolling (0.7 mm)

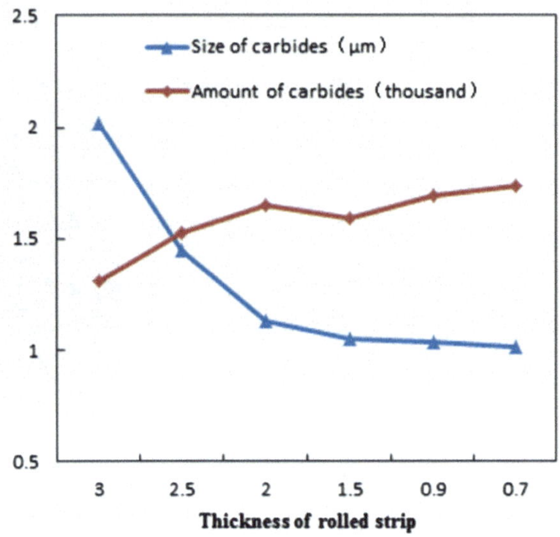

Fig. 3.31 Size and number of carbides particles in rolled 7Cr17MoV steel strips of different thicknesses

strip thickness was reduced. The average size and number of carbides particles in different thickness strips are shown in Fig. 3.31.

As shown in Fig. 3.31, the carbide particles were observed to be rather spherical, and the particles were refined and distributed more uniformly as the cold rolling process continued. This process may be primarily attributed to the distortion and possible breakage of the grains during cold rolling, which benefits the nucleation and precipitation of carbides during recrystallization annealing. Figure 3.32 shows the XRD patterns of carbides in rolled and annealed 7Cr17MoV steel strips of different thicknesses.

The results indicate that the carbide phase mainly consisted of $M_{23}C_6$ in the rolled 7Cr17MoV steel strips, and the carbides mainly consisted of Fe and Cr carbide. The elemental composition of the carbides varied as the cold rolling process continued, whereas the carbide phase remained the same. Moreover, the intensities of the diffraction peaks gradually increased as the strip thickness was reduced. Therefore, the number of carbide particles increased, which improved crystallization.

3.3.4 *Effect of Cold Rolling Slab Thickness on the Mechanical Properties of Steel*

The tensile mechanical properties of 7Cr17MoV strips of different thicknesses obtained after cold rolling and annealing are shown in Fig. 3.33.

It can be seen from Fig. 3.33 that at the beginning of cold rolling, the tensile strength and yield strength decreased. As the strip thickness was reduced to 1.5 mm, the tensile strength and yield strength increased. The tensile strength varied only

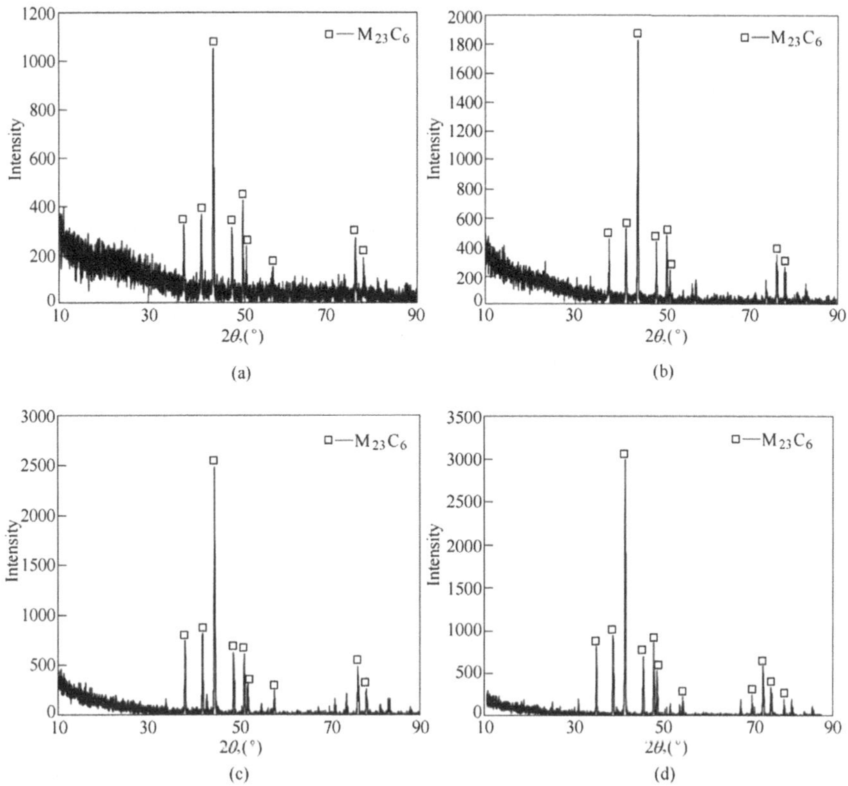

Fig. 3.32 XRD patterns of carbides in rolled 7Cr17MoV steel strips of different thicknesses: **a** Hot rolling (3 mm), **b** Cold rolling (2 mm), **c** Cold rolling (1.5 mm), and **d** Cold rolling (0.7 mm)

slightly, and the strips were markedly elongated throughout the entire cold rolling process. This effect may have been due to the integration of softening annealing and carbide refinement. At the beginning of cold rolling, annealing recrystallization proceeded more thoroughly as the extent of deformation increased. The effect of softening annealing on the material performance was more significant than that of carbide refinement. Thus, the strength decreased and elongation increased. When the deformation reached a given threshold (about 1.5 mm in thickness), annealing recrystallization was almost complete, and the effect of softening annealing was minimal. At this point, the strengthening of carbide refinement played a key role in determining the material's performance, and the strength increased again, and as did the elongation.

The morphologies of the fracture surfaces of the 7Cr17MoV strips are shown in Fig. 3.34.

Figure 3.34 shows that the tensile fracture involved in the formation of mass dimples are the characteristic of plastic fractures. Table 3.10 shows the results of an energy spectrum analysis of the points shown in Fig. 3.34.

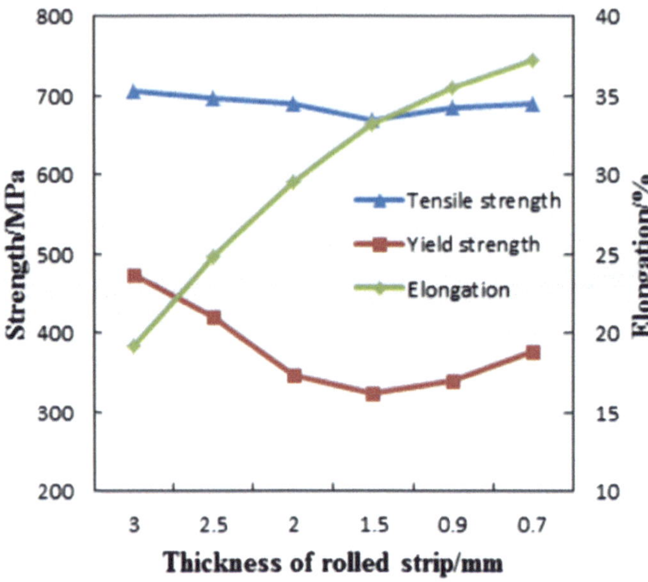

Fig. 3.33 Mechanical properties of 7Cr17MoV steel strips of different thicknesses

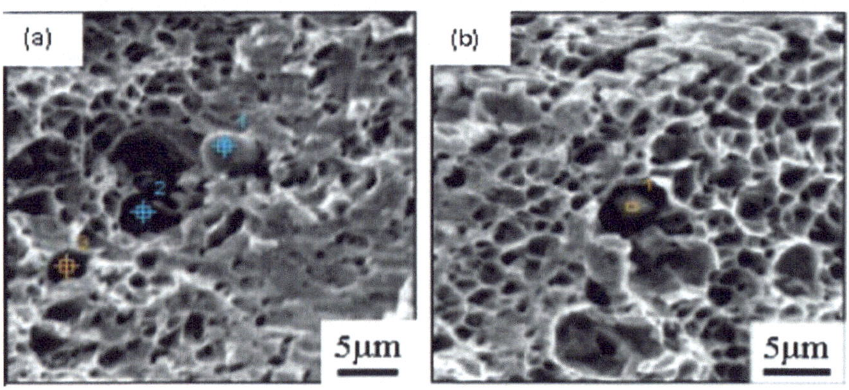

Fig. 3.34 SEM morphologies of fracture surfaces of 7Cr17MoV strips formed be cold rolled

Table 3.10 Compositions of inclusion on fracture surfaces of 7Cr17MoV strips (%)

Number		C	O	Al	Cr	Fe	Mn	V	Ti
(a)	1		38.88	17.02	5.57	5.57	1.85		26.96
	2	0.58			55.01	55.01			
	3	1.63			43.92	43.92		0.37	
(b)	1	3.28		2.68	15.43	15.43	3.23		28.38

It can be seen from Table 3.10 that the particles in the dimple were mainly carbide and oxide inclusions. Controlling the morphology and distribution of carbides during rolling, in addition to the inclusion content, smelting is important in obtaining a 7Cr17MoV strip with good mechanical properties. Generally, uniformly spherical carbides are associated with good performance, whereas plate-like acicular and large carbides are considered harmful. Based on the analysis of the carbides described above, it was observed that both hot and cold rolling resulted in spherical carbides. The distribution and refinement of carbide particles improved as cold rolling proceeded. Thus, the dimples on the fracture surfaces of the cold-rolled strips were more likely to have been equiaxed, and the sizes of the dimples were more uniform, indicating an improvement in the plastic behavior of the cold rolling slab.

References

1. Zhu QT (2018) Effect of the Control of Carbides on the Sharpness of Knives Made of 8Cr13MoV Steel. Dissertation, University of Science and Technology Beijing
2. Yu WT, Li J, Shi CB et al (2016) Effect of electroslag remelting parameters on primary carbides in stainless steel 8Cr13MoV. Mater Trans 57(9):1547–1551
3. Zhu QT, Li J, Shi CB et al (2015) Effect of electroslag remelting on carbides in 8Cr13MoV martensitic stainless steel. Int J Min Met Mater 22(11):1149–1156
4. Chu W, Xie C, Wu XC (2013) Research on controlling eutectic carbides in M2 high speed steel of ESR process. Shanghai Met 35(5):23–26
5. Jiang ZH (2014) Progress and prospect of electroslag metallurgy. Paper presented at the 2014 National special steel annual meeting, Tianjin
6. Zhan LC, Chi HX, Ma DS et al (2013) The as-cast microstructure of ESR-CDS M2 high speed steel. J Mater Eng 7:29–34
7. Jiang ZH, Li ZB (2009) Recent progress on electroslag remelting technology. Spec Steel 30(6):10–13
8. Qi YF, Li J, Shi CB et al (2017) Effect of directional solidification of electroslag remelting on the microstructure and primary carbides in an austenitic hot-work die steel. J Mater Process Tech 249:32–38
9. Chen KX, Wang XY, Wang F (2018) Effects of high temperature diffusion time on microstructure and eutectic carbides of 4Cr5MoSiV1 steel. Hot Work Tech 47(20):239–242
10. Lu ZF, Yuan XX, Wang W et al (2017) Influence of high-temperature diffusion on macrostructure and carbide inhomogeneity of bearing steel. J Iron Steel Res 29(2):144–149
11. Song WX (2011) Metallurgy. Metallurgical Industry Press, Beijing
12. Xiao JM (2006) Metallurgic problems of stainless steel. Metallurgical Industry Press, Beijing
13. Zhang S, Ren Y, Wang S et al (2018) Effect of hot-rolling process on recrystallization and microstructure of X80-grade steel for thick-walled pipeline. Shanghai Met 40(6):55–59
14. Liu M, Feng XM, Lai CB et al (2015) Influence of hot rolling parameters on recrystallization of steel E690 for sea platform. Heat Treat Met 40(10):64–67
15. Zhang J, Lu YL, Zhou DS et al (2018) Effect of hot rolling deformation on microstructures and mechanical properties of aluminum alloy 7085. J Plast Eng 25(4):173–180
16. Yu WT (2017) Study on control technology of carbides in high-carbon martensitic stainless steel 8Cr13MoV used as knives and shears. Dissertation, University of Science and Technology Beijing
17. Yao D, Li J, Li JH et al (2015) Effect of cold rolling on morphology of carbides and properties of 7Cr17MoV stainless steel. Mater Manuf Process 30:111–115

18. Yao D, Li J, Li JH et al (2014) Carbides in high carbon stainless steel used for cutting tools in process of rolling. T Mater Heat Treat 35(11):129–133
19. Yao D (2016) Evolution of the microstructure of high carbon martensitic stainless steel used as a cutlery material during manufacturing process. Dissertation, University of Science and Technology Beijing

Chapter 4
Effect of Heat Treatment on the Carbide in Steel

Abstract Taking 8Cr13MoV steel as an example, the effect of heat treatment process on carbide has been studied. It is found that the optimal austenitizing holding time of spheroidizing annealing and spheroidizing holding time are 90 min and 135 min, respectively. The cooling rate of spheroidizing annealing should be less than 25 °C/h. Increasing the cooling rate of spheroidizing annealing will result in the formation of a large number of fine carbides and sorbite in steel, resulting in a significant increase in hardness and tensile strength. The effects of quenching temperature and holding time on the microstructure, carbide and mechanical properties of special steel are revealed. The effect of tempering temperature at low temperature on the microstructure, carbide, mechanical properties and corrosion resistance of steel is discussed. Finally, the optimal quenching and tempering temperatures are obtained. The roll forging heat treatment process not only refines the grain and primary carbides of the cutting edge, but also reduces the fluctuation range and frequency of sharpness performance in service, generally increases the cutting depth per cut and wear resistance for knife and shear.

Keywords Carbides · Spheroidizing annealing · Quenching · Tempering · Roll forging

8Cr13MoV cold rolling sheet is taken as an example to study the effect of heat treatment on carbide. At room temperature, there are martensite, retained austenite and a small amount of carbides in the forged and rolled 8Cr13MoV steel. At these states, 8Cr13MoV steel has high hardness and poor toughness. If it is directly cold rolled, it is easy to cause plate cracking and scrap. Therefore, spheroidizing annealing is required before cold rolling. The purpose of spheroidizing annealing is to transform martensite into ferrite. Carbon and other alloying elements in steel precipitate in the form of spherical carbides. These evenly distributed carbides in the ferrite matrix reduce the material hardness, improve the plasticity, ensure the steel has good cold working performance, and prevent edge cracks and rolling defects in cold rolling process [1, 2]. Proper holding temperature, holding time and cooling rate should be controlled in spheroidizing annealing process. Recrystallization annealing is also needed for the deformed grains of cold rolled sheet to recrystallize, which refines the grains, removes the processing stress, promotes carbides to precipitation and growth.

J. Li and C. Shi, *Carbide in Special Steel*, Engineering Materials,
https://doi.org/10.1007/978-981-16-1456-9_4

High quality knives and shears require high hardness, wear resistance and corrosion resistance, which are achieved by final quenching and tempering. Reasonable austenitizing temperature and holding time should be controlled during quenching to make proper amount of secondary carbides dissolve into the matrix and play the role of solution strengthening. If there are too many secondary carbides dissolved in the matrix during quenching, the stability of undercooled austenite will increase. There will be more retained austenite in the microstructure after quenching, which will reduce the hardness of steel. During tempering, a large number of fine, uniform and dispersed secondary carbides precipitate on the martensitic matrix, which improves the hardness and wear resistance of the steel. The precipitation of secondary carbides reduces the chromium content in the matrix and form micro cells on the surface of steel, resulting in the decrease of corrosion resistance. Therefore, the heat treatment process system should be reasonably formulated according to the requirements of hardness, wear resistance and corrosion resistance of steel.

4.1 Effect of Spheroidizing Annealing Process on Carbide

4.1.1 Evolution of Carbides During Spheroidizing Annealing

Isothermal spheroidizing annealing is adopted for 8Cr13MoV steel. The process is as follows. Firstly, the material is heated to above Ac_1 point and held for a period of time. Secondly, the material is cooled to a certain temperature below Ac_1 and holding for a period of time. Finally, the material is slowly cooled to a certain temperature then discharged from the furnace for cooling. In this process, carbon and other alloying elements in the matrix precipitate in the form of carbide. In the process of heating above Ac_1 point, the original carbides begin to dissolve and break, and many small granular carbides are obtained. After cooling to a certain temperature below Ac_1 point, carbides begin to spheroidize and grow, which is also called the segregation eutectoid transformation [3, 4]. The transformation of carbides during spheroidizing annealing of martensitic stainless steel with carbon content of about 0.5% has been studied [5], while the change of carbide during spheroidizing annealing of martensitic stainless steel with higher carbon content has not been reported. Therefore, the research on the evolution of carbides during spheroidizing annealing of 8Cr13MoV steel is helpful for optimizing the spheroidizing annealing process of this kind of steel.

8Cr13MoV steel forgings are adopted, and the chemical composition is shown in Table 4.1.

Table 4.1 Chemical composition of 8Cr13MoV steel (wt%)

C	Si	Mn	Cr	Mo	V	S	N	Fe
0.77	0.28	0.45	14.02	0.39	0.45	0.0043	0.011	Bal.

The Ac_1 temperature of the material is 842 °C. Spheroidizing annealing process is shown in Fig. 4.1. The sample is put into the furnace and heated with the furnace. It is held at 800 °C for 30 min, then heated to 860 °C for 90 min, cooled to 750 °C for 90 min, and finally cooled to 600 °C for air cooling at a cooling rate of 25 °C/h. Fourteen samples were prepared, one of which was used as blank sample. The sample taken out was water quenched immediately. The sampling schedule is shown in Table 4.2.

Fig. 4.1 Spheroidizing annealing process

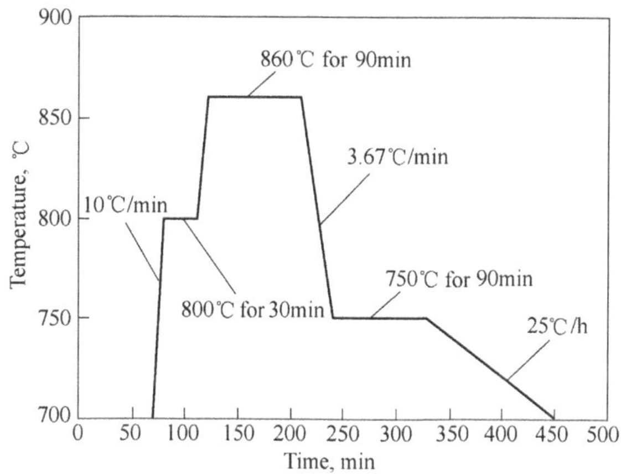

Table 4.2 Heat treatment parameters

Technology	Sequence number	Sampling time, min	Sampling stage	Sampling temperature, °C
Heating	No. 1	0	Unheating	0
	No. 2	30	Heating	330
	No. 3	50	Heating	520
	No. 4	80	Holding at 800°C for 0 min	800
Holding	No. 5	110	Holding at 800°C for 30 min	800
	No. 6	165	Holding at 860°C for 45 min	860
	No. 7	210	Holding at 860°C for 90 min	860
	No. 8	240	Holding at 750°C for 0 min	750
	No. 9	285	Holding at 750°C for 45 min	750
	No. 10	330	Holding at 750°C for 90 min	750
Cooling	No. 11	450	Cooling for 120 min	701
	No. 12	510	Cooling for 180 min	676
	No. 13	570	Cooling for 240 min	651
	No. 14	690	Cooling for 360 min	600

4.1.1.1 Transformation of Microstructure and Carbide During Spheroidizing Annealing

The transformation of structure and carbide during heating is shown in Fig. 4.2.

Figure 4.2a shows the microstructure of the ESR ingot after forging, which is composed of martensite, retained austenite, ferrite, eutectic M_7C_3-type carbides, a few spherical carbides and flake carbides distributed along the grain boundary. The latter two kinds of carbides precipitate during the forging process, and do not exist in the original as-cast structure. The carbides distributed on the grain boundary precipitate along the grain boundary during cooling. During forging, carbides dissolved and the spherical carbides in the grain re-precipitate [6]. When the temperature rises to 330 °C, part of the microstructure changes. The acicular martensite begins to decompose into tempered martensite and carbides, which is more obvious when heated to 520 °C. When heated to 800 °C, the martensite and retained austenite are basically decomposed completely, and the microstructure is composed of ferrite and a large number of carbides in different sizes, but the original martensite region can still be identified. Carbides are easy to precipitate in chain at the edge of original acicular martensite, as shown by the arrow in Fig. 4.2d. The large eutectic M_7C_3-type carbide in the ingot is caused by segregation, and the size of carbide precipitated around

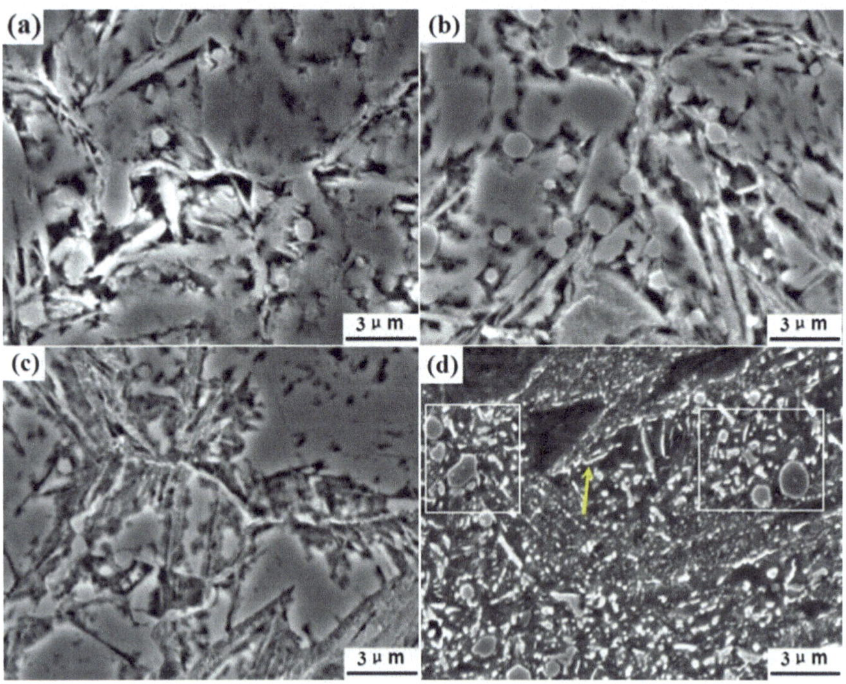

Fig. 4.2 Transformation of microstructure and carbides during heating: **a** Unheated, **b** Heated for 30 min, **c** Heated for 50 min, **d** Heated for 80 min

these eutectic carbide is larger, as shown in the white box position in Fig. 4.2d. The carbides at the grain boundary are divided into chain carbides, which are similar to the carbides in the acicular martensite. The large black area in Fig. 4.2d is the original ferrite region. Due to the low carbon content in this area, less carbides precipitate.

The transformation of microstructure and carbides in the steel at each stage of isothermal process are shown in Fig. 4.3.

Figure 4.3a shows that after holding at 800 °C for 30 min, the original position of martensite and austenite could not be completely distinguished. The number of fine carbides decreases. The carbides further grow up, and the chain carbides coarsen. The statistics of carbide parameters are shown in Table 4.3. R_{max} and R_{min} are the maximum radius and minimum radius of irregular carbide particles, respectively. R_{max}/R_{min} is used to represent the roundness of carbides. Less value of R_{max}/R_{min} mean higher roundness of carbides.

The morphology of carbides in 860 °C steel is more regular than that at 800 °C for 45 min. The number of carbides decreases, the chain carbides start to fuse and

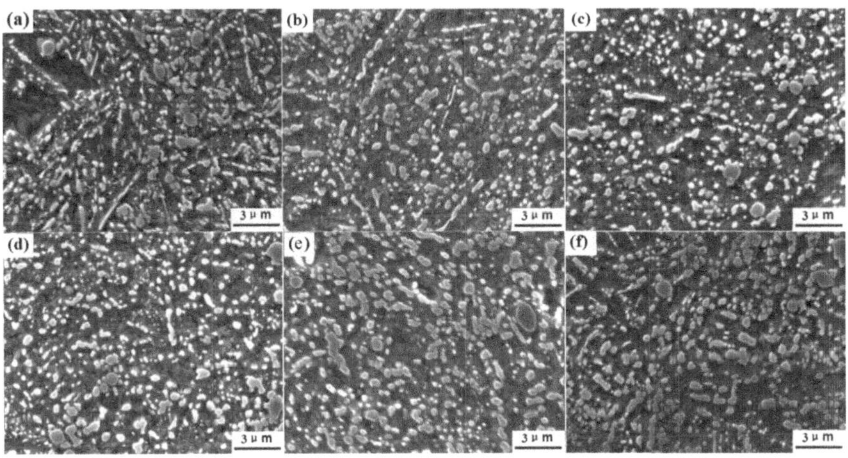

Fig. 4.3 Transformation of microstructure and carbides in holding process: **a** 800 °C for 30 min, **b** 860 °C for 45 min, **c** 860 °C for 90 min, **d** 750 °C for 0 min, **e** 750 °C for 45 min, **f** 750 °C for 90 min

Table 4.3 Statistics of carbide parameters

Sample no.	Amount	Area, μm^2	Average length, μm	Average (R_{max}/R_{min})
No. 5	2008	223.52	0.201	6.91
No. 6	1912	177.28	0.192	4.31
No. 7	1744	162.04	0.179	3.46
No. 8	2140	194.44	0.179	3.14
No. 9	1604	195.72	0.240	4.67
No. 10	1744	200.76	0.231	4.49

the roundness of carbides is improved, as shown in Fig. 4.3b. After holding at 860 °C for 90 min, most of the chain carbides have been melted into particles. Compared with the previous stage, the original large carbides begin to dissolve, and the size of carbides further decreases, as shown in Fig. 4.3c. When the temperature drops from 860 °C to 750 °C, the microstructure enters the spheroidizing stage. Due to the rapid temperature drop, the alloy elements in the matrix do not have sufficient time to diffuse [7]. The original carbides grow and coarsen, and new carbides nucleate and precipitate, which results in a significant increase in the number and size of carbides, as shown in Fig. 4.3d. After holding at 750 °C for 90 min, the coarsening of carbides is more obvious and the size is not uniform and the roundness is poor.

The transformation of microstructure and carbide during cooling are shown in Fig. 4.4. Statistics of carbide parameters are shown in Table 4.4.

Figure 4.4 and Table 4.4 show that the amount of carbides in the microstructure decreases and the average length of carbides increases during cooling for 120 min. After cooling for 180 min, 240 min and 360 min, the parameters of carbides in the microstructure do not change significantly. The results indicate that, after cooling for 180 min, the spheroidizing rate of carbide in the microstructure is very slow due to the decrease of temperature. The spheroidization of carbide is not obvious even

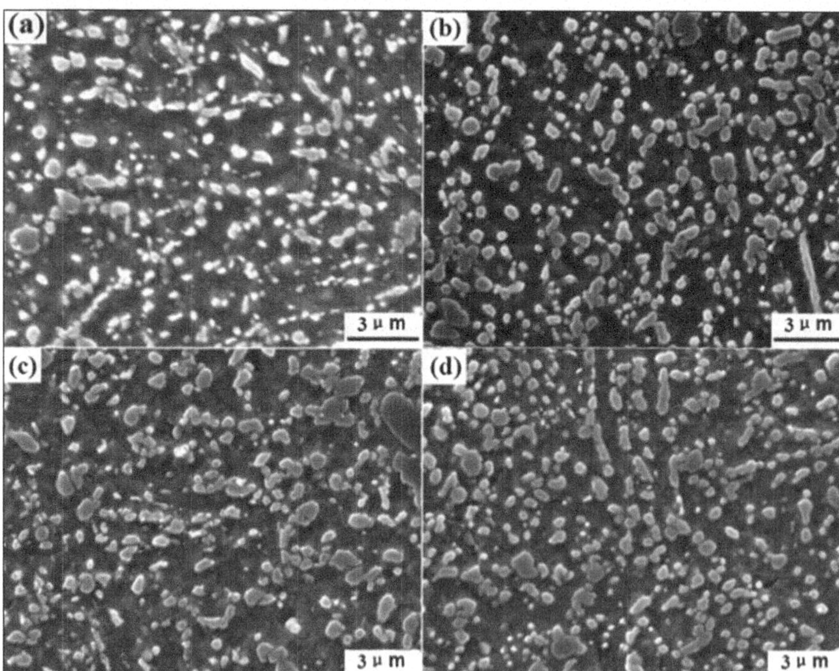

Fig. 4.4 Morphology of carbides during cooling: **a** Cooling for 120 min, **b** Cooling for 180 min, **c** Cooling for 240 min, **d** Cooling for 360 min

Table 4.4 Statistics of carbide parameters

Sample no.	Amount	Area, μm^2	Average length, μm	Average (R_{max}/R_{min})
No. 11	1340	211.43	0.294	4.51
No. 12	1284	198.26	0.285	4.91
No. 13	1332	200.96	0.247	4.69
No. 14	1324	212.44	0.271	4.54

though the temperature keeps decreasing. From the beginning of spheroidization to cooling, the area of carbides basically unchanges, which meets the classical Ostwald ripening theory [8].

4.1.1.2 Change of Carbide Type During Spheroidizing Annealing

The change of carbide composition during spheroidizing annealing is shown in Fig. 4.5.

Figure 4.5a, energy spectrum of typical carbide precipitated during heat treatment, shows that Cr content in precipitated carbides is obviously lower than that in M_7C_3-type carbides. After forging, the eutectic carbides of ESR ingot are broken, but the size of carbides is still larger, generally larger than 1 μm. The content of Cr in M_7C_3-type is significantly higher than that in $M_{23}C_6$-type carbides [9, 10]. According to this feature, it can be found that M_7C_3-type carbide is very stable in the process of spheroidizing heat treatment, and there is no carbide transformation from M_7C_3-type

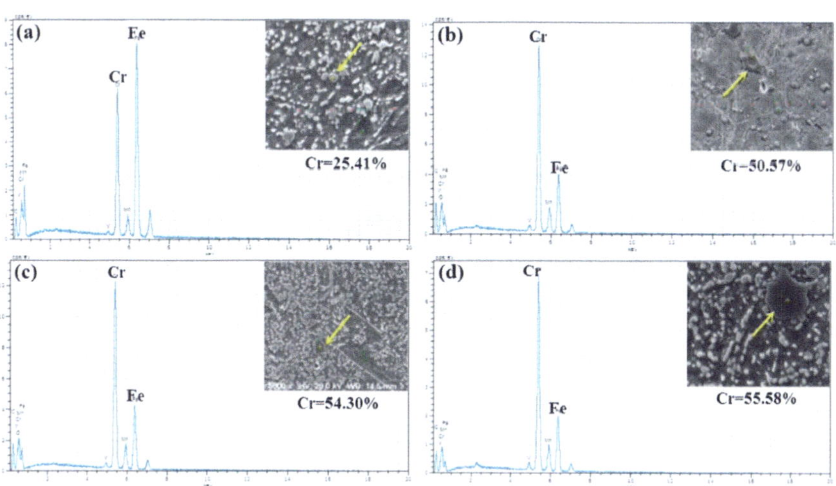

Fig. 4.5 Energy spectrums of carbides: **a** Typical carbide precipitated during heat treatment, **b** Large-sized carbide in sample No. 2, **c** Large carbide in sample No. 6, **d** Large carbide in sample No. 12

to $M_{23}C_6$-type. Figure 4.5b, c and d are energy spectrum of large carbides in Nos. 2, 6 and 12 samples, respectively. The Cr content in these carbides is about 50%, and all of these carbides are M_7C_3-type carbides.

In addition to eutectic carbides, there are some spherical carbides and carbides distributed along grain boundaries in the ESR ingot after forging, as shown in Fig. 4.6.

Figure 4.6a shows the TEM images of carbides in sample No. 1. In Fig. 4.6a, there are flake carbides and some spherical carbides. The lamellar carbides distribute at the grain boundary, while the spherical carbides locate near the grain boundary. The spherical carbides in Fig. 4.6a are hexagonal, as shown in Fig. 4.6b. The diffraction

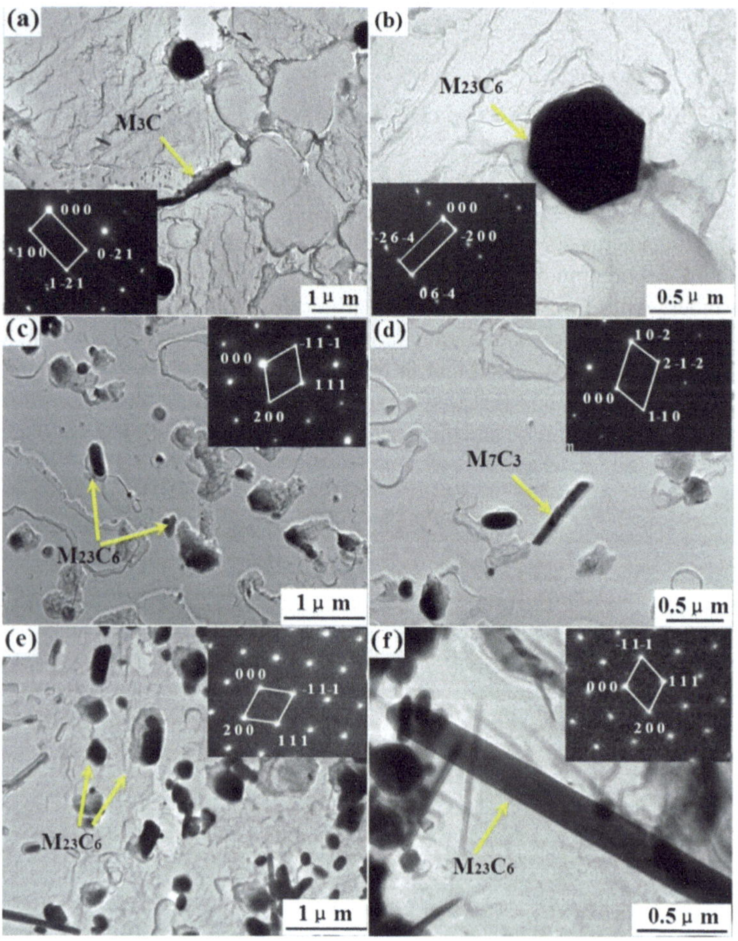

Fig. 4.6 TEM images and diffraction patterns of carbides: **a** Carbides at grain boundary in sample No. 1, **b** Hexagonal carbides in sample No. 1, **c** Granular carbides in sample No. 3, **d** Rod-shaped carbides in sample No. 3, **e** Granular carbides in sample No. 12, **f** Long strip carbides in sample No. 12

patterns of these two carbides are shown in the lower left corner of Fig. 4.6a and b. Results show that the lamellar carbide is M_3C-type carbide, whose crystal belt axis is [012]. The hexagonal carbide is $M_{23}C_6$-type carbide, whose crystal belt axis is [023]. There is no such hexagonal carbide in ESR ingot, which should be precipitated during forging. When the forging temperature is between 800 and 1200 °C, the precipitated carbide should be M_7C_3-type, which will change into $M_{23}C_6$-type during cooling. There are M_3C-type carbides at the grain boundary in the as-forged microstructure, while they are not stably and will transform into $M_{23}C_6$-type carbides during heat treatment.

The TEM image of carbides of No. 3 sample is shown in Fig. 4.6c. After heating for 80 min, the carbides are main $M_{23}C_6$-type. There are few rod-shaped carbides, which are calibrated as M_7C_3-type carbides with a crystal belt axis of [221]. These rod-shaped M_7C_3-type carbides come from the broken eutectic carbides. It is difficult for M_7C_3-type carbides to transform into $M_{23}C_6$-type during heating. According to the calculation of thermodynamic software, the coexistence region of $M_{23}C_6$-type and M_7C_3-type carbides is between 800 and 860 °C. With the increase of temperature, $M_{23}C_6$-type carbides tend to transform into M_7C_3-type carbides, but the transformation takes a long time. After aging at 800 °C for 50 h, the transformation from $M_{23}C_6$-type to M_7C_3-type carbide did not happen in the material that had similar composition to 8Cr13MoV steel [11]. Because the time of holding temperature above 800 °C is short, the spheroidized carbides observed in the samples cooled after holding are $M_{23}C_6$-type. Figure 4.6e shows the TEM image of carbides in No. 12 sample. Some long strip carbides appear in the structure during cooling. These carbides have a large number, regular shape and smooth edge, which are different from the rod-shaped carbides in No. 3 sample. They belong to $M_{23}C_6$-type carbides like granular carbides. The crystal belt axis of the carbide is [0.11], as shown in Fig. 4.6f.

In summary, the carbide transformation from M_7C_3-type to $M_{23}C_6$ type does not occur during the heat treatment process. With the increase of temperature, the M_3C-type carbide at the grain boundary of ESR ingot changes to $M_{23}C_6$-type carbide. Different from other martensitic stainless steels [12], the martensite in 8Cr13MoV steel decomposes and $M_{23}C_6$-type carbide directly precipitates instead of M_3C-type carbide.

4.1.1.3 Microstructure Observation of Annealed Specimen and Tensile Fracture

The original structure of the sample is composed of martensite, retained austenite, primary carbides and a small amount of secondary carbides, as shown in Fig. 4.7.

After spheroidizing annealing, two metallographic samples of 12 mm × 12 mm × 5 mm were obtained. After grinding and polishing, one is used for Rockwell hardness (HRB) test, the average value of each sample is taken as the standard hardness. The other is eroded by $FeCl_3$ hydrochloric acid alcohol solution for microstructure

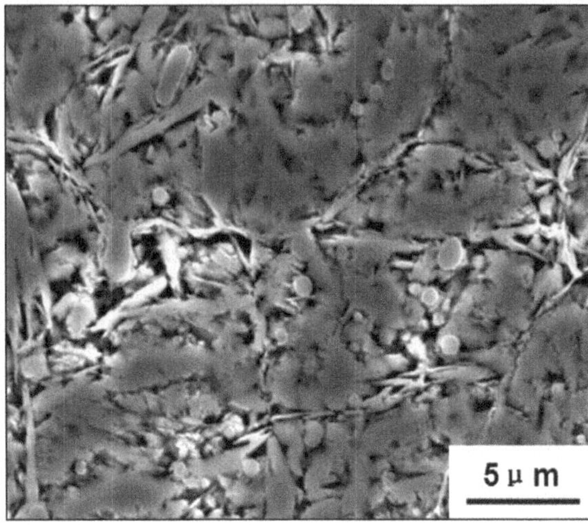

Fig. 4.7 Microstructure of forged sample

observation. The microstructure was observed by SEM, and the carbide was counted by software Image-Pro Plus.

The annealed sample was processed into a round standard tensile specimen with a diameter of 8 mm. The tensile test was carried out on the universal tensile strength testing machine, and the fracture microstructure was observed by scanning electron microscope.

4.1.2 Effect of Austenitizing Time on the Carbides and Mechanical Property of Annealed Steel

Although there are many researches on spheroidizing annealing process, the specific process is quite different due to the difference of alloy element behavior in steel [6, 13]. Based on the evolution of carbide during spheroidizing annealing, the effects of austenitizing holding time (t_1), spheroidizing holding time (t_2) and cooling rate on microstructure and properties of ESR ingot after forging were studied.

Ac_1 of the studied 8Cr13MoV steel is about 840 °C. To investigate the effect of t_1 and the t_2 on microstructure and mechanical properties of steel, the experiments were classified as six groups. 860 °C and 750 °C were selected as the holding temperatures of austenitizing and spheroidizing, respectively. After the holding of spheroidizing, the sample was cooled at the cooling rate of 25 °C/h in each experiment. The experimental scheme is listed in Table 4.5. The test profile for spheroidizing annealing is schematically illustrated in Fig. 4.8.

Table 4.5 Experimental conditions in each experiment

t_2, min	t_1, min		
	D (45)	E (90)	F (135)
A (45)	No. 1	No. 4	No. 7
B (90)	No. 2	No. 5	No. 8
C (135)	No. 3	No. 6	No. 9

Note The holding temperatures of austenitizing and spheroidizing were 860 °C and 750 °C, respectively. t_1 and t_2 were the austenitizing time and spheroidizing time, respectively

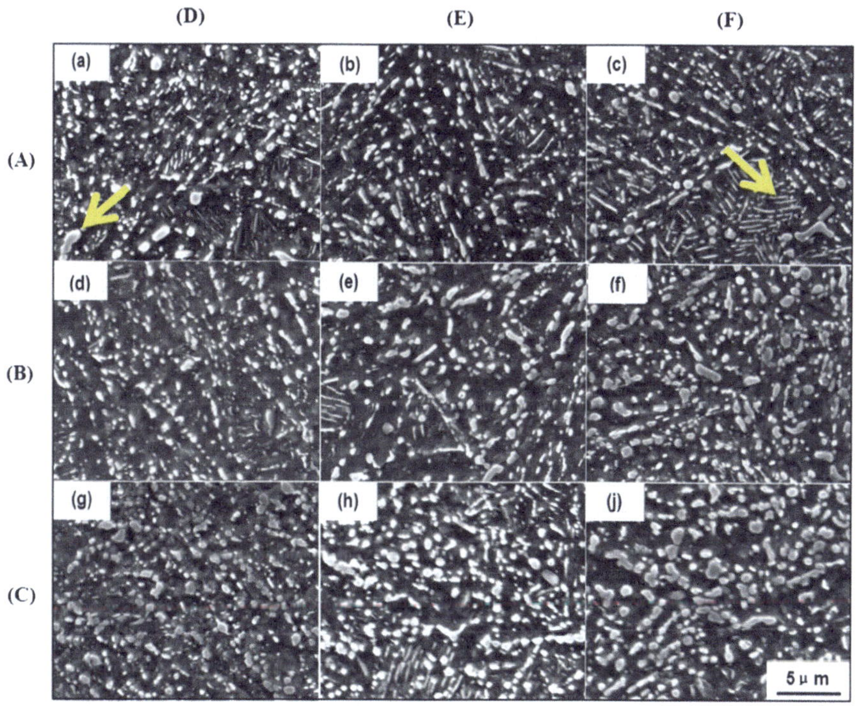

Fig. 4.8 SEM images of specimens holding for different t_1 and t_2: **a** 45 min and 45 min, **b** 90 min and 45 min, **c** 135 min and 45 min, **d** 45 min and 90 min, **e** 90 min and 90 min, **f** 135 min and 90 min, **g** 45 min and 135 min, **h** 90 min and 135 min, **j** 135 min and 135 min. The capital letters before each line and row are their group numbers

4.1.2.1 Effect of Austenitizing Time on the Microstructure and Carbides of Annealed Steel

As shown in Fig. 4.8, the microstructure of all these annealed samples consists of primary carbides, granular, lamellar, short rod-like carbides and ferritic matrix. The primary carbides formed during solidification of liquid steel. These carbides remain in the steel after forging and annealing.

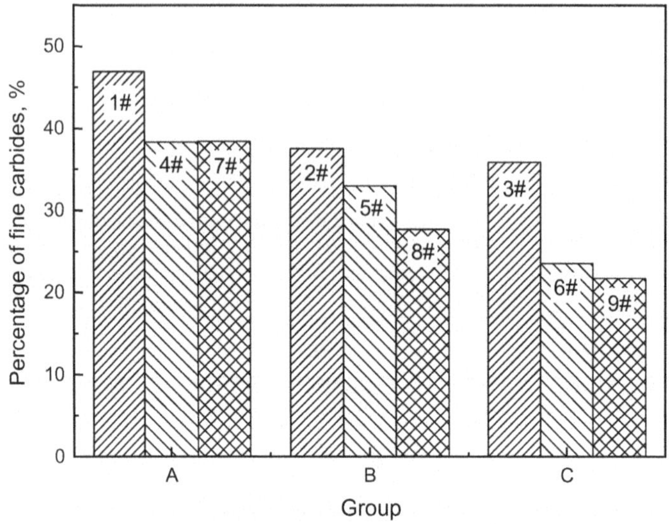

Fig. 4.9 Proportion of fine carbides to the total amount of carbides

As indicated by arrows in Fig. 4.8a, a small amount of carbides show short rod-like or chain-like, indicating incomplete spheroidization of these carbides. In addition, a few carbides are lamellar in morphology, as labeled by arrows in Fig. 4.8c. It can be seen that the characteristics of the microstructure of the steel among various group at the same holding time is identical. Figure 4.9 shows the proportion of fine carbides to the total amount of carbides.

It can be seen from Fig. 4.9 that at the austenitizing holding time of 45 min, the size of carbides is uneven, and there are many fine carbides. The carbides with the size smaller than 0.2 μm are regarded as the fine ones. Taking the total amount of carbides in sample 1# as a base, the ratio of the amount of carbides in other samples to that in sample 1# is shown in Fig. 4.10. At the austenitizing holding time of 90 min, the proportion of fine carbides in the samples of group A decreased from 46.94% to 38.31%, and the amount of carbides decreased to 19.31%. The amount of carbides in sample 7# increased a little with the austenitizing holding time. The amount of lamellar carbides in sample 7# is greater than that in samples 1# and 4#. The amount of carbides in sample 8# increased by 7.41% compared with that in sample 5#. The amount of carbides in sample 9# is least in this group, whereas the size of the carbides is largest.

Although the hardness of samples from group A to group C decreases first and then increases, the microstructure changes of samples in each group are not exactly the same, especially the last sample in each group, according to the microscopic structure. There are more sorbite in sample No. 7 and some fine carbides in sample No. 8. The microstructure of sample No. 9 has little change compared with that of sample No. 6. Samples Nos. 7–9 all belong to group F. Obviously, this large microstructure difference is directly related to spheroidizing holding time.

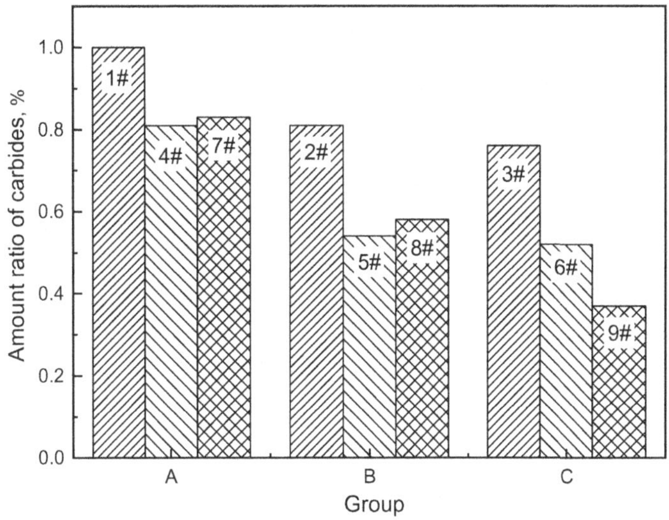

Fig. 4.10 Comparison of carbides amount in different samples

It can be seen from Fig. 4.3 in the previous section that, with the increase of ^{t1}, carbides gradually dissolve, and some positions with far carbide spacing (hereinafter referred to as "interposition") appear in the matrix [14]. The spheroidizing holding time of sample No. 7 is only 45 min, and the alloying elements dissolved in the matrix have no time to diffuse to the carbide core and participate in spheroidization. Therefore, it is very easy to form sorbsite in the "interposition" during the cooling process. The sorbsite also improve the hardness of the matrix, so that the tensile strength increase. The spheroidizing holding time of sample No. 8 is 90 min, and the alloying elements in the matrix diffuse and participate in spheroidization for more time. As a result, there was no large amount of sorbide in Sample No. 8. Compared to sample No. 5, more alloy elements are still in the matrix. Some of these elements are retained in the matrix without the time to migrate to spidization, while others precipitate nearby or in the "interposition" as small particles of carbides, which also strengthen the matrix. This can be proved by the Fig. 4.3c and d that no sorbide appeare with the amount increase of fine carbides. The spheroidizing holding time of sample No. 9 is 135 min. The carbide spheroidization is relatively enough, so there is no obvious precipitation of sorbite and small carbide in the cooling process.

4.1.2.2 Effect of Austenitizing Time on the Mechanical Property of Annealed Steel

The hardness of all annealed steel samples is listed in Table 4.6. The optimum holding times of austenitizing and spheroidizing for obtaining the lowest hardness of steel are 90 and 135 min, respectively.

Table 4.6 Hardness of annealed steel samples (HRB)

t_2, min	t_1, min		
	D (45)	E (90)	F (135)
A (45)	95.56 (No. 1)	91.43 (No. 4)	93.71 (No. 7)
B (90)	95.22 (No. 2)	89.65 (No. 5)	89.95 (No. 8)
C (135)	92.65 (No. 3)	88.05 (No. 6)	89.71 (No. 9)

The hardness variation of annealed samples with the austenitizing holding time is shown in Fig. 4.11. The hardness of annealed samples decreases first, and then increase with the austenitizing holding time. The value of hardness is lowest at the austenitizing holding time of 90 min. The tensile strength of each sample at different holding time is also shown in Fig. 4.11. The tensile strength of each sample exhibits identical trend as its hardness does.

Normally, the plasticity of steel increases with the decrease in the hardness of steel. But the current results show that the relationship between the hardness of the steel and the elongation after fracture is unapparent, as shown in Fig. 4.12. It may result from the presence of non-metallic inclusions and brittle precipitates with large size, especially primary carbides in the steel. It is widely accepted that primary carbides have higher hardness than steel matrix. Therefore, these primary carbides could act as breakage initiation sites during rolling and specific machining of this high-carbon steel.

The heating process of 8Cr13MoV is similar with the high-temperature tempering of high carbon martensitic steel. A large amount of carbides precipitate from martensite, hardness of material decrease, but these carbides also play a hardening role to some extent in the matrix. When the samples were heated up to Ac_1 and hold, matrix

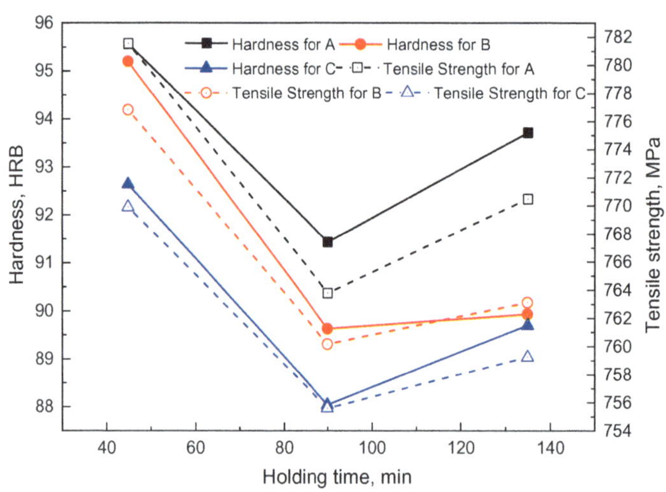

Fig. 4.11 Relationship between the hardness, tensile strength and austenitizing holding time

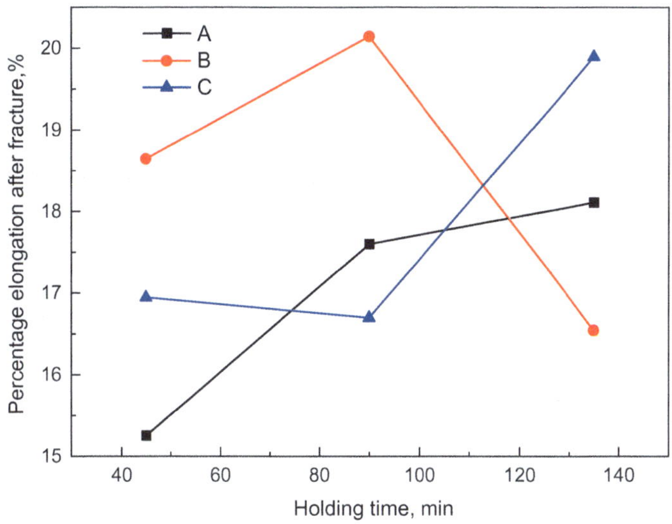

Fig. 4.12 Dependence of austenitizing holding time on elongation after fracture

transformed austenite has higher solubility on carbon [15]. Some small carbides also dissolved, and the remaining carbides would become the nucleation core of newly generated carbides after the heat preservation. When the austenitizing holding time was 45 min, the amount of small carbides was obviously higher than that in other samples. These small carbides were partly precipitated in the heating process directly, and the others were derived from those segmented carbides. When austenitizing holding time was short, the amount of small carbides is greater after the lamelliform and long striped carbides dissolved. The effect of hardening is more significant, so the hardness was higher than that of other samples. With increasing t_l, the small carbides dissolved in the matrix, which resulted in the reduction of reinforcement and decrease of hardness. When austenitizing holding time was 90 min, the hardness of samples decreased to the minimum. When austenitizing holding time was 135 min, more carbides dissolved into the matrix, which resulted in the decrease of nucleation core for carbides and the increase of alloying elements contents in the matrix. Under the same cooling condition, the decrease of spheroidizing degree of carbides could keep more alloying elements in the matrix, which played an important role in strengthening matrix. As a result, the hardness and tensile strength were improved again.

The samples 5# and 6# in group E were selected to observe the tensile fracture. The hardness of sample 6# was the lowest, while the elongation after fracture was lower than that of sample 5#. The SEM photographs of tensile fracture in these two samples were shown in Fig. 4.13. The fracture was typical ductile fracture, which presented a cup-cone shaped feature. The macroscopic fracture of tensile sample was shown in Fig. 4.13a, in which point 1 was crack spreading zone, point 2 was fibrous zone, and point 3 was shear lip aone. he microstructure of 8Cr13MoV after forge and

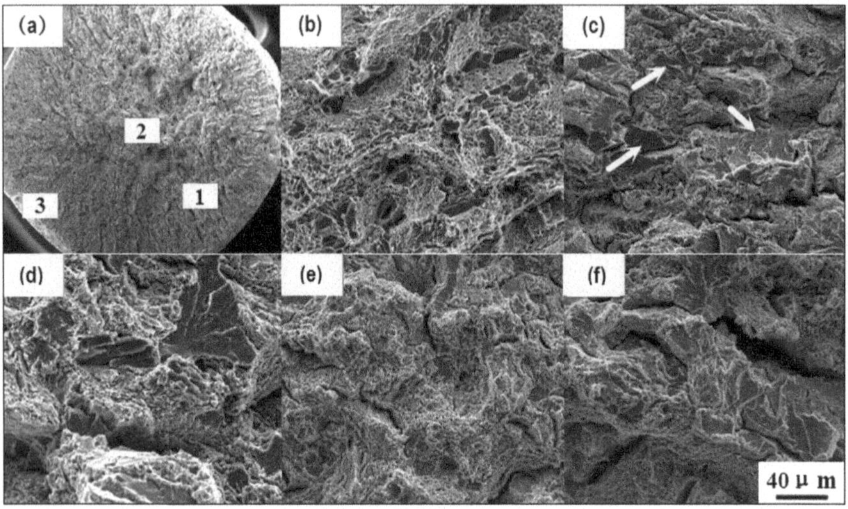

Fig. 4.13 SEM images of tensile fracture morphology: **a** Overall morphology of a specimen, **b** Fiber area, **c** Radiation area, **d** Enlarged view of white arrow pointed location in image (**c**), **e** Radiation area of specimen 5#, **f** Radiation area of specimen 6#

anneal consists of ferrit and carbides. Ferrit has good ductility, while carbides have no ductility. Therefore, in the tensile process, the cracks were easily formed in the position where large primary carbides existed. The enlarged photograph of fibrous zone was shown in Fig. 4.13b. There were many primary carbides and dimples distributed on the fracture. Because of hot working, primary carbides distributed along a certain direction. In the fibrous zone, cracks were produced at the places where these primary carbides existed, and then the cracks resulted in fracture. The enlarged photograph of radiation area is shown in Fig. 4.13c, this was different from that in fibrous zone. Except dimples and large primary carbides, there existed another quasi cleavage fracture in this area. This fracture had the characteristic of river pattern, which was shown in Fig. 4.13c. When it was enlarged partly, some different characteristics compared with quasi cleavage fracture were found. Though it presented river pattern, the interior was not smooth. It presented arborization, which was shown in Fig. 4.13d. Judging from the appearance, there were some secondary carbides precipitated along grain boundary, which presented arborization. This fracture was occurred on the carbides precipitated along grain boundary, which was belong intergranular fracture. The microstructure of crack spreading zone in sample 5# and 6# was shown in Fig. 4.13d and f, respectively. The cracks in sample 5# were less than that in sample 6# obviously. It indicates that that the secondary carbides precipitated at grain boundaries in sample 5# were less than that in sample 6#.

It was concluded that the main factor that affected elongation after fracture remained the precipitation of carbides at grain boundaries. The formation of carbides was directly associated with element segregation [16, 17]. During the electroslag remelting process, the cooling intensity was larger near mold. Therefore, the element segregation is smaller at the position. Consenquently, the amount of primary carbides was smaller, the segregation was more serious at the center of ingot, which resulted in more primary carbides formation [18]. During heat treatment process, the precipitation of secondary carbides is also related to this segregation of alloying elements. More secondary carbides will precipitate at the position where a large amount of primary carbides exist. For this reason, it was found that samples 5# and 9#, which were taken from the edge of ingot, had the maximum percentage elongation after fracture. The sample 6# was obtained from the center of ESR ingot.

4.1.3 Effect of Spheroidizing Time on the Carbides and Mechanical Property of Annealed Steel

4.1.3.1 Effect of Spheroidizing Time on the Microstructure and Carbides of Annealed Steel

It can be seen from Fig. 4.8 that the size of carbides in these samples increases with the increase of spheroidizing time. Taking the samples in group F as an example, the average size of carbides is 0.19 μm, 0.23 μm and 0.31 μm, respectively. The spheroidized degree of carbides is higher with time progresses, whereas the amount of carbides decreases. The volume fraction of carbides is nearly constant (about 30%).

4.1.3.2 Effect of Spheroidizing Time on the Mechanical Property of Annealed Steel

Figure 4.14 presents the hardness and tensile strength of each annealed sample at different spheroidizing time. As can be seen, both the hardness and tensile strength of each sample decreased as the time passed. Combined with the microstructure observations shown in Fig. 4.14, it can be deduced that the hardness of each annealed sample decreases with the increase of spheroidized degree of carbides. The decrease in the hardness of the samples in group E and F became smaller, whereas that in group D increased. If the growth rate of carbides was used to reflect these changes in steel hardness, the growth rate of carbides in the steel samples of group E and F decreases with time goes, whereas the growth rate of carbides in the steel samples of group D increases. From the spheroidizing time of 45–90 min, the growth rate of carbides in the steel samples follows the order: group E > group F > group D. The growth rate of carbides in the steel is in the order of group D > group F > group E when the spheroidizing time increases from 90 to 135 min. This is closely related to the difference in austenitizing time among three groups. Further discussion will be given subsequently.

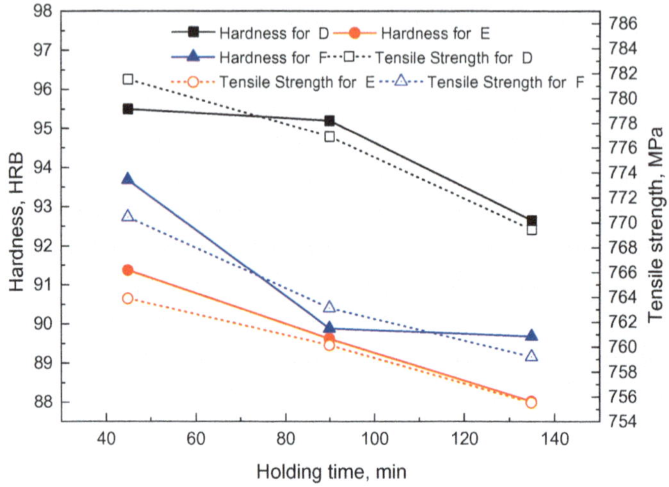

Fig. 4.14 Relationship between spheroidizing time and tensile strength and hardness

In group D, E and F, there is a great difference in the spheroidization rate when spheroidizing time was changed. It can be attributed to the difference in austenitizing time. According to LSW theory [19], the radius of carbides and time during spheroidization process has the following relationship:

$$\frac{dr}{dt} = \frac{2D_M C_0^\alpha \sigma_S V_m^\alpha}{r\left(C^\beta - C_r^\alpha\right)KT}\left(\frac{1}{\bar{r}} - \frac{1}{r}\right) \tag{4.1}$$

where D_M represents the diffusion coefficient of alloying element in matrix, represents the average radius of particals,C^β represents the carbides concentration, K is the gas constant. T is absolute temperature, V_m^α is partial molar volume of precipitates, σ_S is interfacial energy of carbides and matrix.

It can be found from this formula that carbides whose radius is larger than average radius could grow up, while the carbides with the radius smaller than average radius would disappear. Taking the diffusion of multiple component into consideration, the instant growing rate formula of carbides in the spheroidization process was obtained as follows [20]:

$$\frac{dr}{dt} = \frac{2V_m^\alpha \sigma_S}{LKT}\frac{1}{r}\sum_M \frac{D_M}{k_M - k_{Fe}} \tag{4.2}$$

where L is diffusion radius, k_M and k_{Fe} represent the distribution coefficient of alloying element and Fe in carbides and matrix, respectively, namely $k_M = X_M^\beta / X_M^\alpha$ and $k_{Fe} = X_{Fe}^\beta / X_{Fe}^\alpha$.

According to this formula, the instant growing rate of carbides is inversely proportional to L. This means that the greater of the carbides spacing is, the slower of the instant growing rate is. The instant growing rate of carbides is also inversely proportional to $(k_M - k_{Fe})$. The instant growing rate decreases with increasing $(k_M - k_{Fe})$. In addition, it can be obtained that, the instant growing rate increases with the increase in the concentration of the alloying element.

It can be obtained that the austenitizing holding time affects L and $(k_M - k_{Fe})$ directly. In the dissolution process of carbides, with increasing the austenitizing holding time, both carbides spacing and the concentration of alloying element increased. These two parameters had opposite effect on the growing rate. The experimental results showed that the spheroidization rate is in the order of $V_F > V_E > V_D$ in the earlier stage of spheroidization process. It was deduced that the alloying element concentration had a larger effect on the spheroidization rate of carbides. With the increase in spheroidizing time, the concentrations of alloying elements in matrix decreased, and the spheroidization rate also decreased. The difference in the hardness between samples 5# and 6#, and that between samples 8# and 9# decreased, which was coincident with the Eq. (4.2). The hardness change of D group increase, which may be attributed to short of the austenitizing holding time. There were many lamelliform carbides without sectional dissolution and some undissolved small carbides in the microstructure. In the early stage of spheroidizing time, the concentration of alloying element in the matrix is low. With the drive of interfacial free energy, the lamelliform carbides and some undissolved small carbides were still in the process of dissolution. At this stage, the spheroidization rate was small, and the change of steel hardness is not obvious. The instance growing rate formula of lamelliform carbides derived on the principle of dynamics was as follows [21]:

$$\frac{dr}{dt} = \frac{2V_m^\alpha \sigma_S C_0}{l_0 KT}\left(\frac{1}{r_1} - \frac{1}{r_2}\right) \tag{4.3}$$

where l_0 is the length of lamelliform carbides, C_0 is the equilibrium concentration of alloying element at the position of carbides where the radius of curvature is zero. r_l and r_2 are radius when the carbides coarsen to the curvature of ρ_1 and ρ_2, respectively.

According to the formula, the instant growing rate of carbides is inversely proportional to l_0. With the breaking of carbides, l_0 decreased, while the instant growing rate increased gradually. Therefore, the spheroidization rate of samples in group D increased gradually with the increase in spheroidizing time, which is different from those in other two groups.

4.1.4 *Effect of Cooling Rate on the Carbides and Mechanical Property of Annealed Steel*

4.1.4.1 Effect of Cooling Rate on the Microstructure and Carbides of Annealed Steel

The samples with the austenitizing time of 135 min and the spheroidizing time of 45 min were studied under the condition of different cooling rates. The cooling rate was 10 °C/h, 25 °C/h, 50 °C/h, 100 °C/h and 250 °C/h, respectively. SEM images of samples at different cooling rate are shown in Fig. 4.15.

It was observed in Fig. 4.15 that amellar and spherical carbides were observed in each sample under the condition of different cooling rates. With increasing cooling rate, the size of spherical carbides and lamellar carbides became smaller greatly, and the amount of lamellar carbide increased. The thickness and mean spacing in samples with the cooling rate of 250 °C/h decreased by about 50% and the amount of carbide increased by 19.53% compared with that in the samples with the cooling rate of 15 °C/h. Some fine carbides in samples with cooling rate of 250 °C/h appeared between spherical carbides. The minimum size of fine carbides was less than 20 nm. These carbide particles began to precipitate in large quantities when the cooling rate reached 100 °C/h. It was obvious that the cooling rate was directly related to the fine carbide and the lamellar carbide.

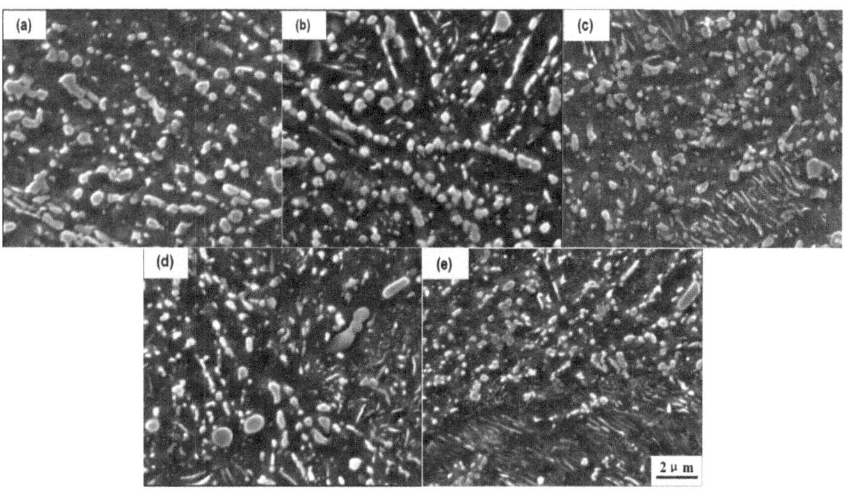

Fig. 4.15 SEM images of the samples cooled at different cooling rate: **a** 25 °C/h, **b** 50 °C/h, **c** 100 °C/h, **d** 250 °C/h

4.1.4.2 Effect of Cooling Rate on the Mechanical Property of Annealed Steel

The hardness, tensile strength and elongation after fracture of samples with different cooling rates are shown in Fig. 4.16. According to the microstructure in Fig. 4.15, the hardness of the sample should increase with the increase of cooling rate, which is consistent with the results in Fig. 4.16.

Figure 4.16 shows that the tensile strength of the sample increases with the increase of cooling rate, especially at the cooling rate of 250°C/h, which reaches 842.57 MPa. The elongation after fracture decreases with the increase of cooling rate, while the change range is small.

The samples with cooling rates of 25 and 250 °C/h were selected for tensile fracture morphology analysis, as shown in Fig. 4.17. There are several dimples on the tensile fracture surface of the specimen at 250 °C/h, which is obviously different from those in other places. As shown in Fig. 4.17b, the dimple is very small and shallow, indicating that the plasticity of the microstructure here is low. When the cooling rate is 25 °C/h, the dimple size is uniform, and there is no small dimple concentration.

Two forms of pearlite formation during spheroidizing annealing are eutectoid transformation and divorced eutectoid transformation. The products are lamellar pearlite and granular pearlite, respectively. The cooling rate is the critical factor of the products. It is found that [6] divorced eutectoid transformation requires a lower undercooling, so a slower cooling rate is required. There is a limit cooling rate in the cooling process of spheroidizing annealing treatment. When the cooling rate exceeds this value, lamellar pearlite will be produced, which is not conducive to the spheroidization of carbides.

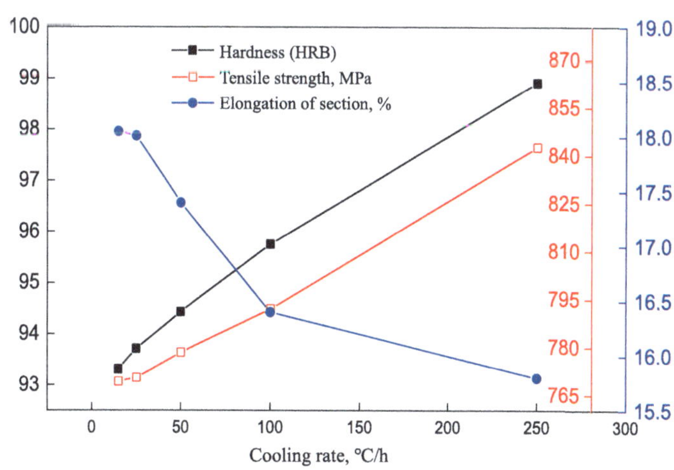

Fig. 4.16 Relationship between cooling rate and hardness, tensile strength and elongation of section

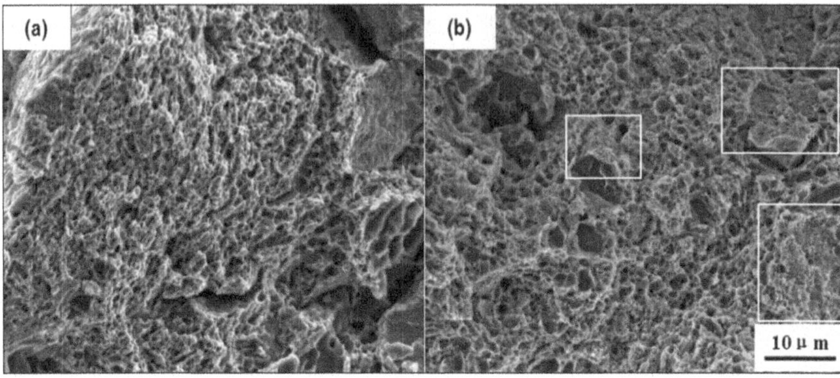

Fig. 4.17 Tensile fracture morphology of samples with different cooling intensity: **a** 25 °C/h, **b** 250 °C/h

When the cooling rate is 15 and 25 °C/h, there is no obvious difference in microstructure and the hardness of the samples changes little, which indicates that the spheroidizing effect is better when the cooling rate is lower than 25 °C/h. When the cooling rate reaches 50 °C/h, the microstructure changes obviously and the cooling rate may exceed the critical cooling rate, which increases the amount of the lamellar pearlite and fine granular carbides. Especially for the sample with cooling rate of 250 °C/h, carbides are generally very small, which indicates that slow cooling process is also an important way for the spheroidization of carbides. High cooling rate obviously restrains the diffusion of alloying elements in the matrix, which makes carbides precipitate nearby. Therefore, a large number of nanoscale carbides precipitate among the large carbides. The sorbitic lamellar spacing is very small and even reaches the nanometer level. This sorbite structure increases the hardness and tensile strength of the matrix. Because the lamellar structure of carbides is very fine, which has limited effect on plasticity. The tensile fracture surface in Fig. 4.17b presents dimple in the white frame, whose size is obviously smaller than other positions.

In summary, there are not only $M_{23}C_6$-type and M_7C_3-type carbides in the forged 8Cr13MoV structure, but also M_3C-type carbides distributed at the grain boundary. During the heating treatment, M_3C-type carbides rapidly convert to $M_{23}C_6$-type carbides, while M_7C_3-type carbides does not change to $M_{23}C_6$-type carbides and remains until the end of heat treatment. Different from other martensitic stainless steels, martensite decomposes and $M_{23}C_6$-type carbides precipitate in 8Cr13MoV structure. The microstructure does not change during slow cooling stage of spheroidizing heat treatment for 180 min.

With the increase of austenitizing holding time from 45 to 135 min, the hardness of the sample first decreases and then increases. When holding time is short, too many fine carbides precipitate and remain in the matrix during heating process, and the strengthening effect still exists. If the holding time is too long, the alloy elements dissolve and distribute evenly in the matrix, which restrains the spheroidization of carbides. In the following cooling process, these alloy elements will not only strengthen the matrix, but also precipitate in the vacancy with sorbite and fine carbides, which increases the hardness of the material [23].

The austenitizing holding time directly affects the growth rate of carbides in the early stage of spheroidization. The longer the holding time is, the faster the growth rate of carbide is. The longer the holding time is, the better the effect of carbide spheroidization is and the lower the hardness of the sample is. When the cooling rate is too fast or the holding time of austenitizing is too long, sorbite and small carbides are easy to form in the microstructure, which need to be eliminated by reducing the cooling rate or prolonging the holding time of spheroidizing stage.

The optimum holding time of austenitizing and spheroidizing is 90 min and 135 min, respectively. In the spheroidizing annealing process of 8Cr13MoV, the cooling rate should be controlled within 25 °C/h. Increasing the cooling rate leads to the precipitation of many fine carbides and sorbite in the microstructure, which results in a substantial increase in hardness and tensile strength [22].

The holding time of austenitizing and spheroidizing has little effect on the elongation after fracture. The main factors affecting the elongation after fracture are precipitation and distribution of carbides in the matrix. Less quantity and uniform distribution of carbides improve the elongation after fracture.

4.2 Effect Quenching Process on the Carbides in Steel

4.2.1 Phase Transformation Temperatures of 8Cr13MoV Steel

The characteristic transformation points of steel mainly include: austenite start transition temperature (A_{c1}), austenite transformation end temperature (A_{c3}), martensite start transformation temperature (M_s) and martensite transformation end temperature (M_f). The transformation point of steel is of great significance to the formulation of heat treatment temperature.

Thermal expansion analyzer (TEA) of DIL805L was employed to measure the A_{c1} and A_{c3} of 8Cr13MoV steel. The results are shown in Fig. 4.18.

As is shown in Fig. 4.18, the expanding length changes as a function of temperature. With the increase of temperature, an approximately linear expansion can be obtained when the temperature is below 841 °C, which is caused by thermal expansion. However, when the temperature exceeds 841 °C, an opposite trend occurs. The Δl begins to decrease, and this phenomenon does not end until 892 °C. After that, the expanding trend returns to normal as before. As is well-known, the specific volume of austenite is smaller than ferrite and martensite. Therefore, the volume shrinkage during 841 and 892 °C can be attributed to the austenite transformation. And the transformation point A_{c1} and A_{c3} are confirmed to be 841 °C and 892 °C, respectively.

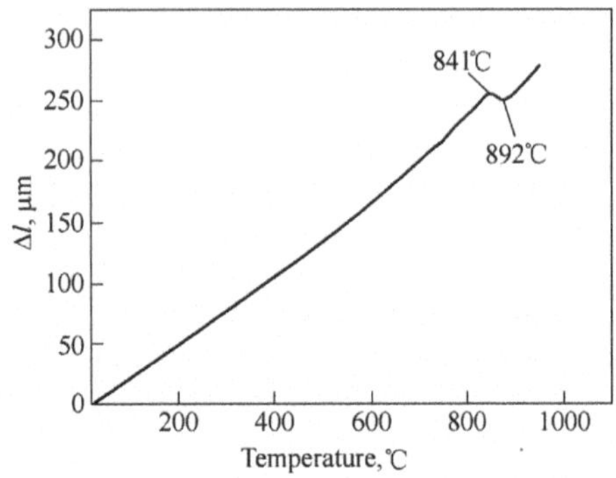

Fig. 4.18 Thermal expansion curve of 8Cr13MoV steel (Δl stands for the expanding quantity of line length)

The M_s and M_f were also measured by TEA. In this experiment, the samples were cooled down to ambient temperature after 10 min heat preservation at 1000 °C. A series of cooling rates, such as 20 °C/s, 10 °C/s, 5 °C/s, 3 °C/s, 1 °C/s, 0.5 °C/s, 0.1 °C/s, 0.03 °C/s, were employed to study the effect of cooling rate on the martensitic transformation. The Schematic view of the measurement of the martensitic transformation is shown in Fig. 4.19.

The volume expansion occurs when the supercooled austenite induces martensitic transformation, because the specific volume of martensite is larger than that of austenite. The M_s and M_f are also confirmed according to this principle. The results are shown in Fig. 4.20. As is shown in Fig. 4.20, with a cooling rate of 0.03 °C/s, the M_s and M_f are 423 °C and 262 °C, respectively. The M_s and M_f continues to decrease with the increasing cooling rate. When the cooing rate is between 0.03 and 1 °C/s, the decreasing tendency is obvious, and after that, this curve gets gradual. With a cooling rate of 20 °C/s, the M_s and M_f are 316 °C and 172 °C, respectively.

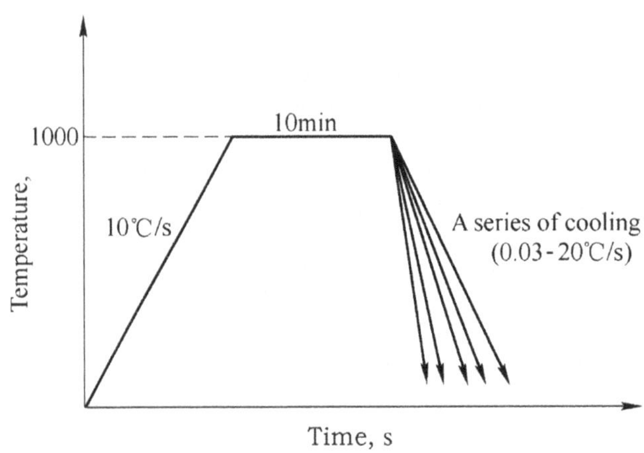

Fig. 4.19 Experimental process for the investigation of effect of cooling rate on martensitic transformation

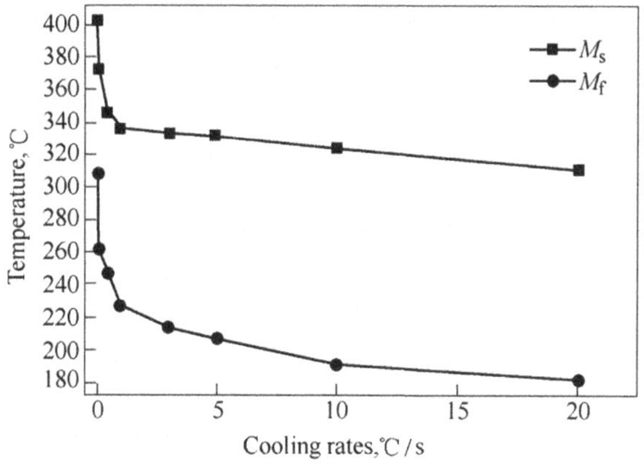

Fig. 4.20 Effect of cooling rate on martensite transformation point

4.2.2 *Effect of Quenching Process on the Carbides of Steel*

The cold rolling annealed samples were heated to different austenitizing temperatures (860–1150 °C) at a speed of 10 °C/min. After holding for 5 min, the samples were cooled by oil or air to room temperature.

Figure 4.21 shows the microstructure of oil-quenched samples at different austenitizing temperatures.

The microstructure of as-annealed steel consists of small, globular carbides in a ferrite matrix, as is demonstrated in Fig. 4.21a. After oil-quenching at low austenitizing temperatures, some small carbides grow up, and the microstructure consists of unequal-sized carbides in a uniform martensite matrix, which can be seen in Fig. 4.21b and c. With the temperature increasing up to 1050 °C, some small carbides dissolve into the matrix, while most larger carbides still keep undissolved. Moreover, some retained austenite appears in the martensite matrix, as is shown in Fig. 4.21d. As the austenitizing temperature continues to rise, the number of carbides decreases rapidly, while the content of retained austenite increases with a high rate. At the temperature of 1150 °C, only coarse martensite and retained austenite can be seen in as-quenched microstructure, and there are almost no residual carbides, as is shown in Fig. 4.21f. The effect of austenitizing temperatures on quenching microstructure can be summarized as follows: with the increase of austenitizing temperature, the amount of carbide in steel decreases. When the amount of carbide decreases to a certain extent, residual austenite appears in the microstructure and the content of residual austenite increases with the increase of austenitizing temperature. In addition, with the increase of austenitizing temperature, the acicular martensite structure gradually coarsened. The quenched samples treated with different austenitizing temperatures were analyzed by XRD, and the results are shown in Fig. 4.22.

The results of phase analysis are shown in Fig. 4.22. $M_{23}C_6$ carbide is detected in as-quenched samples with the austenitizing temperature of 860 and 950 °C. When the temperature reaches 1050,1075 and 1100 °C, diffraction peaks of carbides become less obvious due to their dissolution into the matrix, and they finally disappear at

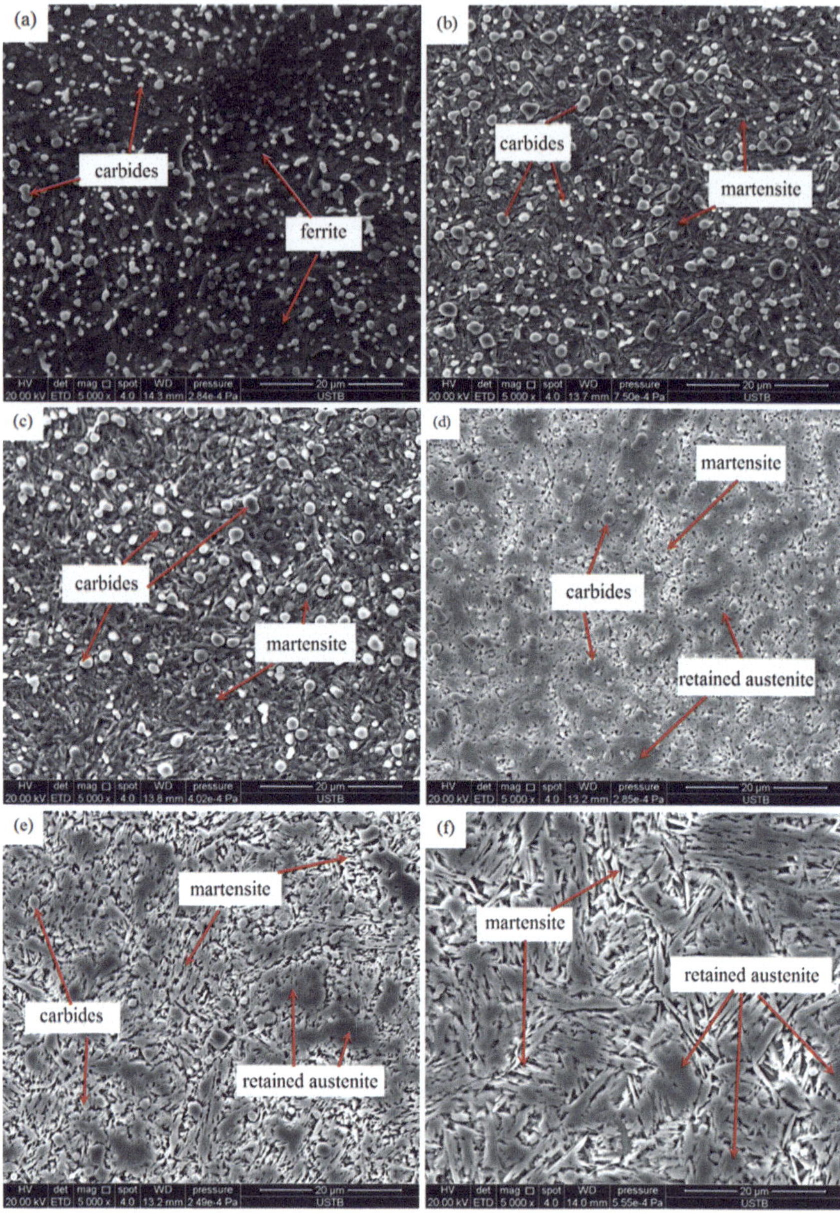

Fig. 4.21 Microstructure of oil-quenched samples with different austenitizing temperatures: **a** As-annealed; **b** 860 °C; **c** 950 °C; **d** 1050 °C; **e** 1100 °C; **f** 1150 °C

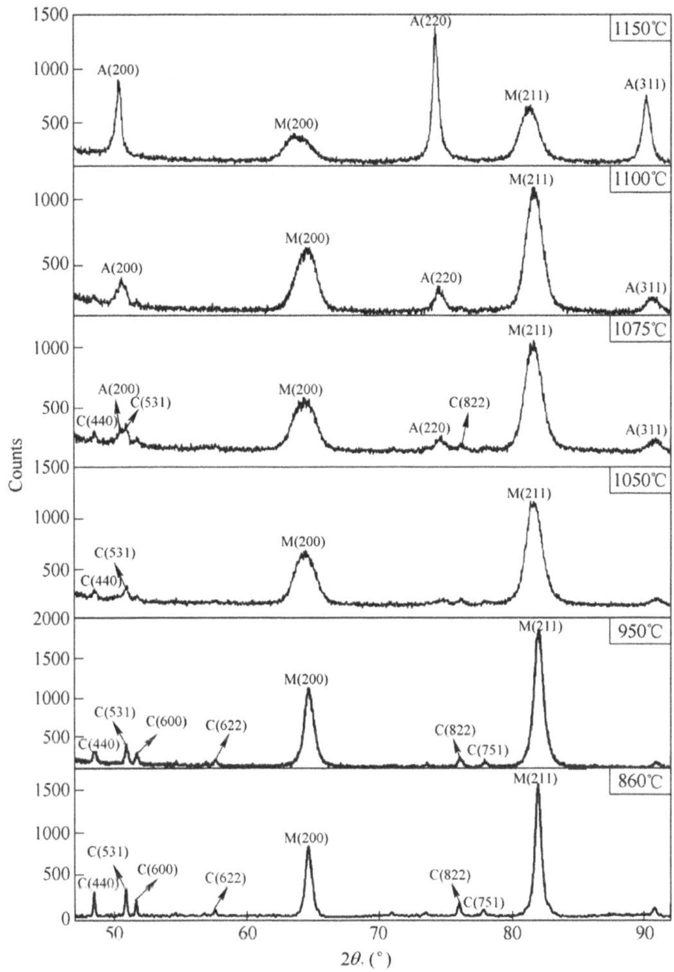

Fig. 4.22 Phase analysis of as-quenched samples under different austenitizing temperatures, where M, A and C stand for martensite, austenite and $M_{23}C_6$ carbides, respectively

1150 °C, which is coincide with the microstructure presented in Fig. 4.22. As is shown in Fig. 4.22, retained austenite first appears at 1075 °C, which is incongruent with the microstructure shown in Fig. 4.21. As can be seen in Fig. 4.21d, retained austenite is first discovered in as-quenched samples at the austenitizing temperature of 1050 °C. This inconformity can be related to the low austenite content at 1050 °C which could not be detected by XRD.

The average size and volume fraction of carbides in oil-quenched samples at different austenitizing temperatures were calculated using image analysis software. The results were reported as the average statistical data of ten SEM photographs for each sample. The results are shown in Fig. 4.23

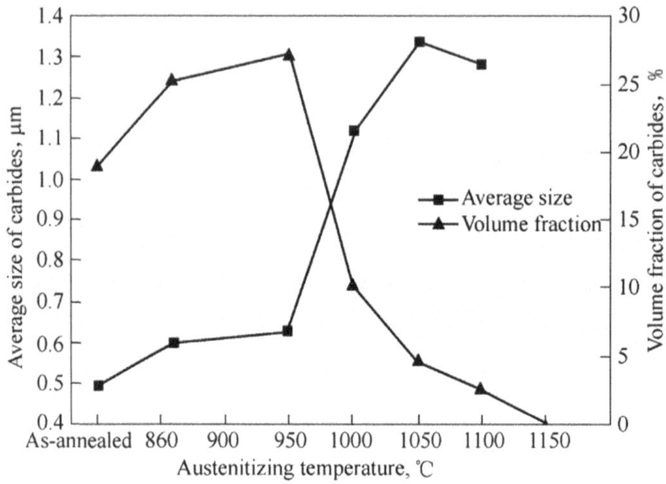

Fig. 4.23 Average size and volume fraction of carbides in oil-quenched samples at different austenitizing temperatures

As is shown in Fig. 4.23, both the average size and volume fraction of carbides increase slowly below 950 °C with increasing austenitizing temperature, It is attributed to the aggregation of alloying elements along the original tiny carbides. When the temperature is higher than 950 °C, some tiny carbides begin to dissolve into the matrix, while many large carbides are still difficult to dissolve. As a result, the volume fraction of carbides decreases, while the average size of carbides increases rapidly. Both the average size and volume fraction of carbides decrease with the increasing temperature above 1050 °C. This is due to the dissolution of larger carbides which is difficult to dissolve at lower temperatures.

The effect of austenitizing temperature on the content of retained austenite in oil-quenched samples is shown in Fig. 4.24.

It can be seen from Fig. 4.24 that the carbides begin to dissolve when austenitizing temperature is higher than 950 °C. When the temperature reaches 1050 °C, the volume fraction of carbides decreases to about 5%. The carbon content around the dissolved carbides is high, which improves the stability of austenite. It may lead to the formation of retained austenite after quenching process. With the increase of austenitizing temperature, especially in the range of 1100–1150 °C, the dissolution of large carbides and the further diffusion of carbon elements lead to the expansion of the high stabe austenite region, resulting in the obvious increase of retained austenite content in the quenched microstructure. When the austenitizing temperature is 1150 °C, the proportion of retained austenite reaches 57.7%. The increase of retained austenite greatly reduces the hardness of steel. Therefore, the austenitizing temperature should be controlled below 1100 °C.

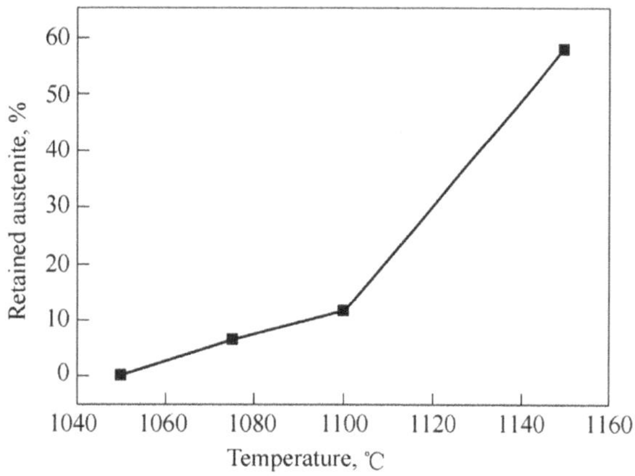

Fig. 4.24 Effect of austenitizing temperature on the content of retained austenite in oil-quenched samples

The 8Cr13MoV steel samples were heated to 1050 °C for 5 min by thermal dilatometer, and then cooled to room temperature at 0.03 and 10 °C/s, respectively. The effect of different cooling intensity on quenching structure is shown in Fig. 4.25. When cooling intensity was 0.03 °C/s, quenching microstructure is still the martensite, this is mainly due to the huge amounts of carbon and alloy elements in 8Cr13MoV steel which increases the stability of the supercooled austenite. The C curve moves to the right, reducing the quenching martensite formed at the critical cooling rate, even make 8Cr13MoV steel under slower cooling rate still can transform into martensite. When the cooling rate increased to 10 °C/s, the content of martensite in quenching microstructure decreased. Studies show that when carbon and alloying elements are dissolved into austenite, the stability of the supercooled austenite can be increased, and the C curve is shifted to the right. If it exists in the form of insoluble carbide, the stability of the supercooled austenite will be decreased instead. When the cooling rate is low, carbides will be precipitated in the austenite during the cooling process, which reduces the stability of the supercooled austenite and is conducive to the transformation from austenite to martensite. However, when

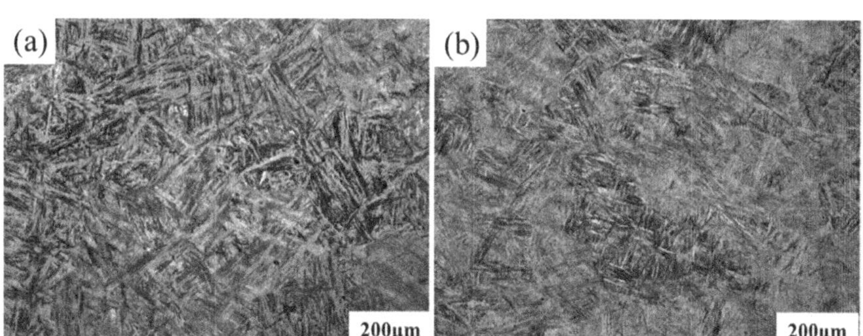

Fig. 4.25 Effect of cooling intensity on quenching structure: **a** 0.03 °C/s, **b** 10 °C/s

the cooling rate is high, it is difficult for carbon and alloy elements to precipitate in the austenite during the cooling process, which improves the stability of the supercooled austenite, thus reducing the martensite content in the quenched microstructure and increasing the residual austenite content. When the cooling rate was increased to 10 °C/s, the length and width of acicular martensite decreased, which was mainly because the increase of cooling intensity which inhibited the growth of martensite and made the martensite structure finer.

4.2.3 Evolution of Microstructure During Quenching Process

An ultra-high temperature confocal microscope was employed to simulate the quenching process in this experiment. The as-annealed sample was heated up to 1050 and 1100 °C, insulated for 15 min and cooled to ambient temperature with a cooling rate of 5 °C/s. The evolution of microstructure is shown in Fig. 4.26.

During the initial stage of heating process, some small carbides emerge from the matrix at 980°C, which can be attributed to the volume shrinkage caused by austenite transformation. With the increasing temperature, the volume fraction of carbides increases obviously, as is shown in Fig. 4.26c, which reveals the alloy elements precipitation and aggregation along some small carbides generated in the spheroidizing annealing process. After heat preservation for 15 min at 1050 °C, the grain boundary can be revealed due to the significant decrease of carbide content, and the grain size is about 20–30 μm. When the temperature of heat preservation is raised to 1100 °C, the austenite grains grow up quickly with the average size of 60 μm, which can be attributed to the dissolving of massive carbides and the weakening role of carbides on pinning grain boundary. There seems no much change when the temperature decreases from 1100 °C to 218 °C with a cooling rate of 5 °C/s, as can be seen in Fig. 4.26g. This phenomenon demonstrates that there is no pearlite or martensitic transformation during this temperature range, which further indicates that this cooling rate is greater than the critical cooling rate of martensitic transformation. The martensitic transformation begins at about 142 °C with a generating mode of skipping in some small area, as is shown in Fig. 4.26h. The grain boundary and undissolved carbides are the preferential positions for the nucleation of martensite. Martensitic transformation is not finished when the temperature cools down to ambient temperature, and there is still much retained austenite in the microstructure, as is shown in Fig. 4.26i. The unfinished transformation may be attributed to the higher dissolved carbon and alloying elements content which depresses the martensitic transformation range and reduces the M_s and M_f. If the M_f is depressed below room temperature, then the retained austenite may be present in the as-quenched microstructure.

The starting point of martensite transformation (M_s) observed in this quenching process is 147 °C, and the ending point of martensite transformation (M_f) is lower than room temperature. When the austenitizing temperature was 1000 °C and the cooling rate was 5 °C/s, Ms and M_f measured by the thermal dilatometer were 339

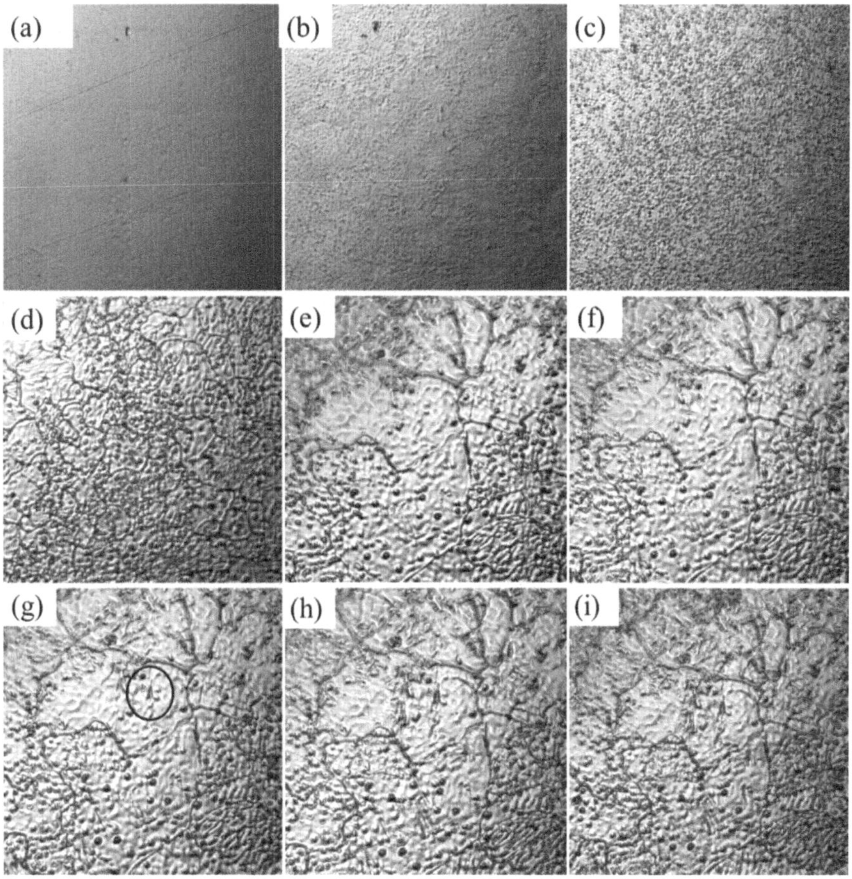

Fig. 4.26 Evolution of microstructure during heat treatment: **a** 100 °C; **b** 980 °C; **c** 1050 °C; **d** heat preservation for 15 min at 1050 °C; **e** heat preservation for 15 min at 1100 °C; f 218 °C; **g** 147 °C; **h** 42 °C; **i** 25 °C

°C and 166 °C, respectively. The difference in martensite transition temperature is mainly due to the fact that the austenitizing temperature in the quenching process is 1100 °C, a large amount of carbides have been dissolved into the austenite and the austenite grains grow significantly. These two factors both improve the stability of the supercooled austenite and make the C curve move to the right, reducing the martensite transition temperature. During the quenching process, parameters such as austenitizing temperature, holding time and cooling speed will affect the martensite transition temperature by affecting the content of carbon and alloy elements in the austenite, and finally affect the martensite content in the quenching structure and the performance of steel.

In general, martensite transformation in low carbon and low alloy steel is almost instantaneously completed by shear, and the continuity between martensite slats is

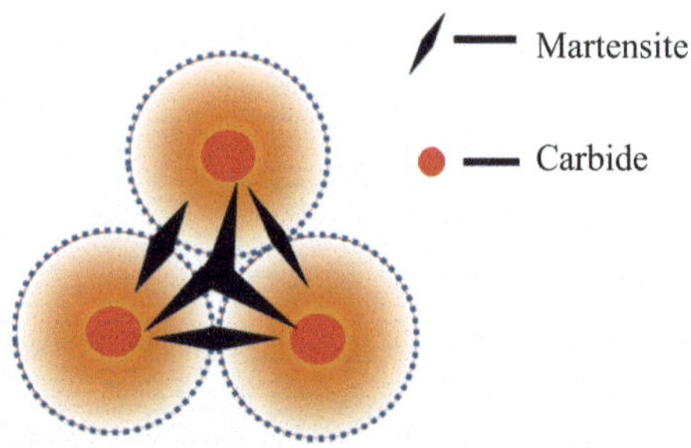

Fig. 4.27 Martensite transformation mechanism of 8Cr13MoV steel in quenching process (red to white represents high to low chromium concentration)

strong. For 8Cr13MoV steel, by observing the microstructure dynamic transformation during quenching, it can be seen that martensite transformation is only carried out in the form of shear in some isolated regions, with a small and relatively dispersed transformation range. This special martensite transformation mode can be explained by Fig. 4.27.

Before quenching, 8Cr13MoV steel contains a large number of secondary carbides in its microstructure. During quenching and heating to austenitizing temperature and heat preservation, some secondary carbides dissolve. The diffusion rate of carbon atoms in steel is faster, while that of chromium atoms is very slow. In the process of quenching and heat preservation, although carbide dissolves, the concentration of chromium element is still uneven. A radiant zone with a chromium concentration gradient is formed around the carbide, with the highest concentration near the original location of the carbide and lower concentrations far from the carbide. The higher chromium content enhanced the stability of the supercooled austenite and inhibited the martensite transformation, which could only occur in the regions with low chromium content. During martensite transformation, areas with high carbon or chromium concentration may be stopped, and areas with low chromium content may easily expand and extend, forming acicular martensite with wide middle and narrow ends.

4.2.4 Effect of Quenching Process on the Mechanical Property of Steel

4.2.4.1 Effect of Quenching Process on the Hardness of Steel

The hardness of experimental steel after oil-quenching and air-quenching at different austenitizing temperatures is shown in Fig. 4.28.

The hardness increases with increasing temperature when the temperature is below 1025 °C for oil-cooled samples, as is shown in Fig. 4.28. The maximum hardness of 62 HRc is observed at 1075 °C. When the temperature is between 1025 and 1075 °C, the hardness value is relatively stable with a slight increase from 61.7 to 62 HRC. Whereafter, it decreases rapidly with a minimum of 40.2 HRC at 1150 °C.

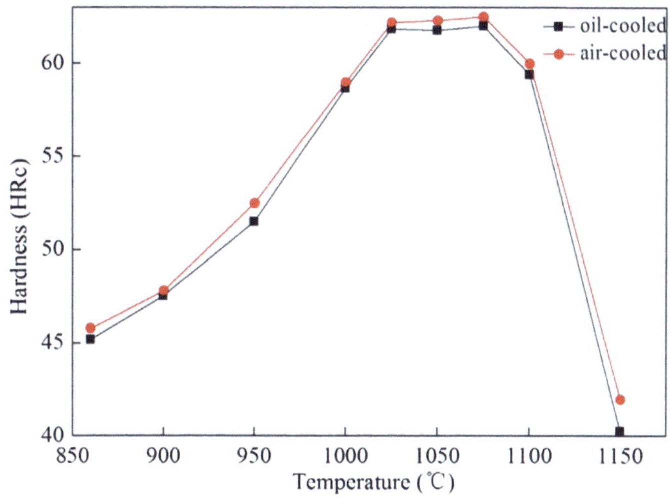

Fig. 4.28 Effect of austenitizing temperature on the hardness of the quenched steel

For air-cooled samples, the hardness is almost same as oil-cooled ones at the same temperature. The difference is that the hardness of air-cooled samples is slightly higher than that of oil-cooled ones at the same austenitizing temperatures.

The hardness of steel is mainly affected by the microstructure, carbide and the content of soluble carbon in the microstructure, especially the microstructure. According to the influence of austenitizing temperature on the volume fraction of carbide shown in Fig. 4.23, when the temperature is below 950 °C, the volume fraction of carbide presents an upward trend with the increase of temperature, resulting in the decrease of carbon solid solubility in martensite after quenching. However, due to the increase of martensite content in steel, the steel hardness increases. According to the XRD pattern of the samples at different austenitizing temperatures in Fig. 4.22, when the temperature is between 950–1025 °C, the steel is mainly composed of martensite and carbide. With the increase of austenitizing temperature, carbide dissolves in large quantities, and the solid solubility of carbon in martensite increases rapidly, so the hardness of steel continues to rise. As can be seen from the quenching structure in Fig. 4.21, when the temperature is 1050 °C, the content of carbon and alloying elements in the matrix increases due to the dissolution of carbide, the stability of supercooled austenite improves, and residual austenite appears in the structure. When the temperature is between 1025 and 1075 °C, with the increase of austenitizing temperature, the residual austenite content in steel increases, and the carbon solid solubility in martensite also increases. Under the joint action of the two, the steel hardness basically remains unchanged. As the austenitizing temperature continues to increase, the residual austenite content increases rapidly, the austenite grain grows, and the hardness of steel decreases rapidly. According to Fig. 4.28, the hardness of samples under air cooling condition is higher than that under oil cooling condition, on

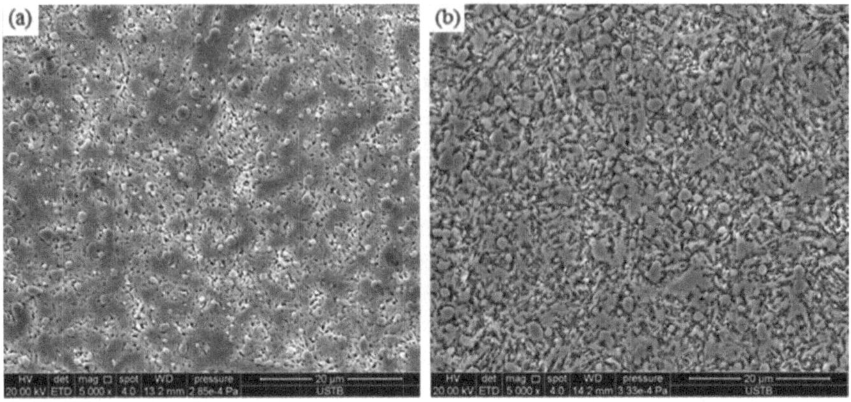

Fig. 4.29 Microstructure of as-quenched samples: **a** Oil-cooled, **b** Air-cooled

the premise that other quenching processes are kept unchanged. The microstructure of steel under oil and air cooling conditions is shown in Fig. 4.29.

As shown in Fig. 4.29, there is no great change in the size and amount of carbides. However, the retained austenite content in air-cooled sample is less than that of oil-cooled sample. In addition, the microstructure of martensite is more compact in air-cooling sample. Hardness tests show that the hardness values of oil-cooled and air-cooled samples are 61.7 and 62.3 HRC, respectively. Therefore, this experiment gives an evidence that high cooling rate can depress the martensitic transformation and increase the content of retained austenite. In conclusion, air-cooling can be selected to obtain less retained austenite, higher hardness and more compact microstructure of steel.

4.2.4.2 Effect of Quenching Process on Wear Resistance of Cutting Knives

The wear resistance of knives at different austenitizing temperatures of 1000, 1025, 1050 and 1100 °C were analyzed. As shown in Fig. 4.30, before the sharpness test, the distance between the cutting edge plane and the vertex of the edge wrapping Angle is L_0; after the sharpness test, the distance between the cutting edge plane and the vertex of the edge wrapping Angle is L_1, L_2 is the amount of wear on the cutting edge after the sharpness test.

Considering that there are differences in the thickness of the cutting sandpaper during the sharpness test, in order to more accurately represent the abrasion resistance of the tool itself, the thickness of the cutting sandpaper of the tool is added to the calculation method of the abrasion resistance of the tool, as shown in Eq. (4.4):

$$N = \frac{H}{L_2 \times 10^3} \tag{4.4}$$

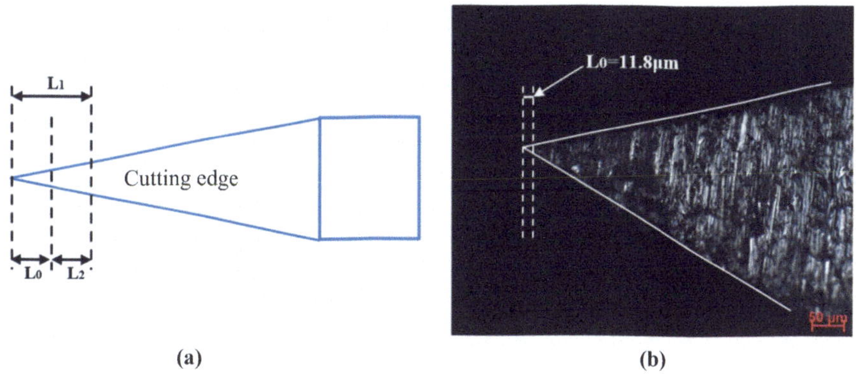

Fig. 4.30 Schematic view of cutting edge parameters and blade morphology without sharpness test: **a** Diagram of cutting edge of cutting tool; **b** Blade morphology not tested for sharpness

where N is the wear resistance of the tool; H is the thickness (i.e. sharpness durability) of the cutting sandpaper after the sharpness test, mm; L_2 is the amount of blade wear in the sharpness test, mm.

The cutting edge Angle is 37°, and the distance (L_0) between the cutting edge plane and the vertex of the cutting edge Angle is 11.8 μm.

After different austenitizing temperatures are adopted in the quenching process and cutting tools are tested for sharpness, the blade morphology is shown in Fig. 4.31.

It can be seen from Fig. 4.31 that when the austenitizing temperature is 1100 °C, the cutting edge has the best retention ability and less wear, followed by 1000 °C.When the austenitizing temperature is 1025 and 1050 °C, the wear degree of the cutting edge is more obvious, and the thickness of the cutting edge also increases obviously. The effect of austenitizing temperature on the blade morphology after sharpness test is determined by the factors such as the sharpness durability and wear resistance of the tool itself. The influence of austenitizing temperature on the wear amount (L_2) and wear resistance (N) of the cutting edge during quenching is shown in Fig. 4.32.

It can be seen from Fig. 4.32 that when the austenitizing temperature increases from 1000 to 1025 °C, the wear resistance of the tool does not change much, but the wear amount of the edge increases. As the austenitizing temperature continues to rise to 1050 °C, the abrasion resistance of the tool and the wear amount of the edge increase simultaneously. When the austenitizing temperature rises to 1100 °C, the abrasion resistance of the tool is greatly improved, and the edge wear is rapidly reduced. In general, the wear resistance of cutting tools will increase with the increase of austenitizing temperature, while the wear of cutting edge will increase first and then decrease with the increase of austenitizing temperature. According to the analysis in Fig. 4.23, the volume fraction of carbide in steel decreases with the increase of austenitizing temperature. The carbide on the blade surface is easy to fall off in the process of using. After the carbide falls off, many pits will be formed on the steel matrix, which improves the friction coefficient of the blade surface, reduces the

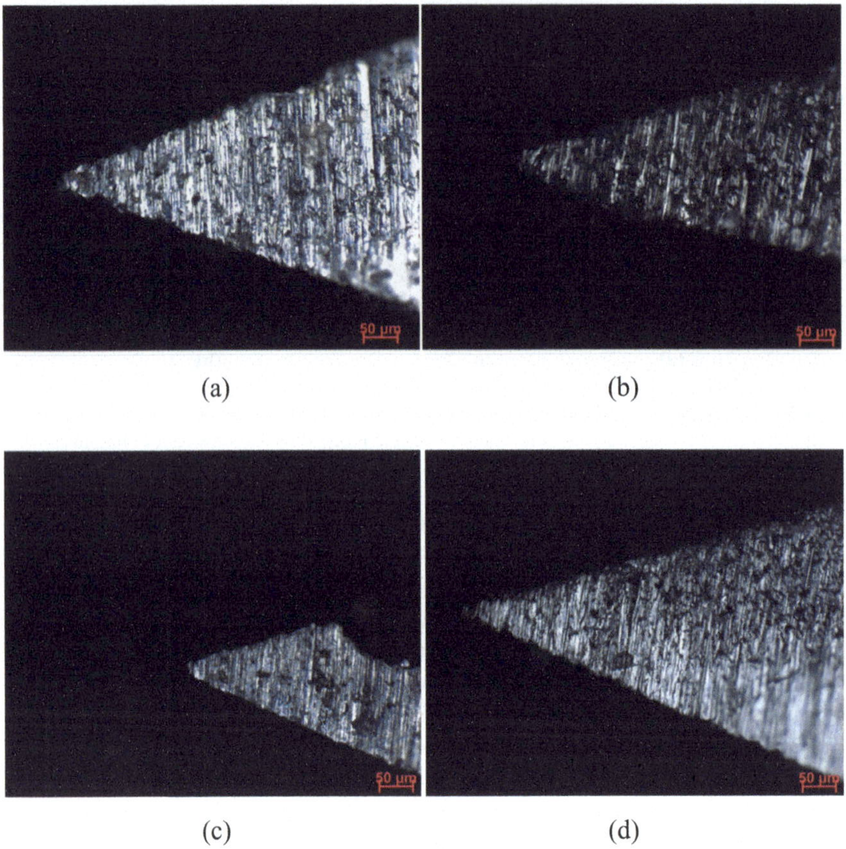

Fig. 4.31 Blade morphology after cutting tool sharpness test at different austenitizing temperatures during quenching: **a** 1000 °C, **b** 1025 °C, **c** 1050 °C, and **d** 1100 °C

sharpness of the tool and aggravates the wear of the tool. Therefore, with the increase of austenitizing temperature, the volume fraction of carbide in steel decreases, and the wear resistance of steel increases gradually.

After quenching at 1000, 1025, 1050 and 1100 °C, the sharpness of the cutting tool was tested. During the test, the thickness of the cumulative cutting sandpaper was 220 mm, 250 mm, 300 mm and 427 mm, respectively. When the austenitizing temperature increases from 1000 to 1050 °C, although the abrasion resistance of the tool does not change much, the cumulative cutting sandpaper thickness of the tool increases, resulting in the wear amount gradually increases. When the temperature rises to 1100 °C, although the cumulative cutting thickness of the tool still increases greatly, the wear amount of the tool decreases rapidly due to the obvious improvement of the wear resistance of the steel. This also explains why the blade morphology in Fig. 4.32 maintained better at 1000 and 1100 °C, while the wear was severe when the austenitizing temperature was 1025 and 1050 °C.

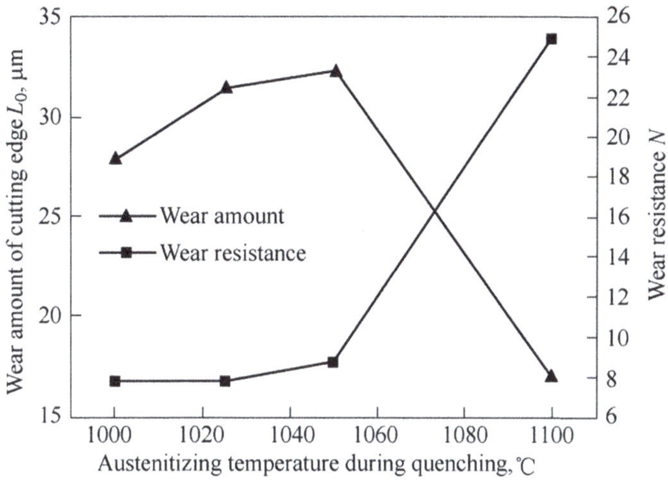

Fig. 4.32 Effect of austenitizing temperature on the wear amount and wear resistance of cutting edge during quenching

4.2.4.3 Effect of Quenching Process on the Sharpness of Cutting Knives

The sharpness test curve of cutting knives at different austenitizing temperatures in the quenching process, shown in Fig. 4.33.

As can be seen in Fig. 4.33, the amount of per cutting generally fluctuates and the cumulative amount of cutting (sharp durability) of the tool increases with the increase of austenitizing temperature. When the austenitizing temperature is 1000 °C, the cutting quantity of each knife is basically below 15 mm, and the sharpness of the tool is poor. When the temperature rises to 1025 °C, the cutting amount of each

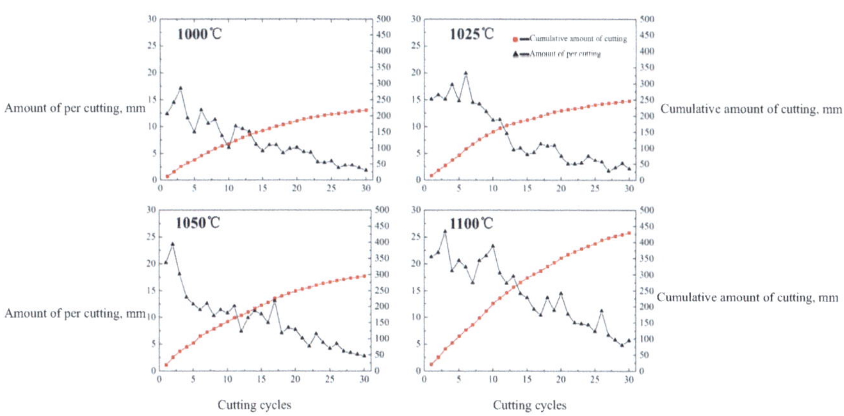

Fig. 4.33 Effect of austenitizing temperature on the sharpness of a knives during quenching

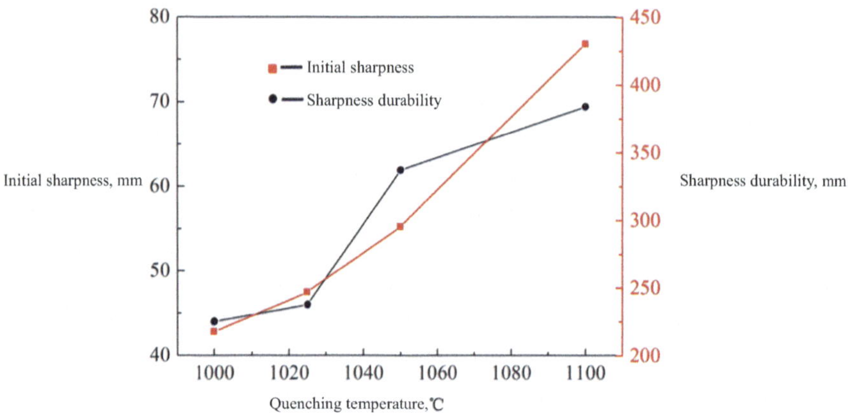

Fig. 4.34 Effect of quenching temperature on initial sharpness and sharpness durability of knife

cutter in the first 7 cutting cycles is greater than 15 mm, and the sharpness of the cutter is improved. When the austenitizing temperature rises to 1050 °C, the sharpness of the cutter is good in the first 3 cutting cycles. In the 5–15 cutting cycles, the cutting amount of each cutter fluctuates around 10 mm, and then decreases gradually. When the austenitizing temperature was 1100 °C, the per cutting amount in the first 12 cutting cycles was all greater than 15 mm, and then the per cutting amount showed a decrease in fluctuation. The effect of austenitizing temperature on initial sharpness and sharpness durability is shown in Fig. 4.34.

According to Fig. 4.34, when the quenching temperature increases from 1000 to 1025 °C, the initial sharpness increases slightly, while when the quenching temperature increases to 1050 °C, the initial sharpness is greatly improved. According to Figs. 4.23 and 4.28, when the quenching temperature increases from 1000 to 1025 °C, the steel hardness increases from HRC59 to HRC 62.2, and the carbide volume fraction decreases from 10% to 7.5%. When the quenching temperature increased from 1025 to 1050 °C, the steel hardness increased from HRC62.2 to HRC 62.3, and the carbide volume fraction decreased from 7.5% to 5%. It can be seen that the hardness of steel has little effect on the initial sharpness, and the increase of initial sharpness is mainly due to the reduction of the carbide volume fraction in steel. When the volume fraction of carbide is large, more pits are formed on the blade, which leads to a large friction coefficient on the blade surface and reduces the initial sharpness of the tool. In addition, combined with Figs. 4.32 and 4.34, when the austenitizing temperature increased from 1000 to 1050 °C, the wear resistance of the tool did not change much, and the initial sharpness gradually increased, indicating that the initial sharpness of the tool has little relationship with the wear resistance of the tool. When the quenching temperature increases from 1050 to 1100 °C, the carbide volume fraction in steel decreased to around 3%, while the steel hardness decreased, but initial sharpness tool is still in a rising state, which again indicated that when the blade geometry morphology was consistent, the main factors influencing the cutting tool initial sharpness is the volume fraction of carbides in the steel, tool initial sharpness increased along with the decline of carbide volume fraction.

It can be seen from Fig. 4.34 that the sharpness durability is always on the rise with the increase of austenitizing temperature, which is similar to the change trend of the initial sharpness, indicating that the reduction of the volume fraction of carbide can improve the initial sharpness and the sharpness durability at the same time. According to Figs. 4.23, 4.28 and 4.32, when the austenitizing temperature increases from 1000 to 1050 °C, the volume fraction of carbide in steel gradually decreases from 10 to 5%, the hardness of steel increases from HRC 59 to HRC 62.3, and the wear resistance of steel remains basically unchanged. When the austenitizing temperature increased from 1050 to 1100 °C, the volume fraction of carbide in steel decreased to 3%, and the hardness of steel decreased from HRC 62.3 to HRC 60.3, but the wear resistance of steel increased greatly. It shows that the volume fraction of carbide in steel is the decisive factor affecting the wear resistance of steel. When the carbide volume fraction continues to decrease, the wear resistance of steel increases rapidly. Knife sharp durability and wear resistance were positively correlated, when austenitizing temperature increased from 1000 to 1050 °C, although wearability remain unchanged, but the carbide volume fraction decreases, sharp and durable degree increased, when the austenitizing temperature continues to rise to 1100 °C, due to the cutting tool abrasion resistance increased significantly, carbide volume fraction in steel also decreases, both role led to sharp durability increased significantly.

From the above analysis, it can be concluded that the change of austenitizing temperature in the quenching process mainly affects the sharpness of steel through the way of affecting the volume fraction of carbide in steel, the hardness of steel itself has no obvious effect on the sharpness. With the decrease of the volume fraction of carbide in steel, the initial sharpness and sharpness durability of the cutting tools are on the rise. Combined with the analysis of blade surface morphology in the process of sharpness test, the main reason for the decrease of sharpness performance is the increase of friction coefficient of blade surface caused by carbide falling off. In general, carbide hardness is higher than martensite, and the uniform distribution of carbide in steel matrix can improve the wear resistance of steel. During the sharpness test, it was found that the carbides distributed on the cutting edge which fell off easily. After the carbide fell off, it would not only lose the function of protecting the steel matrix, but also accelerate the wear of the cutting tool and reduce the sharpness of the cutting tool.

Carbide shedding is related to its size. When the austenitizing temperature is between 1000 °C and 1100°C, the average size of carbide is greater than 1 μm. The depth of the spherical carbide embedded in the matrix is not large, and it is easy to be cut off by the shear force of abrasive particles in the friction process. To reduce the large size, spherical carbide of tool steel, obtain a large number of nanoscale, irregular shape carbide, will effectively improve the wear resistance of steel and the ability to maintain the edge of the tool, and further improve the sharpness of the tool.

4.3 Effect of Tempering on the Carbides of Steel

4.3.1 Effect of Tempering Temperature on Carbides and Microstructure

The microstructure after quenching is very unstable due to the high supersaturation of carbon in the as-quenched martensite, the high strain energy and interface energy of the martensite, and the existence of a certain amount of retained austenite in the martensite. The difference of free energy between unequilibrium state and equilibrium state of martensite and retained austenite provides the driving force for tempering transformation. The transformation will be spontaneous once the dynamic conditions are met. This dynamic condition is that the atom have enough activity ability. Tempering treatment could increase the activity ability of the atom by heating, so that the transformation can be carried out and reach the required degree. Tempering of martensitic stainless steel usually includes the following processes: carbon atom segregation in martensite; martensite decomposition; transformation of retained austenite; precipitation and transformation of carbide; the aggregation and growth of carbide and the recovery and recrystallization of ferrite. These five processes are different from each other and overlapped with each other, and controlled by diffusion factors. Therefore, their transformation depends on tempering temperature and holding time, of which temperature is one of the most important factors. The alloy elements (Cr, etc.) play a role in the microstructure transformation in the tempering process, which generally hinder the transformation, so that each stage temperature of tempering transformation moves to high temperature.

7Cr17MoV could obtain better mechanical properties and corrosion resistance, when the quenching temperature is 1050–1100 °C and the quenching holding time is 15–30 min. Therefore, this quenching process is selected for the further tempering research. After being quenched at 1060 °C, 7Cr17MoV is tempered. The tempered microstructure under different tempering temperatures is shown in Fig. 4.35. During tempering, due to the precipitation of carbide, the supersaturation of carbon in α-Fe decreases continuously. At the same time, the residual austenite will also decompose and change into supersaturated α-Fe and carbides, which are equivalent to tempering martensite. Therefore, the tempering structure of 7Cr17MoV martensitic stainless steel is annealing martensite + carbides + a small amount of residual austenite.

Figure 4.35 shows that when the tempering temperature is 100 °C, the martensitic structure keeps its morphology during quenching with a small amount of carbide precipitation. The increasing tempering temperature enhances the diffusion ability of alloy elements. The carbon atoms keep precipitating from supersaturated solid solution. In the meanwhile, the martensite reverts, dislocation moves, carbide precipitates and grows up. The residual austenite obviously decomposes when the tempering temperature is increased to 200 °C. The morphology of martensite is relatively obvious, and there is no recrystallization of martensite, when the tempering temperature is below 300 °C. The total precipitation amount of alloy carbide also changes

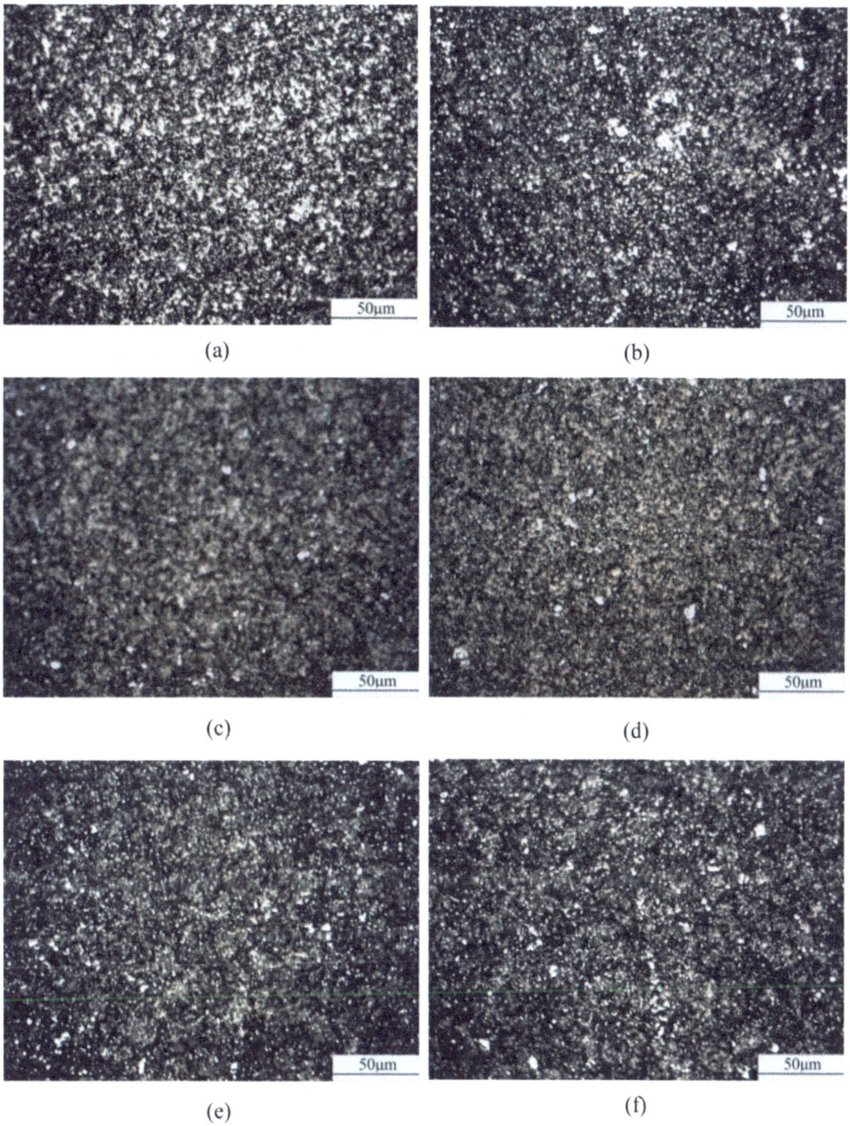

Fig. 4.35 Metallography of 7Cr17MoV at different tempering temperatures: **a** As-quenched, **b** 100 °C; **c** 150 °C, **d** 200 °C, **e** 250 °C, **f** 300 °C

little. The XRD patterns of 7Cr17MoV at different tempering temperatures are shown in Fig. 4.36.

Figure 4.36 shows that the phases stay the same at low-temperature tempering due to the less precipitation quantity of carbide. The type of carbide is mainly $M_{23}C_6$.

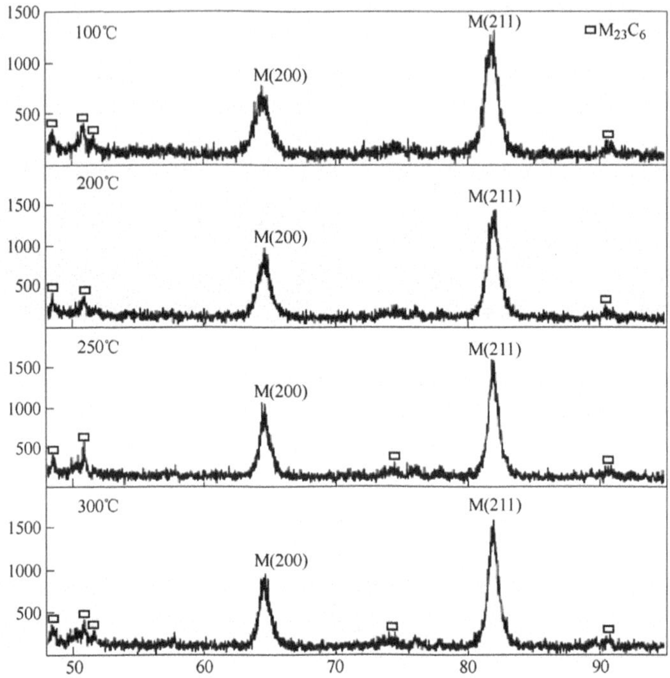

Fig. 4.36 XRD patterns of 7Cr17MoV at different tempering temperatures

The microstructure and carbide of 7Cr17MoV at different tempering temperatures are shown in Fig. 4.37.

Figure 4.37 shows that carbide precipitates from the matrix, the quantity increases and the average size decreases due to the increase of tempering temperature. When the tempering temperature is 100 °C, only a small amount of carbide precipitates, and the average size of carbide is about 1.0 μm. When the tempering temperature is between 150 and 200 °C, the quantity of carbide increases, the average size of carbide is about 0.9 μm. When the tempering temperature is between 250 and 300 °C, the quantity of carbide in the structure further increases, and the average size of carbide is about 0.77 μm. Although the size of the original carbide will increase due to the increase of tempering temperature, the average size of carbides generally decreases as the new fine secondary carbides precipitate. With the increase of tempering temperature, M_3C-type and M_7C_3-type carbides precipitate first, then $M_{23}C_6$-type carbides precipitate. M_7C_3-type carbides precipitate at 300 °C, $M_{23}C_6$-type carbides precipitate at 500 °C [23, 24]. M_3C-type and M_7C_3-type carbides are the main secondary carbide precipitated at a low tempering temperature which is below 300 °C, and $M_{23}C_6$-type carbides are mainly the undissolved carbide during quenching process.

Quantitative analysis of carbide and retained austenite content at different tempering temperatures is shown in Fig. 4.38.

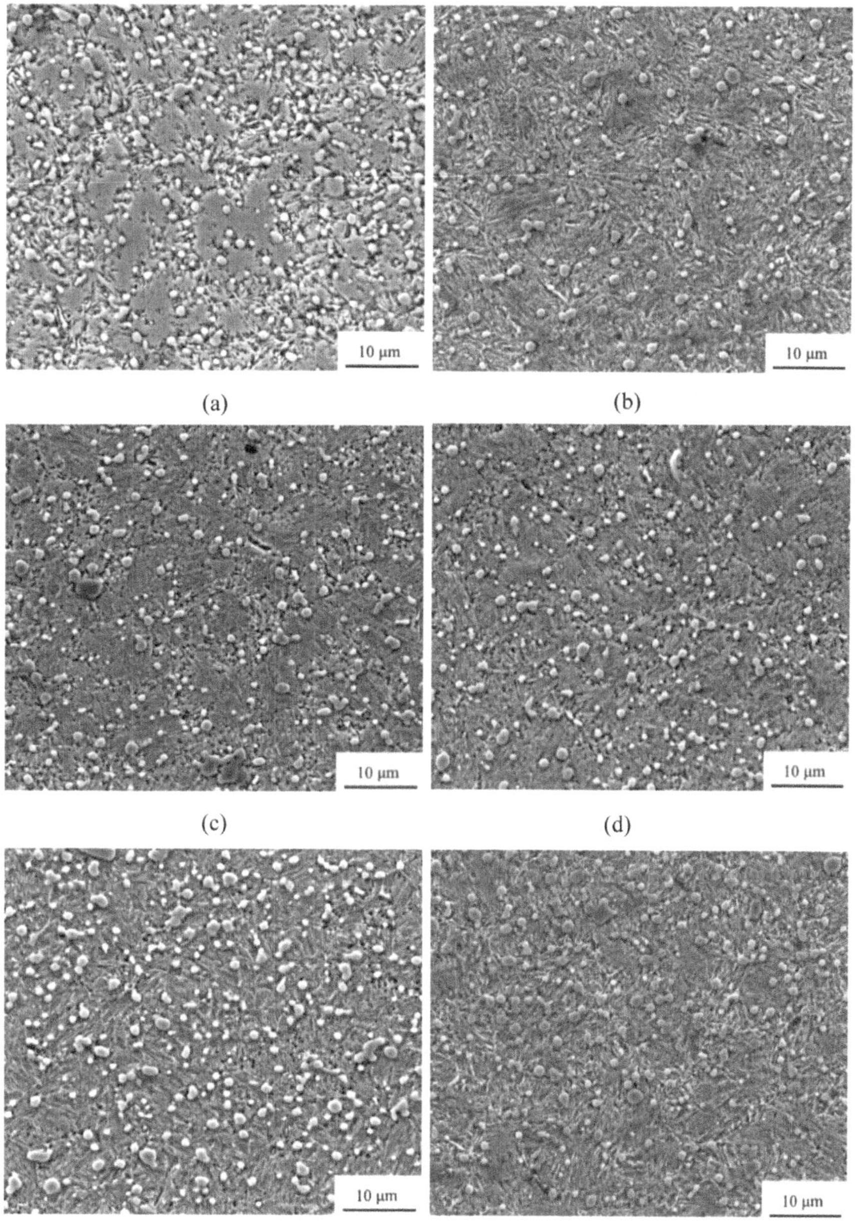

Fig. 4.37 Microstructures of 7Cr17MoV at different tempering temperatures: **a** As-quenched; **b** 100 °C; **c** 150 °C; **d** 200 °C; **e** 250 °C; **f** 300 °C

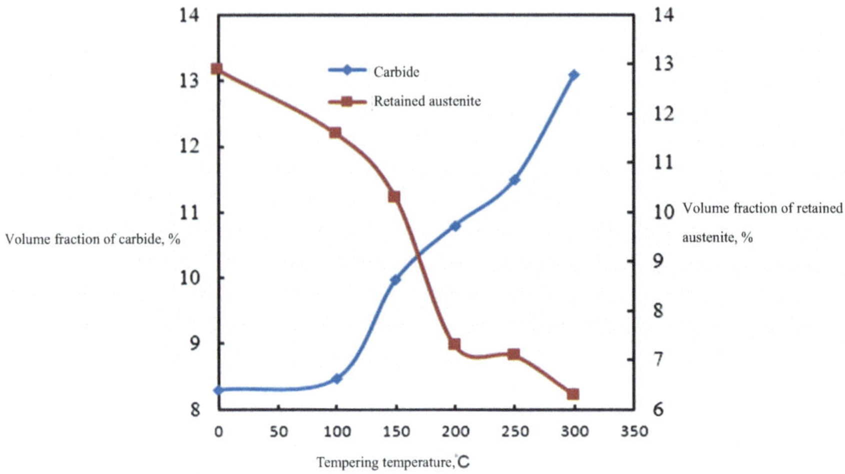

Fig. 4.38 Effect of different s on the content of carbide and retained austenite in 7Cr17MoV

Figure 4.38 shows that carbon atoms redistributed when tempering temperature is 100 °C. A small amount of ε-carbide precipitates, and the residual austenite stays the same [25]. At this time, the volume fraction of carbide is 8.47%, and the content of retained austenite is 11.6%, which are similar with the content of carbide and retained austenite in as-quenched structure. When the tempering temperature is 150 °C, ε-carbide precipitates further. The volume fraction of carbide increases to 9.97%, and the content of retained austenite gradually decreases to 10.3%. When the tempering temperature reaches 200 °C, ε-carbide converts to M_3C carbide. In the meantime, the retained austenite begins to decompose and form ferrite and M_3C carbide. The volume fraction of carbide rises to 10.8%, and the content of retained austenite rapidly reduces to 7.3%. When the tempering temperature reaches 250 °C, the retained austenite decomposes continuously and M_3C carbide further precipitates. The volume fraction of carbide is 11.5% and the content of retained austenite is 7.1%. When the temperature reaches 300 °C, M_3C carbide precipitates continuously. At the same time, M_3C carbide continuously converts to M_7C_3 carbide. The volume fraction of carbides is 13.1%, and the content of retained austenite further reduces to 6.3%.

The line scan energy spectrum of micro-region of 7Cr17MoV steel at tempering temperature of 150 °C is shown in Fig. 4.39.

Figure 4.39 shows that, similar to the linear scanning pattern of as-quenched sample, the chromium content in the carbide is obviously higher than other areas. The molybdenum content in the carbide is slightly higher than other areas. And the iron content in the carbide is obviously lower than other areas. From the edge to the center of the carbide, the chromium content gradually increases and the iron content shows an opposite tendency. There is no obvious chromium depleted zone in the matrix. But the chromium content of the matrix that is far away from the carbide is

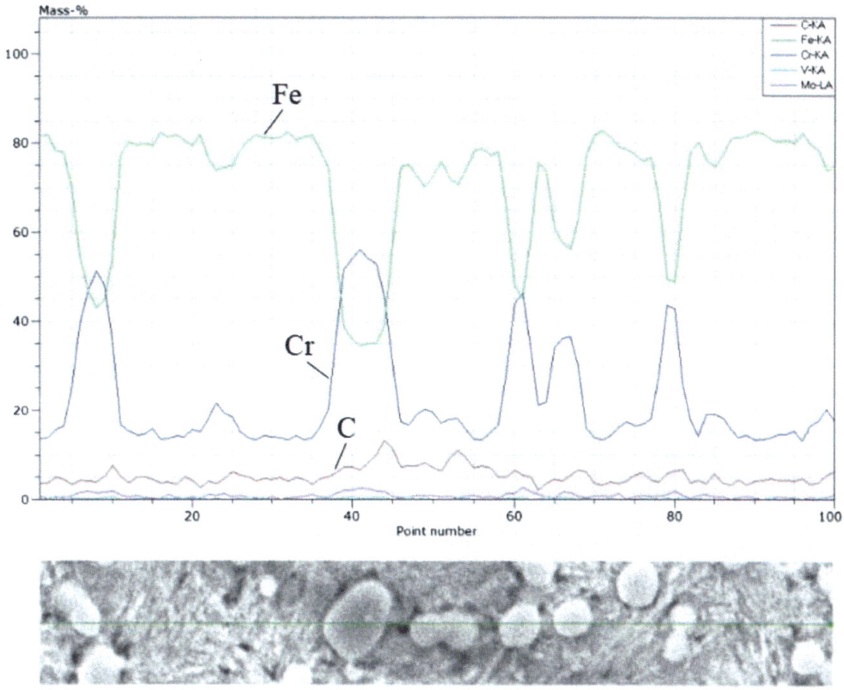

Fig. 4.39 Line scanning energy spectrum of micro-region of 7Cr17MoV steel at tempering temperature of 150 °C

the lowest, which indicates that most of the carbides are insoluble carbides and the post-precipitated carbides are few. Figure 4.40 is the area scan energy spectrum of tempered 7Cr17MoV at 150 °C.

Figure 4.40 shows the area scan energy spectrums have good correspondence to the line scan energy spectrum. There is obvious deficiency of iron element, obvious enrichment of chromium and carbon elements and slight enrichment of molybdenum element in carbide. The vanadium element shows relatively uniform distribution in the matrix and has a slight accumulation in some parts, which is due to the low tempering temperature. The vanadium and molybdenum elements dissolve in the matrix and could not precipitate at such a low temperature. Compared with the area scan energy spectrums of the as-quenched samples, the alloy elements in the matrix uniformly distribute after tempering, and the microstructure recovers well.◦

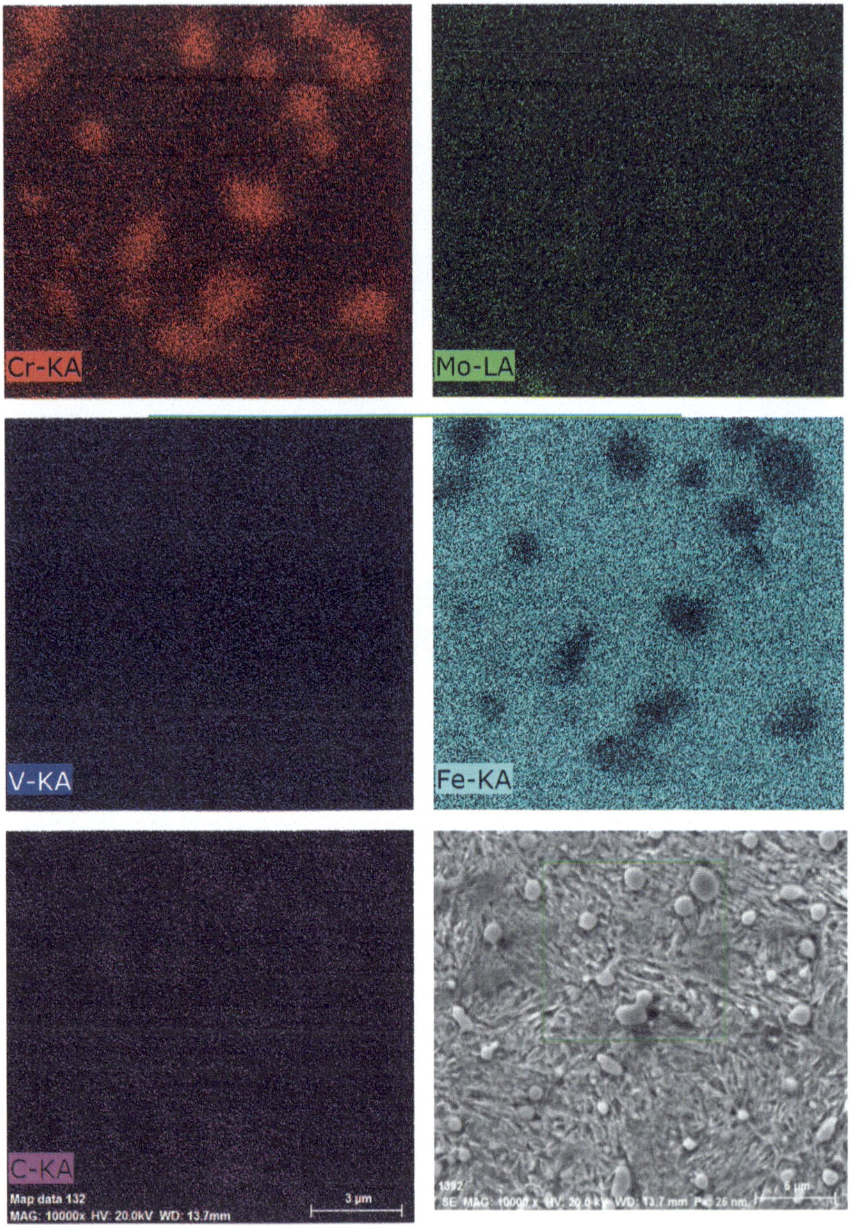

Fig. 4.40 Area scan energy spectrums of tempered 7Cr17MoV at 150 °C

4.3.2 Effect of Tempering Temperature on Mechanical Properties

Tempering of steel includes two processes, which are softening and hardening, respectively. The recovery of martensite and dislocation structure lead to the softening process. The decomposition of austenite, the desolvation of supersaturated carbon and the precipitation of the second phase lead to the hardening process [26]. Figure 4.41 is the effect of different tempering temperature on 7Cr17MoV hardness.

Figure 4.41 shows that the hardness of the material decreases with the increase of tempering temperature. When the tempering temperature is 100 °C, the hardness of the material is HRC 60.5, which is even higher than HRC 59.2 of the as-quenched samples. During tempering, the high chromium and other alloy contents in the steel improve the tempering resistance, keep the morphology of martensitic structure, promote the precipitation of fine carbides and finally increase the steel hardness. With the increase of tempering temperature, the diffusion ability of carbon and alloy elements increases, which leads to the continuously precipitation of carbides from supersaturated solid solution and the recovery of martensite and the movement of dislocation. In the meanwhile, the undissolved carbides grow and coarsen, which softens the material [27]. When the tempering temperature reaches 200 °C, the retained austenite decomposes obviously and produces tempered martensite and carbides, which strengthens the hardness of the material. The hardness decline trend slightly slows down when tempering at 200–250 °C. The hardening effect of carbide is not obvious due to lacking of the dispersed carbides, and the hardness of the material shows a downward trend with the increase of tempering temperature.

Figure 4.42 is the effect of tempering temperature on tensile properties of 7Cr17MoV.

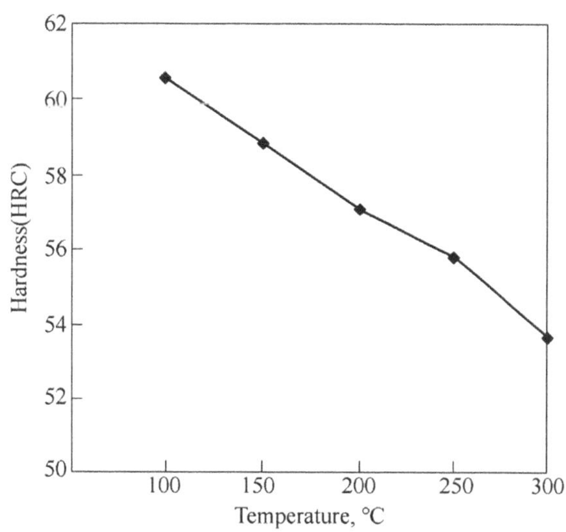

Fig. 4.41 Effect of different tempering temperature on 7Cr17MoV hardness

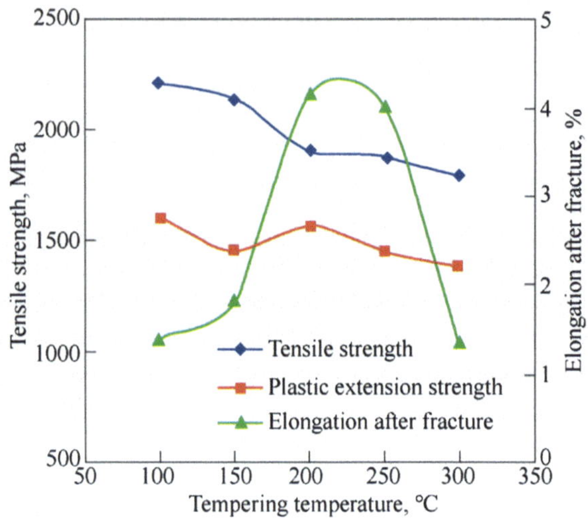

Fig. 4.42 Effect of tempering temperature on tensile properties of 7Cr17MoV

Figure 4.42 shows that the tensile strength and plastic extension strength of the material generally decrease with the increase of tempering temperature, which is related to the tempering softening of the material. The trends of the tensile strength and plastic extension strength are basically the same as the hardness while different with the elongation after fracture. When the tempering temperature is 100 °C, the elongation after fracture of the material is as lower as 1.28%, which is due to limited tempering effect on the improvement of the plasticity of the material at such low tempering temperature. With the increase of tempering temperature, the elongation of the material increases. When the tempering temperature is between 200 and 250 °C, the elongation of the material increases obviously, which is about 4%. This is because the increase of tempering temperature reduces the supersaturation of martensitic structure, promotes the carbide precipitation and the recovery of the martensitic structure and improves the homogeneity of the material structure. When the tempering temperature rise to 300 °C, the elongation decreases significantly, which is related to the low temperature tempering brittleness.

Figure 4.43 is the tensile fracture morphology of 7Cr17MoV at different tempering temperatures.

Figure 4.43 shows that the fracture of material contains the characteristics of plastic fracture and brittle fracture. There are dimples, micropores and a few cleavage surfaces on the fracture surface. When the tempering temperature is 100°C, the cleavage surface is more obvious. With the increase of tempering temperature, dimple become more obvious. Dimple is an typical feature of plastic deformation of material, which indicates that the plastic deformation of material is more obvious with the increase of tempering temperature. Especially when the tempering temperature is between 200 and 250 °C, the fracture surface is rougher, the distribution of dimples is more uniform, and the plasticity of the material is relative better than the other

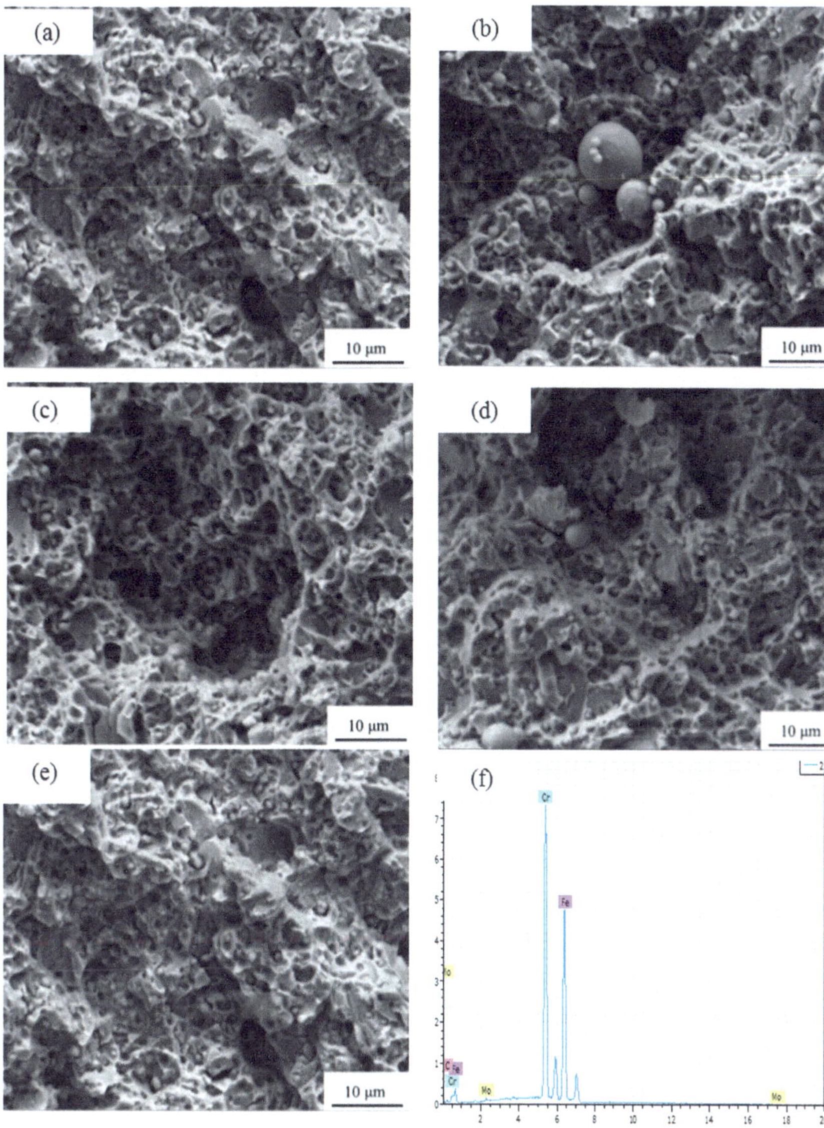

Fig. 4.43 Tensile fracture morphology of 7Cr17MoV at different tempering temperatures. **a** 100 °C, **b** 150 °C, **c** 200 °C, **d** 250 °C, **e** 300 °C and **f** energy spectrum of carbide

samples. When the tempering temperature is 300 °C, the cleavage surface becomes obvious and the plasticity of the material decreases. In the meantime, there are obvious second phase particles in the dimples. EDS analysis is carried out for these particles and the result is shown in Fig. 4.43f. Most of the second phase particles are carbides, which indicates that the formation and fracture of dimple are closely related to the second phase particles such as carbide. Therefore, it is helpful to improve the plasticity of the material through controlling the precipitation, distribution and size of the second phases.

4.3.3 Effect of Tempering Temperature on Corrosion Performance

Figure 4.44 shows the curve of open circuit potential with time of 7Cr17MoV soaked in 3.5% (mass fraction) NaCl solution at different tempering temperature. The open circuit voltage slowly decreases with time then reaches a relatively stable state when tempering at 100, 150 and 250 °C. The open circuit voltage gradually increases with time then reaches a stable state when tempering at 200 and 300 °C. The curve changes with small fluctuations, which should be attributed to the occurrence and repair of metastable corrosion at some active particles (carbides and inclusions). The open circuit voltage of samples tempering above 200 °C is significantly higher than that tempering at 100 and 150 °C.

Figure 4.45 shows the polarization curve of 7Cr17MoV at different tempering temperatures. Table 4.7 shows the corresponding self-corrosion potential and self-corrosion current. The increase of tempering temperature raises the self-corrosion potential, reduces the self-corrosion current and improves the corrosion resistance of the material. Tempering at 100 °C has similar electrochemical corrosion process

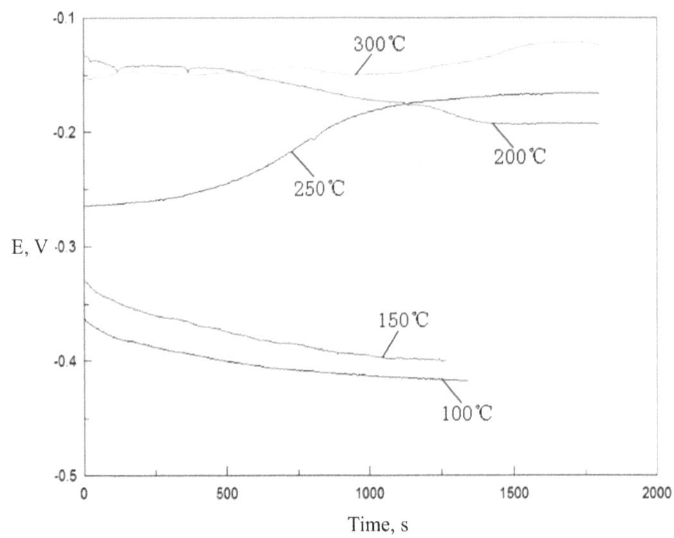

Fig. 4.44 Effect of tempering temperature and time on the open circuit potential of 7Cr17MoV

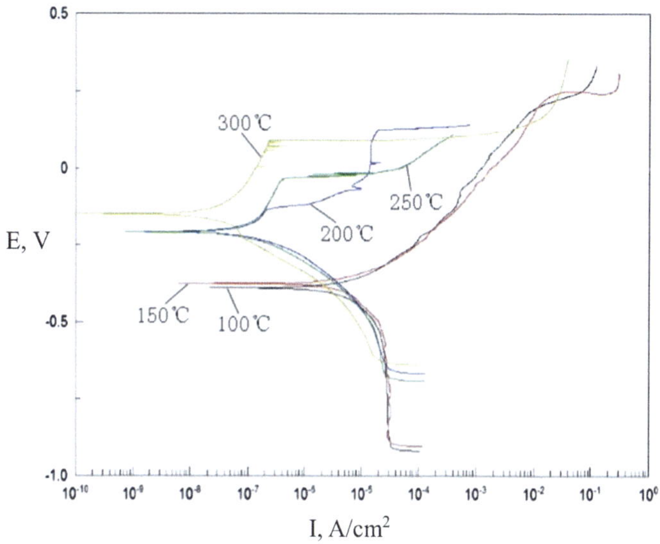

Fig. 4.45 Dynamic polarization curves of 7Cr17MoV at different tempering temperatures

Table 4.7 Self-corrosion potential and current parameters of 7Cr17MoV at different tempering temperatures

Temperature, °C	100	150	200	250	300	1050
E, V	−0.39	−0.38	−0.21	−0.21	−0.15	−0.28
I, A/cm^2	9×10^{-6}	8.18×10^{-6}	1.93×10^{-7}	1.98×10^{-7}	5.02×10^{-8}	6.7×10^{-7}

with tempering at 150 °C. Tempering at 200 °C has similar electrochemical corrosion process with tempering at 250 °C. Carbides precipitate at the higher tempering temperature and may damage the corrosion resistance of the material. However, the precipitated carbides are mainly M_3C-type and M_7C_3-type, which occupy less chromium element. The quantity of precipitated carbides is also small. Hence, low temperature tempering has limited effect on the corrosion resistance of the material, while the undissolved $M_{23}C_6$-type carbides have negative effect on the corrosion performance of the material. In the meantime, when the tempering temperature is low, the diffusion speed of elements in the structure is relatively slow, which generates chromium depleted region, leads to the weak areas of corrosion resistance in the structure of the material, forms micro electrochemical corrosion and declines the corrosion resistance of the material. When the tempering temperature increases, the diffusion speed of chromium and other elements rise, the material structure is more uniform, the internal stress is further reduced and the corrosion resistance of the material is increased. The dispersed carbides which precipitate during tempering is also conducive to the improvement of the pitting resistance of the material.

Figure 4.46 shows the pitting morphology of 7Cr17MoV quenched at 1060 °C and tempered at different temperatures after electrochemical test. A number of pitting holes can be observed in the quenched sample. Serious corrosion occurs when the sample are tempered at 100 and 150 °C, especially at 150 °C. A large number of large corrosion holes appear on the surface, and the corrosion holes gradually expand and connect each other to the long strip corrosion area. Some long strips are even through the whole sample surface, and the expansion of corrosion holes has directionality. When tempering at 200 °C, individual pitting areas can be seen on the sample surface. When the tempering temperature rises to 250 and 300 °C, it is difficult to find pitting areas on the sample surface, which is consistent with the test results of polarization curve.

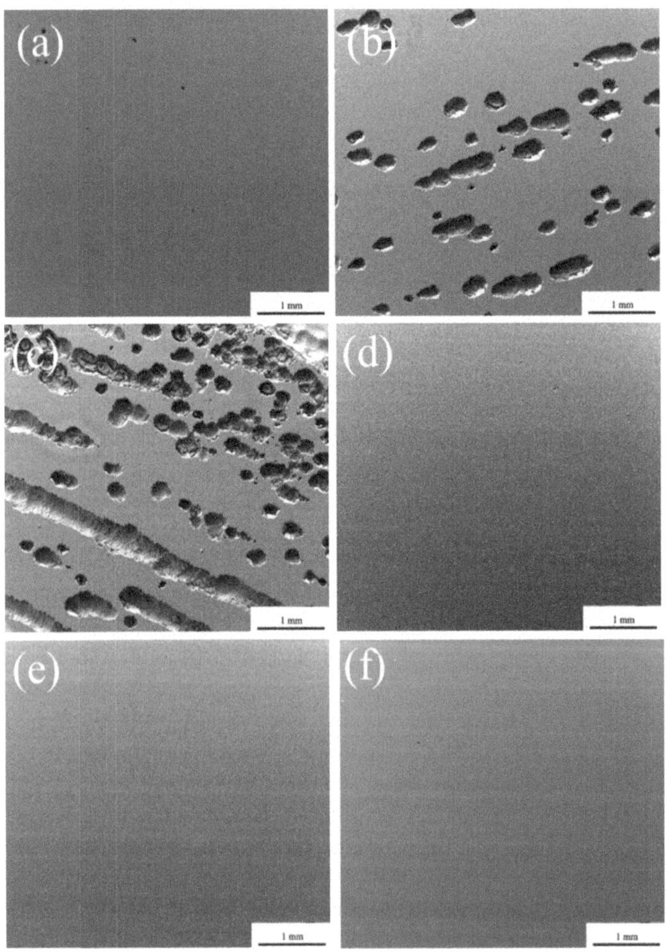

Fig. 4.46 Morphology of pitting at different tempering temperatures. **a** As-quenched; **b** 100 °C; **c** 150 °C; **d** 200 °C; **e** 250 °C; **f** 300 °C

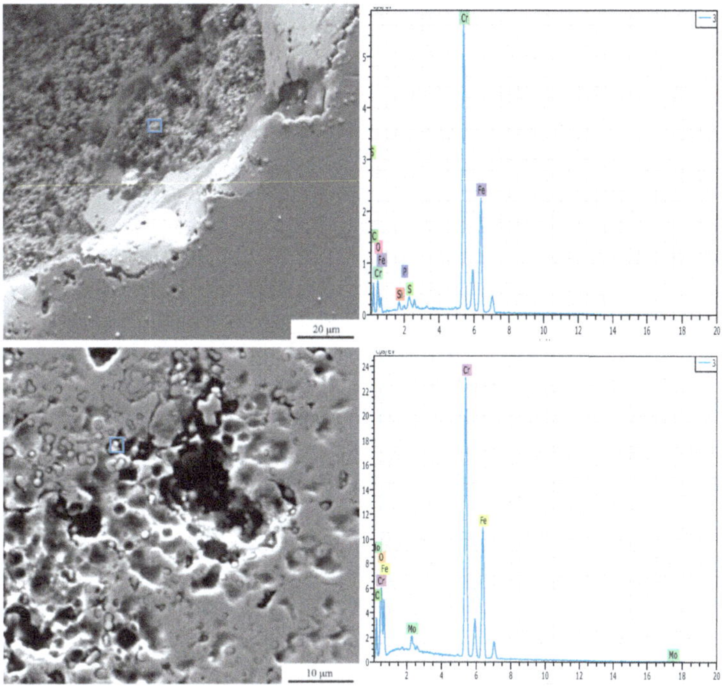

(a) (b) (c) (d)

Fig. 4.47 Microstructure and EDS analysis of pitting holes during tempering at 150 and 200 °C: **a** Tempering microstructure at 150 °C; **b** Corresponding energy spectrum of carbides in (**a**); **c** Tempering microstructure at 200 °C; **d** Corresponding energy spectrum of carbides in (**c**)

Figure 4.47 shows the morphology of pitting holes and EDS analysis of samples tempered at 150 and 200 °C. It can be seen from Fig. 4.47c that white globular particles are usually seen in the pitting holes. EDS analysis in Fig. 4.47d indicates that these white particles are $M_{23}C_6$-type carbides and pitting usually occurs around $M_{23}C_6$-type carbides and develops into a stable pitting area with the falling of carbide particles. The chromium depleted zone around $M_{23}C_6$-type carbide forms a positive and negative electrode in the metal and is prone to corrosion. Figure 4.47a shows the local micro morphology of the pitting hole of the sample during tempering at 150 °C. Figure 4.47b shows that the metal in the corrosion hole dissolves, the remaining carbides stay in the hole and the surrounding metal of the corrosion hole collapses.

Therefore, the corrosion mechanism of high carbon martensitic stainless steel can be divided into the following processes. (1) The pitting source is formed around the carbide due to chromium deficiency. (2) The pitting source develops into a stable pitting area, forming pitting holes. (3) The metal in the pitting hole is continuously dissolved, the holes become deeper, and grow lengthwise. (4) The surrounding metal collapses and the holes expand, connect and grow laterally.

4.4 Effect of Roll Forging Heat Treatment on Carbide

In the traditional knife making process, the blade is directly ground to the specified thickness after quenching and tempering. This grinding process will increase the temperature of the cutting edge, reduce the hardness of the cutting edge. The following grinding process is time-consuming and laborious. The wear residue is mixed with the grinding wheel material, and the recovery of the residual material is difficult and the utilization rate is low. A roll forging heat treatment process is proposed to solve these problems. The process steps are as follows. Firstly, heat the cutting edge to austenitizing temperature and preserve heat for a period of time. Secondly, take the blade out of the furnace and roll forging the cutting edge of the blade for multiple times. Thirdly, after roll forging, the thickness of the cutting edge is reduced from 2.5 mm to 1.0–1.5 mm, and then air-cooled to room temperature. Fourthly, recrystallization annealing is carried out for, followed by secondary quenching and tempering. The hot deformation of the blade during roll forging could refine the grain and primary carbide of the blade, which will improve the sharpness of the tool. In addition, the roll forging process reduces the labor burden of the craftsman and improves the labor efficiency. The surplus material can be directly recycled, which raises the utilization rate of the surplus material.

Following are the particulars of heat treatment process of roll forging. Firstly, heat the blade to 1050 °C for 5 min, and then roll forging. Secondly, the initial roll forging temperature is 650 °C, and the blade thickness is reduced from 2.5 mm to 1.5 mm. Thirdly, the blade after roll forging is kept at 800 °C for 2 h for recrystallization annealing. Fourthly, heat the blade to 1050 °C for 15 min and then air cooling. The traditional heat treatment process for blade is that heat to 1050 °C for 15 min and then air cool. After the two heat treatment processes, all the blades are tempered at 180 °C for 3 h.

4.4.1 Effect of Roll Forging Heat Treatment Process on Grain Size

The grain morphology of the blade edges after heat treatment is shown in Fig. 4.48.

Figure 4.48 shows that the grain is mixed in the blade without roll forging heat treatment, and the size of the grain is larger than 2 mm. The largest grain size in the blade without roll forging reaches 40.03 mm and the average grain size is 5.26 mm. After roll forging heat treatment, the grain size is relatively uniform. The largest grain size is 10.84 mm and the average grain size is 3.37 mm. Therefore, the roll forging process can effectively refine the grain size and make the grain size more uniform. In general, the thickness of the cutting edge is only about several microns. The cutting edge sharpness is achieved by the excellent mechanical properties provided by the slim cutting edge. Therefore, the microstructure of the material needs to be

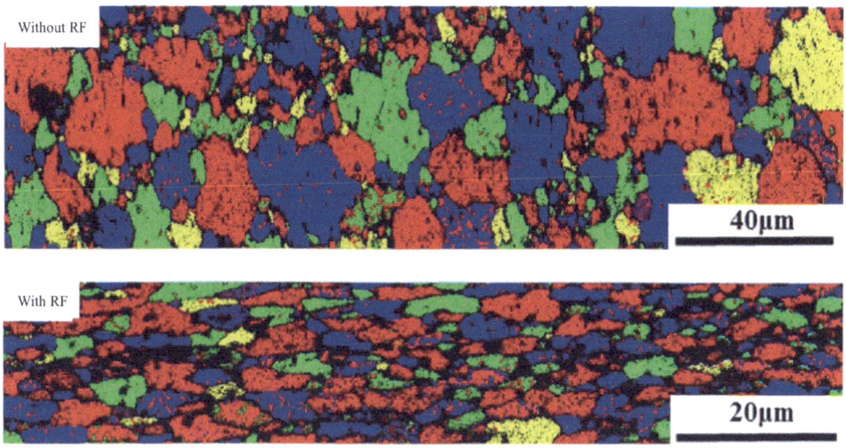

Fig. 4.48 Effect of roll forging heat treatment process on grain size

very uniform and small to improve cutting edge sharpness. Refining the grain of the cutting edge can effectively improve the sharpness of the blades.

The reasons for recrystallization annealing in roll forging heat treatment process are as follows. Due to the small thickness of the blades (about 2.5 mm), the temperature drop of the blade from the heating furnace to roll forging equipment is large, and the actual temperature (650 °C) during roll forging is lower than the recrystallization temperature of steel. When the blade is forged at this temperature, the grains of the blade will not recrystallize after being broken or deformed. The binding force between grains is small, which make the grains fall off and leading to a blunt cutting edge. When the blade is kept at 750–800°C for 2–3 h, the deformed or broken grains recrystallize at the interface of a large number of dislocations, defects, precipitations and martensite, which refines the blade grains and enhances the binding force between grains.

4.4.2 *Effect of Roll Forging Heat Treatment on Carbide*

The sharpness was seriously affected by the resistance force during cutting process and the friction coefficient had an important influence on the resistance force. Once the primary carbides fell off, the friction coefficient of cutting edge will increase and the sharpness of knives would decrease consequently.

The statistics of the primary carbides whose size is bigger than 8 μm^2 indicate that the average size of large primary carbides decreased from 21.88 μm^2 to 16.44 μm^2 after the roll forging heat treatment. The roll forging process can refine the large primary carbides. The total area and quantity of primary carbides with an area of 3–8 μm^2 in the sample after different roll forging passes are counted, and the results are shown in Fig. 4.49.

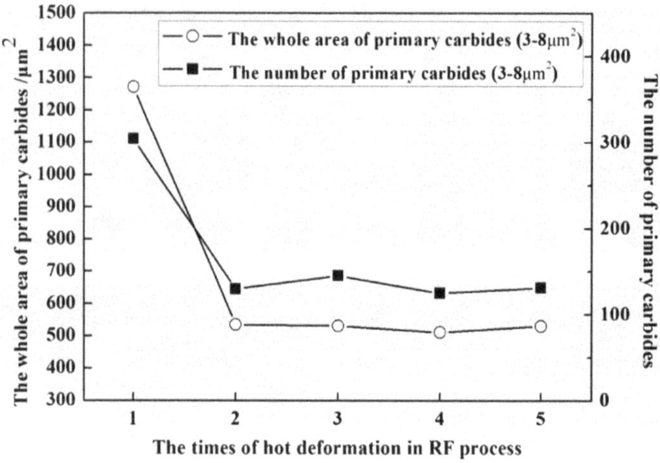

Fig. 4.49 Whole area and number of primary carbides during RF process

It is clear from Fig. 4.50 that both the whole area and number of primary carbides decreased to a value less than half after the second deformation of RF process. These results indicated that more than half of the primary carbides in the range of 3–8 μm^2 were broken into smaller ones less than 3 μm^2 after RF process, and this was the ultimate reason for the different cutting performance between traditional process and RF process. Small primary carbides may result in slight decrease in sharpness, while large primary carbides may result in serious decrease in sharpness. The following several roll forging deformation is small, whose main purpose is to make the thickness of the blade uniform at different positions, but have no effect on crushing primary carbide with an area of 3–8 μm^2. Therefore, the heat treatment process of roll forging has a better thinning effect on the primary carbide with the blade area greater than 3 μm^2.

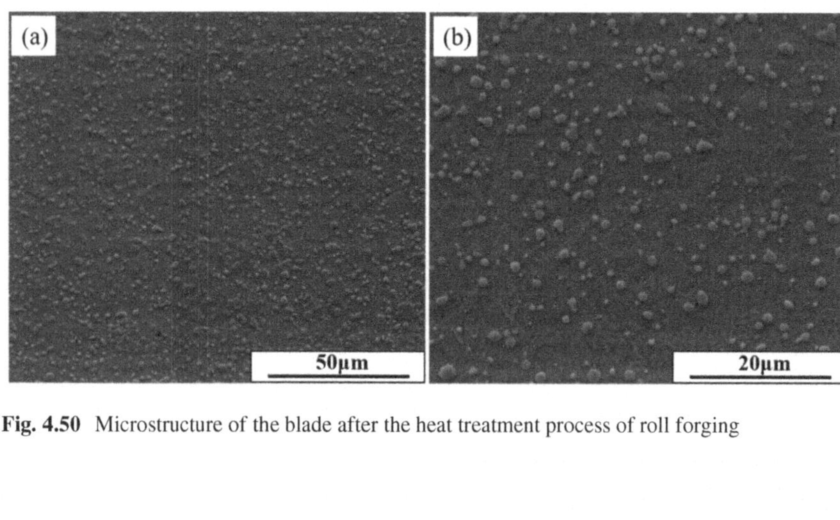

Fig. 4.50 Microstructure of the blade after the heat treatment process of roll forging

Figure 4.50 shows the microstructure of the blade after the heat treatment process of roll forging. After heat treatment of roll forging, the matrix structure is mainly tempered martensite, and there are some spherical secondary carbides of micron scale and nanometer scale. In addition, a small number of irregularly shaped primary carbides with an area of less than 3 μm^2 could be observed.

4.4.3 Effect of RF Process on the Sharpness of Steel

Before the sharpness test, the initial distance from the edge plane to the apex of the edge wrap angle is 11.8 μm. The profile of the cutting edge after sharpness test is shown in Fig. 4.51. The wear of the tool without roll forging heat treatment is 30.5 μm, while the wear of the tool after roll forging heat treatment is 17 μm. According to the calculation Eq. (4.4) of blade wear resistance, the wear resistance of the tool without roll forging heat treatment is 7, and the wear resistance of the tool after roll forging heat treatment is 16. It can be seen that the roll forging heat treatment process can effectively improve the wear resistance of the tool.

The sharpness test curve of the tool without roll forging heat treatment is shown in Fig. 4.52, and the tool edge wrap angle is 39°.

Figure 4.52 shows that during the test process, the single cutting thickness has been in a state of fluctuation, but the overall trend is gradually decreasing. When the number of cutting cycles exceeds 26, the blade is blunt, and the cumulative total cutting amount curve also shows a flat tendency. The initial cutting capacity of the tool without roll forging heat treatment is 40 mm, and the total cutting capacity is 214 mm.

The falling off of fine secondary carbides at the edge only causes a slow decline of single cutting thickness, while the falling off of large primary carbides causes a sharp

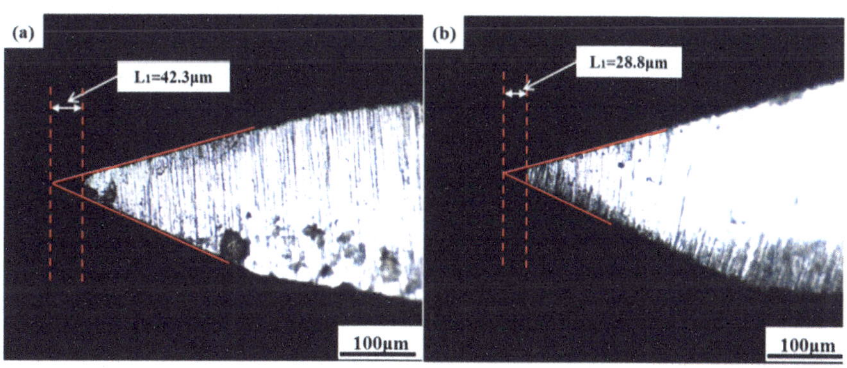

Fig. 4.51 Morphologies of cutting edge after sharpness test: **a** Traditional process, **b** RF process. *Note* L_1 is the distance from intersection line of two planes along cutting edge to the tip of cutting edge after sharpness test

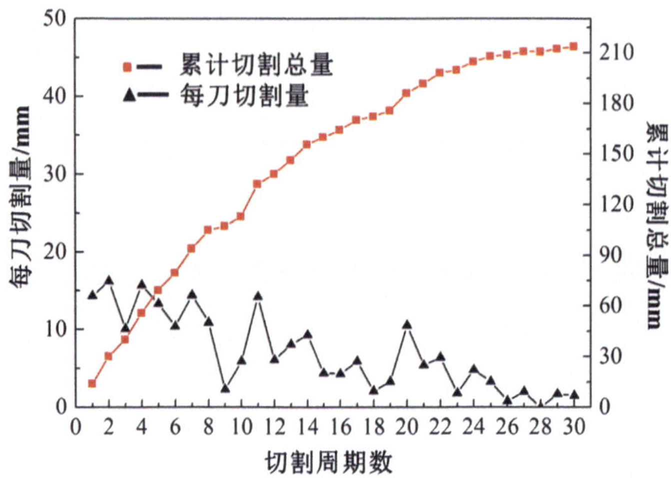

Fig. 4.52 Sharpness test curve of 8Cr13MoV tool without roll forging

decline of single cutting thickness. Figure 4.52 shows that in 30 cutting cycles of the blade without roll forging heat treatment, the curve of single cutting thickness has experienced 10 times of descending process, and the descending speed is fast. The descending process only includes 1–2 cutting cycles, which is due to the falling off of the large primary carbides in the blade. The decrease of single cutting thickness represents the damage of the geometric shape of the blade. Therefore, the more times single cutting thickness decreases, the faster the blade becomes blunt.

The sharpness test curve of the tool after the roll forging heat treatment is shown in Fig. 4.53, and the tool edge wrap angle is also 39°.

Figure 4.53 shows that after the roll forging heat treatment, the sharpness test curve also shows a trend of fluctuation and decline. Compared with the blade without roll forging heat treatment, the decline of sharpness test curve is relatively gentle. After roll forging heat treatment, each falling section generally contains 2–4 cutting cycles, and the first 30 cutting cycles have 8 falling sections. The blade without roll forging heat treatment has only one or two cutting cycles in each falling section, and the first 30 cycles have experienced 10 falling sections. The results show that the initial cutting capacity and total cutting capacity of the tool are 63.3 mm and 273.8 mm respectively. Compared with the blade without roll forging heat treatment, the initial cutting capacity increases by 56%, and the total cutting capacity increases by 28%, which indicates that the roll forging heat treatment process can effectively improve the sharpness of the tool.

After the first cycle of sharpness test for the blades, the microstructure of the cutting edge is shown in Fig. 4.54.

Figure 4.54 shows that after the first cycle of sharpness test, the primary carbides of the edge are large, and the number of primary carbides with an area of 3–8 μm^2 is large, while most of the primary carbide of the blade after the roll forging heat

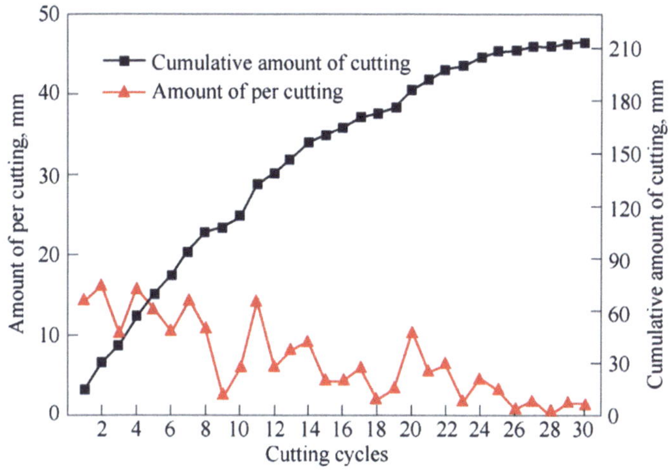

Fig. 4.53 Sharpness test curve of 8Cr13MoV blade after roll forging heat treatment

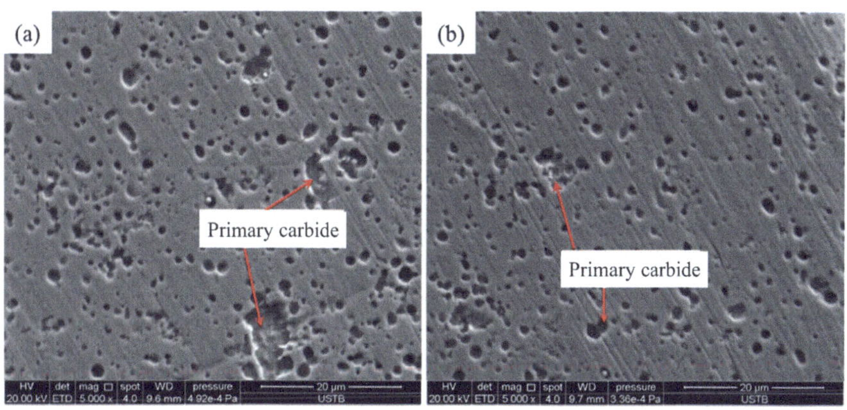

Fig. 4.54 Microstructure of cutting edge after the first cycle of sharpness test: **a** Blade after traditional heat treatment process, **b** Blade after roll forging heat treatment

treatment is less than 3 μm^2. Due to the large size and irregular morphology, these primary carbides are easy to fall off during the wear process, which leads to the formation of pits, increases the friction coefficient on the surface of the cutting edge, and reduces the sharpness of the blades. The falling off of the primary carbides in the blades without roll forging heat treatment greatly reduces the single cutting thickness, while falling off of the primary carbide in the blades with roll forging heat treatment has little effect on the single cutting thickness. Therefore, the refinement of primary carbide in the blade after roll forging heat treatment is an critical factor for the improvement of the initial sharpness.

The reasons for the improvement of initial cutting capacity (ICC) and retention ability of cutting edge (RACE) the blade after the roll forging heat treatment are as follows:

(1) The tensile strength of the blade after the traditional heat treatment is 1280 MPa, while it increases to 1608 MPa after roll forging heat treatment. The roll forging heat treatment process promotes the grains from serious mixed grain state to uniform state and reduces the average grain size, which improves the strength and toughness of the blade and the binding force between grains, and protects the grains from falling off in service.

(2) Figure 4.54 shows that under the grinding effect of abrasive paper, the primary carbides in the blade are easy to fall off. The falling off of the large carbides obviously reduce the blade sharpness. However, roll forging heat treatment effectively refines the primary carbides whose area is bigger than 3 μm^2. Compared with the falling off of the primary carbides in the blades without roll forging heat treatment, they only slowly decrease the sharpness of the blades with roll forging heat treatment. The roll forging heat treatment process reduces the fluctuation range and frequency of the blade sharpness.

(3) The roll forging heat treatment process improves the wear resistance and the retention ability of the cutting edge shape of the blade, and further improves the retention ability of cutting edge.

References

1. Zhao NQ, Yang ZG, Feng YL (2008) Solid phase transition of alloy. Central South University Press, Chang sha
2. Li DL, Ma DS, Chen ZZ et al (2010) Spheroidizing annealing process of 7CrMn2Mo steel. Heat Treat Met 35(11):57–61
3. Verhoeven JD, Gibson ED (1998) The divorced eutectoid transformation in steel. Metall Mater Trans A 29(4):1181–1189
4. Verhoeven JD (2000) The role of the divorced eutectoid transformation in the spheroidization of 52100 steel. Metall Mater Trans A 31(10):2431–2438
5. Wu X, Zhao ZY, Xue RD (2014) Carbide behavior in 5Cr15MoV steel during the isothermal and slow cooling process of spheroidization annealing. Trans Mater Heat Treat 10:98–102
6. Chen W, Li LF, Yang WY et al (2009) Microstructure evolution of hypereutectoid steels during warm deformation II. Cementite Spheroidization and Effects of Al. Acta Metall Sin 45(2):156–160
7. Yu WT, Li J, Shi CB et al (2016) Evolution of carbides in high carbon martensite stainless steel Cr13MoV during spheroidizing annealing process. Heat Treat Metals 9:25–31
8. Xiao JM (1999) Material energetics. Shanghai Science and Technology Press, Shang hai
9. Bjärbo A, Hättestrand M (2000) Complex carbide growth, dissolution, and coarsening in a modified 12 pct chromium steel-an experimental and theoretical study. Metall Mater Trans A 32(1):19–27
10. Thomson RC, Bhadeshia HKDH. Carbide precipitation in 12Cr1MoV power plant steel. Metall Mater Trans A 23(4):1171–1179
11. He YL, Zhu NQ, Wu XY et al (2011) Thermodynamic and kinetic calculation on precipitation behavior of chromium carbide. Trans Mater Heat Treat 32(1):134–137

12. Deng FY, Ma DS, Liu JH et al (2010) Effect of heat treatment process on structure and properties of new martensite stainless steel 6Cr15Mo for tool. Special steel 6:53–55
13. Song ZL, Du XD, Chen LQ et al (2011) Microstructure and impact toughness of 7Cr17Mo martensitic stainless steel. Trans Mater Heat Treat 5:95–99
14. Wang QS, Jiao ZG, Fan HS (1982) Study on spheroidizing annealing mechanism of steel. Special steel 3:3–12
15. Luzginova NV, Zhao L, Sietsma J (2008) The cementite spheroidization process in high-carbon steels with different chromium contents. Metall Mater Trans A 39(3):513–521
16. Zhou XF, Fang F, Li G et al (2010) Morphology and properties of M2C eutectic carbides in AISI M2 steel. ISIJ Int 50(8):1151–1157
17. Chumanov VI, Chumanov IV (2011) Control of the carbide structure of tool steel during electroslag remelting: Part I. Russ Metall (Metally) 6:515–521
18. Zhu QT, Li J, Shi CB et al (2015) Effect of quenching process on the microstructure and hardness of high-carbon martensitic stainless steel. J Mater Eng Perform 24(11):1–9
19. Xu ZY (1999) Thermodynamics of Materials. Science press, Beijing
20. Hu XB, Li L, Wu XC (2005) Kinetic analysis of carbide coarsening during thermal fatigue in 4Cr5MoSiV1 hot work die steel. Trans Mater Heat Treat 26(1):57–61
21. Di HS, Zhang XM, Wang GD et al (2005) Spheroidizing kinetics of eutectic carbide in the twin roll-casting of M2 high-speed steel. J Mater Process Tech 166(3):359–363
22. Yu WT, Li J, Shi CB et al (2017) Effect of spheroidizing annealing on microstructure and mechanical properties of high-carbon martensitic stainless steel 8Cr13MoV. J Mater Eng Perform 26(2):478–487
23. Qin B, Wang ZY, Sun QS (2008) Effect of tempering temperature on properties of 00Cr16Ni5Mo stainless stee. Mater Charact 9:1096–1100
24. Lin YL, Lin CC, Tsai TH et al (2010) Microstructure and mechanical properties of 0.63C-12.7Cr martensitic stainless steel during various tempering treatments. Mater Manuf Process 25:246–248
25. Hu GL, Xie XW. (1985) Heat treatment of steel (principle and process). Northwestern Polytechnical University Press, Xi'an
26. Seol J, Jung JE, Jang YW et al (2013) Influence of carbon content on the microstructure, martensitic transformation and mechanical properties in austenite/ε-martensite dual-phase Fe-Mn-C steels. Acta Mater 61:558–578
27. Tkalcec I, Azctia C, Crevoiserat S et al (2004) Tempering effects on a martensitic high carbon steel. Mater Sci Eng A 387(36):352–356

Chapter 5
Effect of Magnesium on the Carbide in H13 Steel

Abstract Magnesium can refine non-metallic inclusions, change the morphology of carbides in steel, and act as the nucleation core of austenite to induce the nucleation of austenite. The effect of magnesium on microstructure, inclusions, carbides and mechanical properties of H13 steel is ascertained in this chapter. The results show that the number of large inclusions and the size of carbides in H13 steel decrease significantly after magnesium treatment of the steel. Magnesium treatment of H13 steel inhibits the chain growth of carbides and promotes the spheroidization of carbides. $MgAl_2O_4$ inclusions formed after magnesium treatment are more conducive to the nucleation of γ-Fe, M(C, N) and M_6(C, N) than that of Al_2O_3 inclusions. Magnesium could improve the stability of tempered martensite, and give rise to a large number of tempered troostite after tempering, increasing the strength and hardness of the steel. Magnesium treatment also improves the toughness of the steel and reduces the friction coefficient and wear ratio of H13 steel.

Keywords Magnesium · Inclusions · Carbides · Heat treatment · Mechanical property

H13 hot work die steel is commonly used in die casting, hot forging and hot extrusion. Compared with foreign H13 steel, the domestic H13 steel has a larger gap in the control of inclusions and carbides, especially the large amount of primary carbides, and the carbides in the annealed structure are obviously distributed along the grain boundaries.

Magnesium is a strong deoxidant, which can make the dissolved oxygen in molten steel reduce to an extremely low level [1, 2]. Al_2O_3 cluster inclusions can be modified to tiny spinel after Mg treatment in Al killed molten steel [3–5], which can improve the fatigue performance of steel [6, 7]. Magnesium can also improve the shape of carbide in steel from strip to spherical or nearly spherical and refine annealed carbides [8]. Magnesium bearing inclusions can be used as the core of deformed austenite in the process of deformable austenite rolling refinement and recrystallization [9].

J. Li and C. Shi, *Carbide in Special Steel*, Engineering Materials,
https://doi.org/10.1007/978-981-16-1456-9_5

5.1 Formation and Removal of Mg Bearing Inclusions in H13 Steel

5.1.1 Physical Properties of Mg Bearing Inclusions

5.1.1.1 Density of Mg Bearing Inclusions

The inclusions are mainly MgO and $MgO{\cdot}Al_2O_3$ after Mg treated H13 steel, Mg bearing inclusions are represented by $m\mathrm{MgO}{\cdot}n\mathrm{Al_2O_3}$.

To analyze the density of Mg bearing inclusions ($m\mathrm{MgO}{\cdot}n\mathrm{Al_2O_3}$), the molar ratio of MgO and Al_2O_3 m/n are 0.41, 1.24, 2.16 and 2.75, which were measured by EDS. The spinel with different m/n ratio were prepared by chemical reagent.

The spinel powder was ground to form a fine and even powder, the MgO and Al_2O_3 molar ratio m/n were taken to be 1:1. The measured masses of the powder were 0.615, 0.942 and 1.527 g, respectively, and the measured volumes were 0.16, 0.25 and 0.40 cm^3, respectively. The density is 3.85, 3.71 and 3.82 g/cm^3 by calculation, and the average density is 3.79 g/cm^3.

The density of spinel can be calculated by Eq. (5.1):

$$\rho = x\mathrm{Mg}O \cdot \rho\mathrm{MgO} + x\mathrm{Al_2O_3} \cdot \rho\mathrm{Al_2O_3} \tag{5.1}$$

x_{MgO}: The mass fraction of MgO in $MgO{\cdot}Al_2O_3$,
$x_{\mathrm{Al_2O_3}}$: The mass fraction of Al_2O_3 in $MgO{\cdot}Al_2O_3$,
ρ_{MgO}: MgO density, g/cm^3,
$\rho_{\mathrm{Al_2O_3}}$: Al_2O_3 density, g/cm^3.

Substituting $\rho_{\mathrm{MgO}} = 3.63\ \mathrm{g/cm^3}$ and $\rho_{\mathrm{Al_2O_3}} = 3.97\ \mathrm{g/cm^3}$ into Eq. (5.1), $\rho = 3.88\ \mathrm{g/cm^3}$, there is a certain error of 2.3% between the actual measured result and the theoretical calculated value, that is

$$\Delta = \frac{\left|\rho - \rho_{\mathrm{MgO\cdot Al_2O_3}}\right|}{\rho_{\mathrm{MgO\cdot Al_2O_3}}} \times 100\% = 2.3\% \tag{5.2}$$

The $MgO{\cdot}Al_2O_3$ spinel with different molar ratios of MgO and Al_2O_3 were calculated by the same calculation method, the results are shown in Table 5.1.

As can be seen in Table 5.1, the density of spinel is lower than that of Al_2O_3 (3.97 g/cm^3), but it is larger than that of MgO. The spinel density decreases with the increasing of MgO content.

Table 5.1 Density of different kinds of $MgO{\cdot}Al_2O_3$ spinel (g/cm^3)

Content	$MgO{\cdot}2Al_2O_3$	$MgO{\cdot}Al_2O_3$	$1.5MgO{\cdot}Al_2O_3$	$2MgO{\cdot}Al_2O_3$	MgO
Density	3.92	3.79	3.72	3.66	3.55

5.1.1.2 Effect of Inclusion Density on Removal Rate

The collision, agglomeration and growth of inclusions occur in molten steel. The inclusions in molten steel float up in three ways during this process: the inclusions float by their own buoyancy, the inclusion adheres to the bubble surface and floats by the buoyancy of the bubble, the inclusions are brought to the steel slag interface by the turbulent flow of molten steel.

The floating speed of the inclusions follows the Stokes formula when the molten steel is at rest [10]:

$$v_{\mathrm{p}} = \frac{2(\rho_m - \rho_p)gd_p^2}{9\mu}. \tag{5.3}$$

where v_p is the removal rate of inclusions, m/s. ρ_m is the density of molten steel, kg/m^3. ρ_p is the density of inclusion, kg/m^3. g is the gravitational acceleration, m^2/s. d_p is the diameter of inclusions, m. u is the viscosity of molten steel, kg/(m·s). g is 9.8m^2/s, μ is 0.002 kg/(m·s), and ρ_m is 7.8×10^3 kg/m^3 at 1873 K [11].

According to Eq. (5.3), the floating rate of the inclusions increase with the decreasing of inclusion density, that is to say, the inclusions are easier to be removed through the floating process.

d_p value is 100×10^{-6} m, and the density values of $MgO \cdot Al_2O_3$ spinel are listed in Table 5.1. v_p value is 4.17×10^{-2} m/s, 4.23×10^{-2} m/s, 4.37×10^{-2} m/s, 4.45×10^{-2} m/s and 4.51×10^{-2} m/s respectively after substituting different density value into Eq. (5.3).

The relation between floating rate of inclusion in liquid steel and the composition of $mMgO \cdot nAl_2O_3$ inclusion is shown in Fig. 5.1. As shown in Fig. 5.1 and Table 5.1,

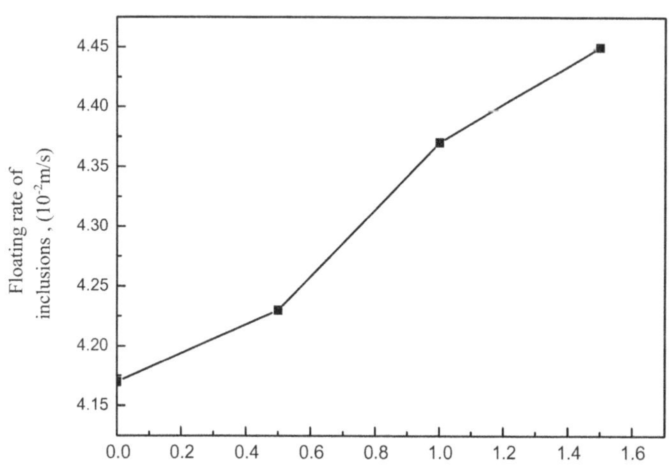

Fig. 5.1 Relation between floating rate and the composition of mMgO·nAl$_2$O$_3$ inclusion

the density of Mg-bearing inclusions increase with the increasing of Mg content of $MgO{\cdot}Al_2O_3$ spinel, and the floating rate of inclusions increase.

5.1.2 Collision and Agglomeration of Al_2O_3 and $MgO{\cdot}Al_2O_3$ Particles

The confocal scanning laser microscope (CLSM) was adopted to investigate the effect of magnesium addition on the collision, agglomeration, and growth behavior of inclusions. The schematic view of CLSM is shown in Fig. 5.2.

The chemical composition of experimental material H13 steel, that is, steel A and steel B, are shown in Table 5.2.

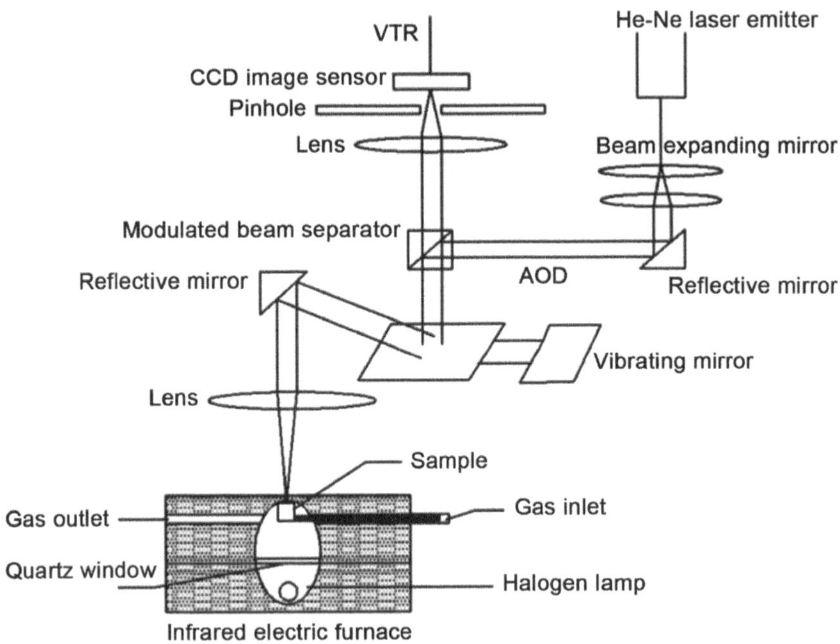

Fig. 5.2 Schematic view of CLSM

Table 5.2 Chemical composition of experimental steels (wt%)

Steel	C	Si	Mn	Cr	Mo	V	Ni	P	Al	Mg
A	0.40	0.97	0.29	5.04	1.21	0.90	0.14	0.028	0.024	0
B	0.41	0.99	0.28	5.01	1.21	0.91	0.13	0.028	0.017	0.0010

The heating rate can be effectively controlled by setting the temperature program (typically 200 ℃/min–250 ℃, and then heating to 1350 ℃ with heating rate of 150 ℃/min, and finally 100 ℃/min to steel melting temperature).

5.1.2.1 Collision and Agglomeration of Al_2O_3

Before the melting of sample A, the δ ferrite was preheated. Meanwhile grain boundaries of δ ferrite slightly expanded and formed grooves. Steel in these grooves (soluteenriched region) first melted to form liquid. The area of partially melted region of steel gradually became larger with increasing temperature, and there were particles and particulate aggregates emerged on steel surface in the process. These particles mainly came from the inside of liquid metal pool, and particulate aggregates were formed by the collision and coalescing of small single particles.

In order to determine the chemical compositions of particles and particulate aggregates, sample surface was scanned and detected by the observation using scanning electron microscopy and energy dispersive spectroscopy (SEM–EDS) after experiment, as shown in Fig. 5.3.

The EDS spectrum of the inclusion particles in Fig. 5.3 was presented in Fig. 5.4, indicating that the inclusion particles were Al_2O_3.

The shape of alumina particles on the surface of steel liquid is irregular and there are many sharp tips in their periphery, as shown in Fig. 5.5b. The total number of fine alumina particles and their particulate aggregates decreased with time after the

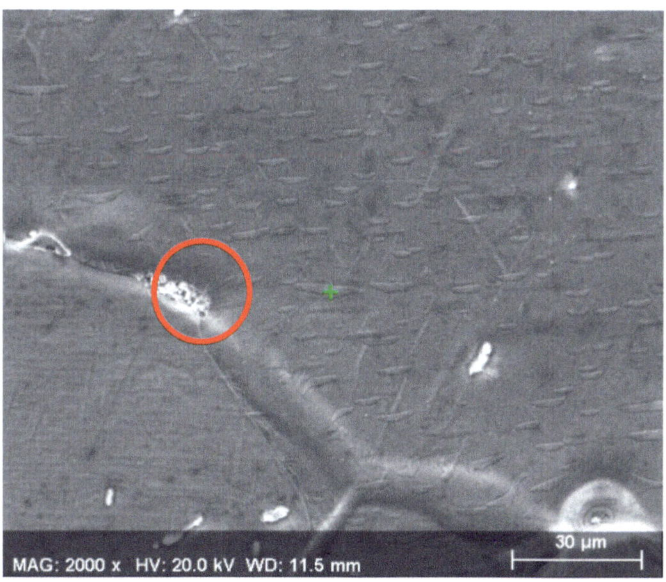

Fig. 5.3 Particles on the surface of sample A

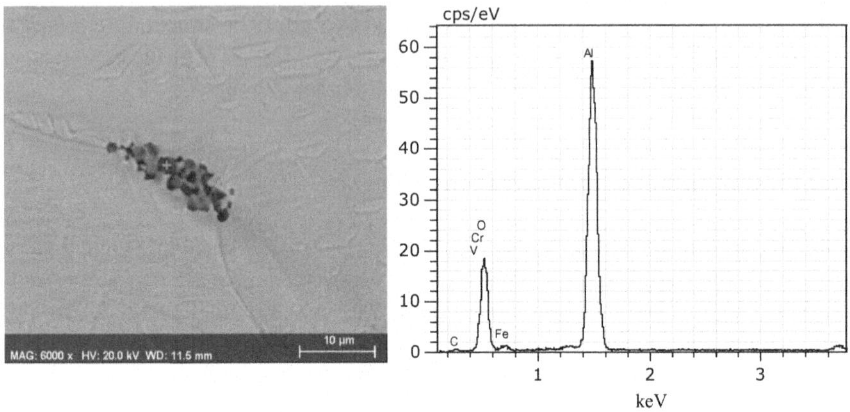

Fig. 5.4 EDS spectrum of the inclusion particles

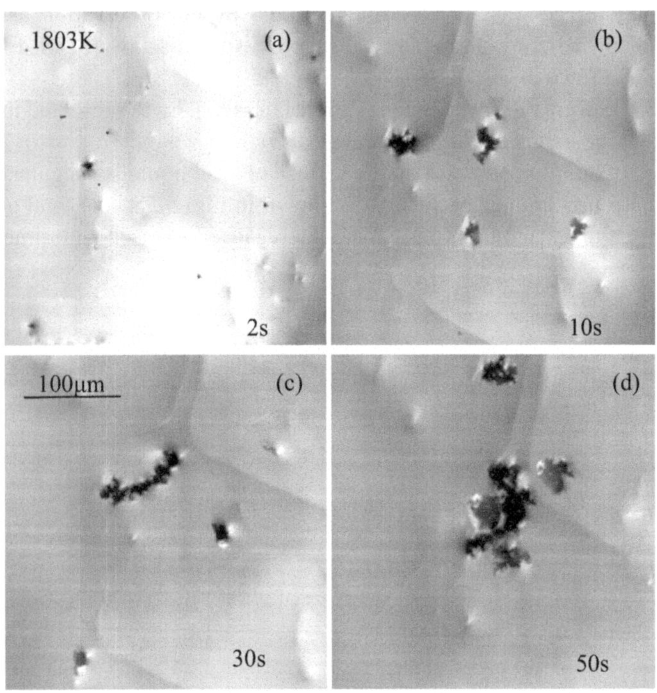

Fig. 5.5 Formation process of an alumina cluster on the surface of liquid H13 steel without magnesium at 1803 K

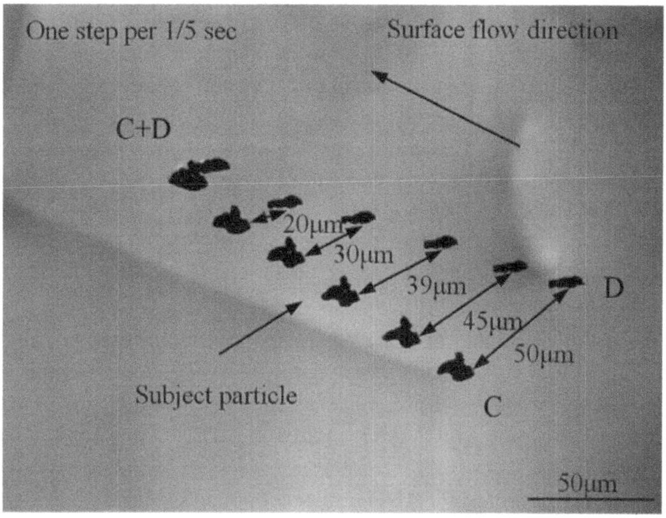

Fig. 5.6 Forming process of particulate aggregates caused by the long-range attractive force between two alumina particles on the surface of liquid steel A at 1803 K

surface of sample A melted. However, their size rapidly increased to the extent of nearly one hundred micrometers, as shown in Fig. 5.5c, d.

To estimate the movement law of inclusions on melt surface, the trajectories of inclusions were measured and analyzed within the time interval between continuous frames. The attraction process between two particles (C and D) is shown in Fig. 5.6 and the intervals of two sequential steps are 1/5 s. It can be seen from Fig. 5.6 that particles A and B drifted along with the surface flow of steel liquid and the distance between them was reduced from 50 to 20 μm with the elapse of experimental time. The two particles attracted each other, then collided and combined into an integrated whole within 1 s. Thus, it can be inferred that there must be a strong mutual attractive force between particles C and D which make them attract and collide each other. This kind of long-range attractive force existed between two particles of a pair at the melt surface (it should be noted here that there is a certain difference between the behavior of inclusions at the melt surface and in the bulk, regarding acting forces for example) is obviously not influenced by the surface flow of steel liquid.

Clusters formed by the collision between alumina particles were usually large and loose. At the beginning of particle agglomeration, the nearest floating particles on the surface of liquid steel attracted each other and collided, and then formed particulate aggregates with a size of 5–10 μm caused by long-range attractive force. At the initial stage of melting, the distribution of single particles on the surface of steel liquid was homogeneous. With the rapid decrease of particles number on steel surface, the distance between particulate aggregates was becoming progressively larger, and finally exceeded the sphere of action of long-range attractive force between two alumina particles. Because of the existence of surface flow, the distribution of particulate aggregates on steel surface was not homogeneous。In this case, clusters formed by particulate aggregates will grow in a certain direction leading to the formation of inhomogeneous distribution of clusters, and sometimes forming a chain. Clusters of 50 μm or even hundreds of microns can be formed about 1 min after melting on the steel surface. The initial clusters are dendritic, with many nodes and branches, and the apparent density is very small.

5.1.2.2 Collision and Agglomeration of $MgO{\cdot}Al_2O_3$ Spinel

The observed particulate matter in the experiment above mentioned was also found in experiment of sample B by the observation using scanning electron microscopy and energy dispersive spectroscopy (SEM–EDS) afterexperiments. The SEM micrograph and EDS analysis results of $MgO{\cdot}Al_2O_3$ inclusion particles in sample B are shown in Fig. 5.7.

The formation process of $MgO{\cdot}Al_2O_3$ clusters on melt surface is shown in Fig. 5.8.

It can be seen from Fig. 5.8 that the appearance of $MgO{\cdot}Al_2O_3$ particles looks a little bit smoother than that of alumina particles. The formation sequence of $MgO{\cdot}Al_2O_3$ cluster was the same as that of alumina cluster, but rapid deformation and densification occurred at the early stage of its formation process as shown in Fig. 5.8d. Thus, the deformation and densification rate of $MgO{\cdot}Al_2O_3$ clusters was more quickly than that of alumina clusters. It can be concluded that, unlike Al_2O_3 clusters, $MgO{\cdot}Al_2O_3$ clusters cannot form the outside branch and attract more distant particles and consequently the agglomeration and growth of $MgO{\cdot}Al_2O_3$ clusters were slower and the formed clusters were also much denser than Al_2O_3 clusters as shown in Figs. 5.5d and 5.8d. It also can be seen from Fig. 5.8 that the formation process of $MgO{\cdot}Al_2O_3$ clusters took much shorter time than that of Al_2O_3 clusters.

The rapid densification of $MgO{\cdot}Al_2O_3$ clusters shows that the sintering process between $MgO{\cdot}Al_2O_3$ particles in the cluster conducted faster than that between Al_2O_3 particles, possibly due to higher specific surface free energy at the particle boundaries for the larger interfacial area and formation of lower melting point substance at the boundaries. It can be concluded that the long-range attractive force between $MgO{\cdot}Al_2O_3$ particles is also exist, but is smaller than that between Al_2O_3 particles. It is important to note here that the long-range attractive force between liquid spherical inclusions ($CaO–Al_2O_3$, $Al_2O_3–SiO_2$, $CaO–Al_2O_3–SiO_2$ and all silicoaluminate with a mass fraction of less than 60% Al_2O_3) particles has never been found. The agglomeration takes place in a completely different way between these liquid inclusions in molten steel [12].

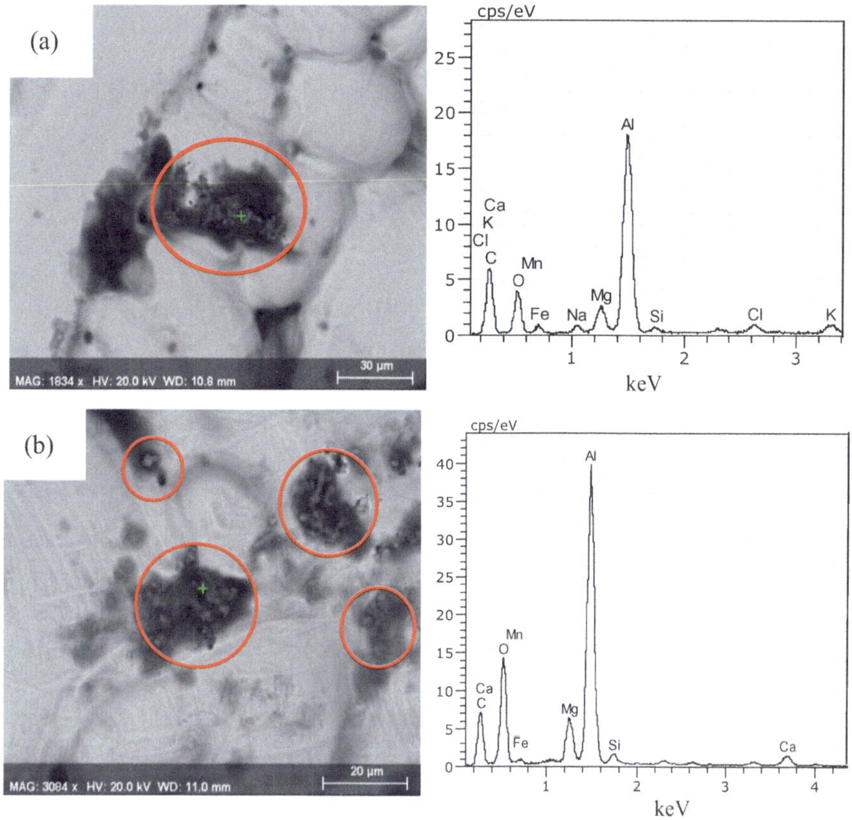

Fig. 5.7 SEM micrograph and EDS analysis results of $MgO{\cdot}Al_2O_3$ inclusion particles in sample B

5.1.3 Factors Affecting the Magnitude and Action Radius of Long-Range Attractive Force

By measuring the acceleration and weight of particles, the magnitude of long-range attractive force can be roughly calculated. Assuming the particle is approximately semi-cylindrical shape. The diameter of particles can be observed, and the height of particle was taken as 2 mm. In a system of two particles of a pair, the smaller particle moved to another larger particle due to long-range attractive force, while the larger particle still maintained a static state. The viscous resistance on the movement of particles caused by liquid viscosity was not considered in calculation [13].

The Schematic view of calculation model for acceleration of disk-like particle according to its moving distance is shown in Fig. 5.9. In this model, the smaller particle 1 in the system of two particles of a pair is defined as object particle, and the other larger particle 2 is labeled as subject particle accordingly. The acceleration

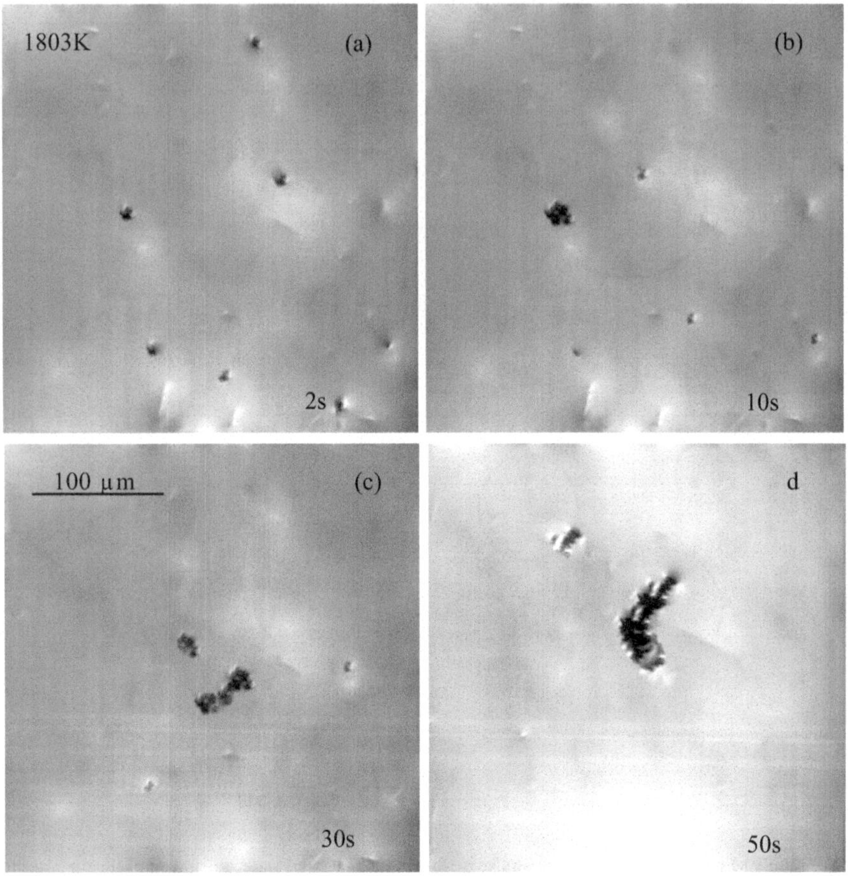

Fig. 5.8 Formation process of $MgO{\cdot}Al_2O_3$ cluster on the surface of liquid H13 steel containing magnesium at 1803 K

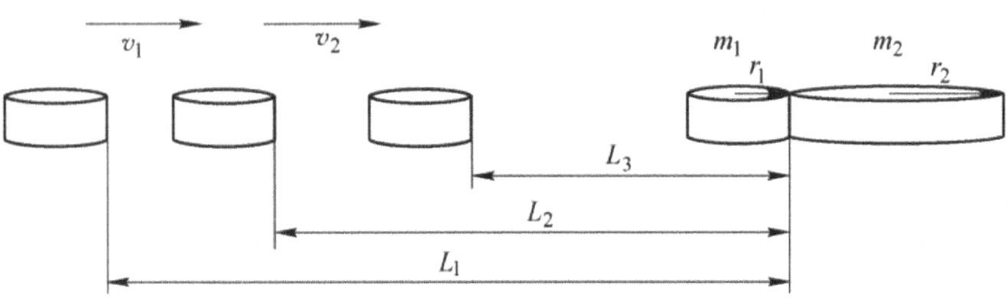

Fig. 5.9 Schematic view of calculation for acceleration of disk-like particle according to its moving distance

of object particle can be calculated by this model according to its moving distance within the 1/15 s when the subjectparticle remain static, as shown in Fig. 5.9. The longrange attractive force, F, is shown as the following Equations

$$F = m_1\dot{a}_1 \tag{5.4}$$

$$v_1 = (L_1 - L_2)/t \tag{5.5}$$

$$v_2 = (L_2 - L_3)/t \tag{5.6}$$

$$a_1 = (v_2 - v_1)/t \tag{5.7}$$

where, m_1 is the mass of the object particle (the corrected value of m1 can be expressed as $m_1 \cdot m_2/(m_1 + m_2)$ when two particles moved simultaneously), kg; m_2 is the mass of the subject particle, kg; a_1 is the acceleration of particle, $m \cdot s^{-2}$; F is the long-range attractive force between two particles; v_1 is the velocity of particle within the first interval of 1/15 s, $m \cdot s^{-1}$; v_2 is the velocity of particle within the second interval of 1/15 s, $m \cdot s^{-1}$; t is the duration of movement of particle between two sequential steps, s; L_1 is the moving distance of particle within the interval of 3/15 s, m; L_2 is the moving distance of particle within the interval of 2/15 s, m; L_3 is the moving distance of particle within the interval of 1/15 s, m.

The viscous resistance of particles to liquid steel is ignored in this calculation.

5.1.3.1 Effect of Particle Size of Inclusion on the Magnitude of Long-Range Attractive Force

The derived magnitude of long-range attractive force L_2 and inter-particle distance between particles of a pair with different sizes are shown in Fig. 5.10.

It can be seen from Fig. 5.10 that under the same conditions of the distance between inclusions, the larger the diameter of smaller particle, the bigger the attractive force it is subject to. For example, when the distance between particles in a pair is in the neighborhood of 32 μm, the attractive force exerted on the smaller particle in a pair with the diameter less than 5 μm was 8.1×10^{-16} N, while the attractive force exerted on the smaller particle in a pair with the diameter in the range of 5–10 μm was 5.3 $\times$ 10^{-15} N and the attractive force exertedon the smaller particle in a pair with the diameter larger than 10 mm was 9.8×10^{-15} N.

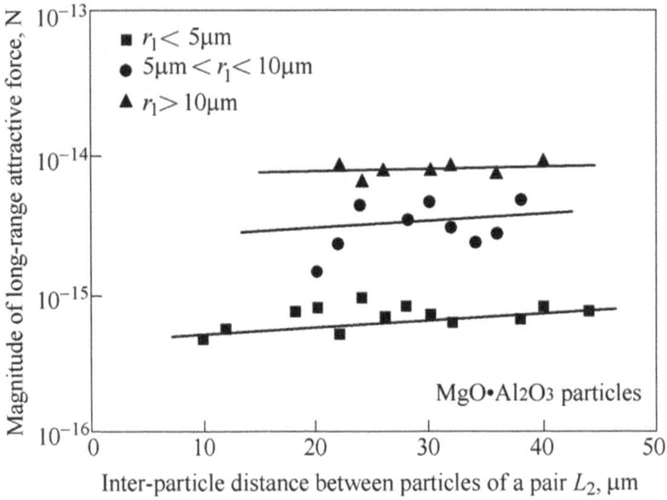

Fig. 5.10 Influence of particle size of inclusion on the magnitude of long-range attractive force between particles of a pair on the surface of liquid steel

5.1.3.2 Effect of Particle Size of Inclusion on the Action Radius of Long-Range Attractive Force

The relationship between action radius of long-range attractive force L_1 and radius of inclusion particle with larger size in a pair is shown in Fig. 5.11.

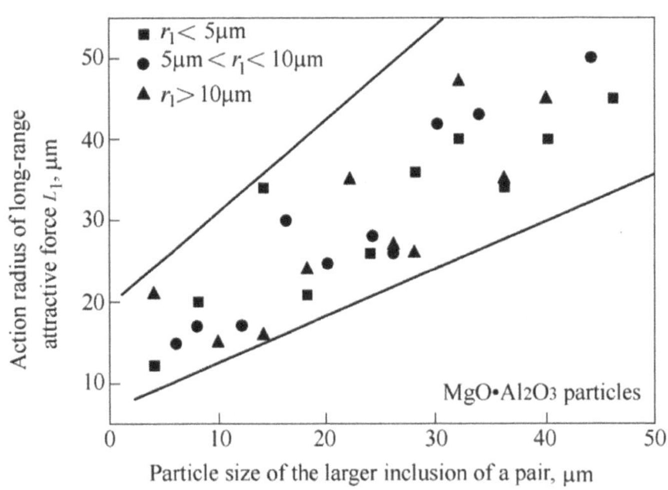

Fig. 5.11 Relationship between action radius of long-range attractive force and particle size of the larger inclusion of a pair

It can be seen from Fig. 5.11 that action radius of long-range attractive force increases with the increase of the radius of larger particle in a pair. When particle size of larger inclusion of a pair is larger than 4 μm, action radius of long-range attraction is greater than 10 μm. It also can be seen in this case, that unlike magnitude of long-range attractive force, being proportional to particle size of the smaller inclusion of a pair, action radius of long-range attractive force has no obvious relationship with particle size of the smaller inclusion of a pair. However, it is determined by the article size of the larger inclusion of a pair.

5.1.3.3 Effect of Particle Type of Inclusion on the Magnitude and Action Radius of Long-Range Attractive Force

The derived attractive forces between alumina particles of a pair and between $MgO{\cdot}Al_2O_3$ particles of a pair are shown in Fig. 5.12 as a function of the inter-particle distance L_2 between particles of a pair.

It can be seen from Fig. 5.12 that under the conditions of the same size range of the smaller inclusion particle of a pair, the average magnitude of long-range attractive force between $MgO{\cdot}Al_2O_3$ particles was less than that between alumina particles. The maximum action radius of long-range attractive force between $MgO{\cdot}Al_2O_3$ particles was observed to be in the neighborhood of 46 μm, while that between alumina particles was observed to be in the region of 56 μm. It means that $MgO{\cdot}Al_2O_3$ particles are more difficult to coagulate than alumina particles.

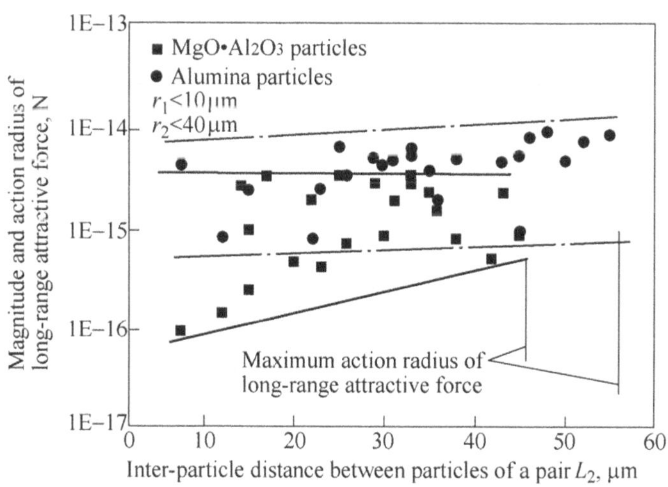

Fig. 5.12 Attractive force measured between $MgO{\cdot}Al_2O_3$ particles and between alumina particles on the surface of liquid steel

5.1.4 *Effect of Magnesium on Inclusions in H13 Steel During ESR*

5.1.4.1 Effect of Magnesium on Inclusions in H13 Steel

Figure 5.13 shows the size distribution of inclusions in H13 die steels with Mg content of 0, 0.0006, 0.0027 and 0.0032%.

As can be seen in Fig. 5.13a, the number of observed inclusions increased after Mg treatment. And with the increasing of Mg content in H13 die steel, the number of observed inclusions increased slightly. The size distribution of inclusions are shown in Fig. 5.13b, it can be seen that the inclusions smaller than 1 μm in size take up more than 60% of the total inclusions after Mg treatment, and the number of the inclusions smaller than 1 μm in size increased with the increasing of Mg content. It can be concluded that the inclusions become tinier and more dispersed after Mg treatment. The morphology and compositions of inclusions in each heat is shown in Fig. 5.14.

Figure 5.14a–c show the typical oxide inclusions are Al_2O_3 in the Mg-free sample. Most of the observed Al_2O_3 are about 2–5 μm in size. It is clear from Fig. 5.14b, c that a different structure can be distinguished in all these precipitates by their colour. It appears that the Al_2O_3 acts as the nucleation core of V(C, N). Figure 5.14d–l show that the oxide inclusions change from Al_2O_3 to $MgO{\cdot}Al_2O_3$ after Mg treatment. No cluster like Al_2O_3 inclusions were found. V(C, N) precipitated around $MgO{\cdot}Al_2O_3$ inclusions. $MgO{\cdot}Al_2O_3$ and a small amount of MgO exist in the steel with 0.0032% Mg. Figure 5.15 shows the variation of Mg, Al and O content of inclusions.

As can be seen in Fig. 5.15, Mg content of inclusions increases with the increasing of Mg content of steel. Meanwhile, Al and O content decrease with the increasing of Mg content of steel. It indicates that the component of inclusions would change with

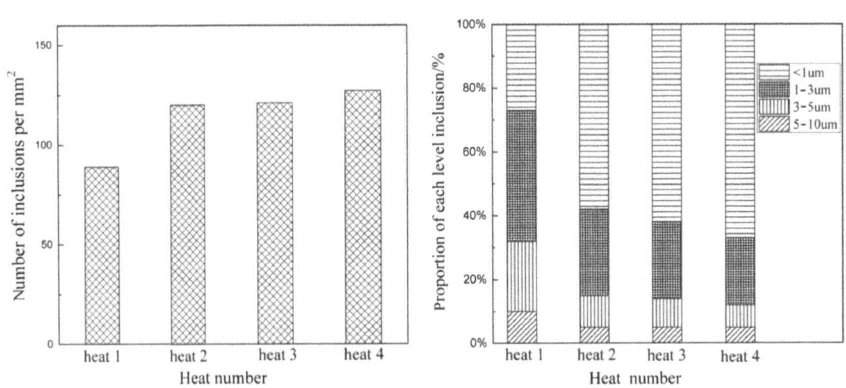

(a) (b)

Fig. 5.13 Statistical results of inclusions: **a** Number of inclusion; **b** Proportion of each level inclusion

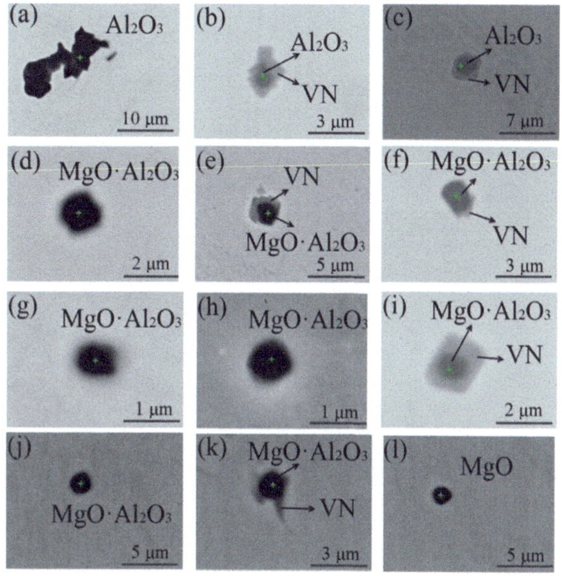

Fig. 5.14 SEM images and EDS analysis results of typical inclusions: **a–c** [Mg] = 0; **d–f** [Mg] = 0.0006%; **g–i** [Mg] = 0.0027%; **j–l** [Mg] = 0.0032%

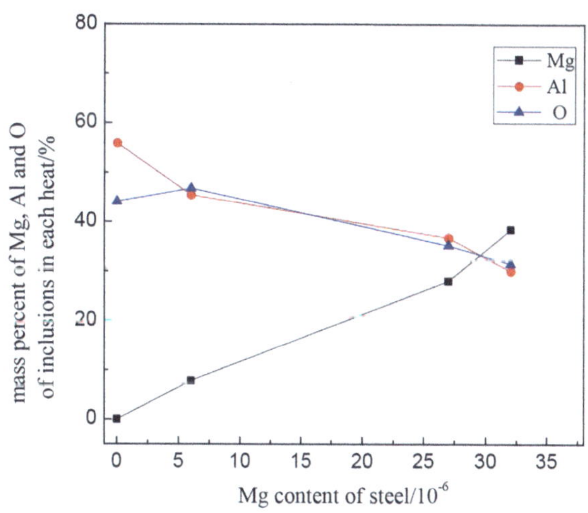

Fig. 5.15 Relation between the Mg content of steel and Mg, Al and O content of inclusions

different Mg content of steel. Mg content of inclusions in the steel with 0.0032% Mg. Element line scanning results of the MgO·Al_2O_3 inclusion in the steel with 0.0006 and 0.0032% Mg are shown in Fig. 5.16.

As can be seen in Fig. 5.16a, Al_2O_3 exist as the core of the inclusion, and MgO exist as the outside layer around the Al_2O_3. As can be seen in Fig. 5.16b, the inclusions contain MgO at its core was detected and at its edges of the inclusions is MgO·Al_2O_3. The SEM line scanning of MgO·Al_2O_3 in the steel with 0.0032% Mg is shown in Fig. 5.16c. Its core is mainly composed of MgO and a significant amount of Al_2O_3

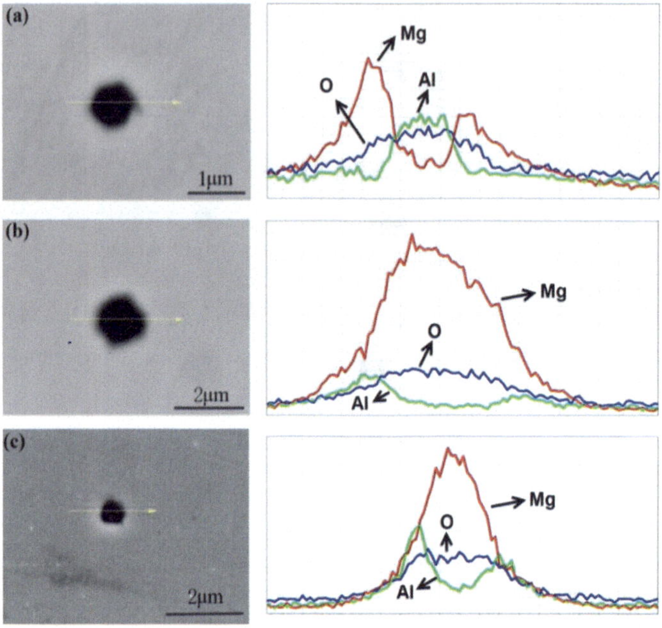

Fig. 5.16 Element line scanning of $MgO{\cdot}Al_2O_3$ inclusion after Mg treatment: **a** 0; **b** 0.0006%; **c** 0.0032%

around its periphery. It deduces that two kinds of $MgO{\cdot}Al_2O_3$ can be formed after Mg treatment by different reaction and mechanisms [14].

The evolution of Al_2O_3 by Mg treatment in H13 die steel takes place in three stages.

At first, the injected Mg (as vapour or dissolved magnesium) reacts with the Al_2O_3 and then $MgO{\cdot}Al_2O_3$ with the core of Al_2O_3 would be formed. It is considered that the decreasing of Al_2O_3 inclusions is due to Eq. (5.8):

$$3[Mg]/Mg_{vapor} + 4Al_2O_3 = 3MgO \cdot Al_2O_3 + 2[Al] \tag{5.8}$$

At the second stage, Mg reacts with the dissolved oxygen for its strong thermodynamic affinities with oxygen. Al reacts with MgO and then $MgO{\cdot}Al_2O_3$ with the core of MgO would be formed. The increasing in $MgO{\cdot}Al_2O_3$ after Al_2O_3 inclusions disappear can be attribute to Eqs. (5.9) and (5.10):

$$[Mg] + [O] = MgO \tag{5.9}$$

$$2[Al] + 4MgO = MgO \cdot Al_2O_3 + 3[Mg] \tag{5.10}$$

At the last stage, Mg reacts with $MgO{\cdot}Al_2O_3$, $MgO{\cdot}Al_2O_3$ is further reduced by Mg and reduction product MgO is formed. The appearance of MgO phase isdue to Eq. (5.11):

$$3[Mg] + MgO \cdot Al_2O_3 = 4MgO + 2[Al] \quad (5.11)$$

It can be seen clearly that the Eqs. (5.10) and (5.11) are the same reaction with different direction. The variation of Mg content will result in the transformation of $MgO{\cdot}Al_2O_3$ and MgO in the molten steel, and that is also the major reason of transformation of the direction of Eqs. (5.10) and (5.11).

5.1.4.2 Modification of Mg Bearing Inclusions in H13 Steel During ESR Process

Steel ingots with Mg contents of 0, 0.0027 and 0.0032% were forged to consumable electrodes, and ESR process was carried out. The Mg contents of ESR ingots were 0, 0.0005 and 0.0006%, respectively. The inclusions size distribution of ESR ingots is shown in Fig. 5.17.

As shown in Fig. 5.17, most of the inclusions in ESR ingots were in the range of 0–1 and 1–2 μm after ESR process, inclusions size larger than 5 μm were not detected. The amount of inclusions in ESR ingots of Mg contained consumable electrode after ESR process is lower than Mg free electrode, it can be inferred that the removal rate

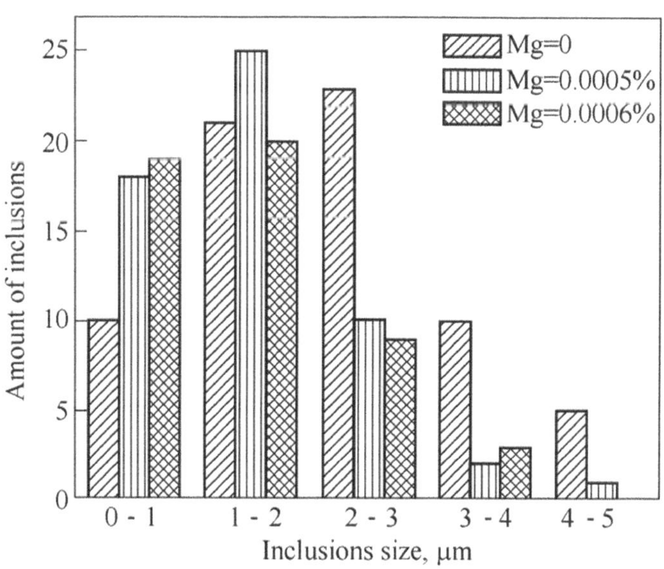

Fig. 5.17 Size distribution of inclusions in ESR ingots

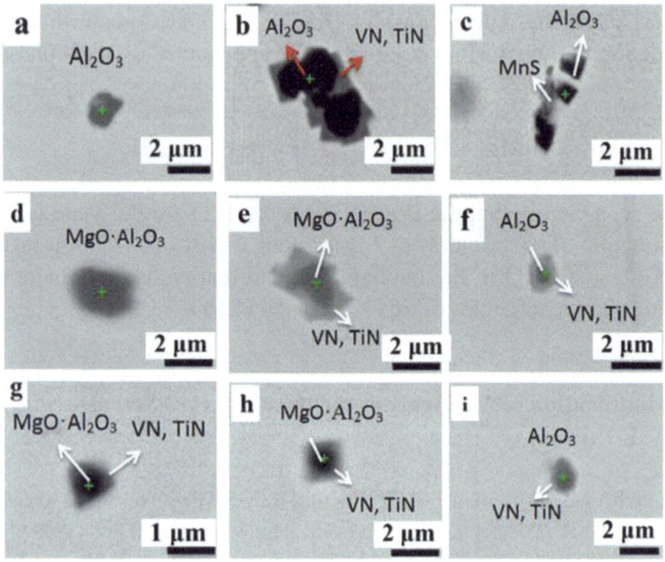

Fig. 5.18 Typical inclusions in ESR ingots: **a–c** Mg free; **d–f** Mg content 0.0005%; **g–i**: Mg content 0.0006%

of inclusions in Mg contained electrode is much higher. The typical inclusions after ESR process are shown in Fig. 5.18.

As can be seen in Fig. 5.18a–c, the inclusions in Mg free ESR ingot are Al_2O_3 and (Ti,V)N formed on the edge of Al_2O_3. The morphology of Al_2O_3 inclusions are irregular and some present as cluster. As can be seen in Fig. 5.18d–f, g–i, the inclusions in Mg contained ESR ingots are mainly $MgO{\cdot}Al_2O_3$ and (Ti,V)N precipitated surround $MgO{\cdot}Al_2O_3$. Al_2O_3 can also be detected in the ESR ingots, but MgO cannot be detected. The relationship between the Mg content of ESR ingots and the Mg content of inclusions is shown in Fig. 5.19.

As shown in Fig. 5.19, both of the Mg content in ESR ingots and inclusions decrease after ESR process, the mean Mg content of inclusions is merely around 5%.

5.1.4.3 Analyzation of Modification of Mg Bearing Inclusions in H13 Steel During ESR Process

The Mg content of steel decrease sharply after ESR process. The relation between total Mg content of steel and dissolved Mg content is calculated by FactSage7.0, the result is shown in Fig. 5.20.

As shown in Fig. 5.20, the dissolved Mg existed in the molten steel when the Mg content in electrodes are 0.0027 and 0.0032%. The Mg exists in the form of oxide inclusion when the Mg content of the ESR ingots is 0.0005%. A large amount Mg bearing inclusions were removal by adsorption of slag during ESR process, which

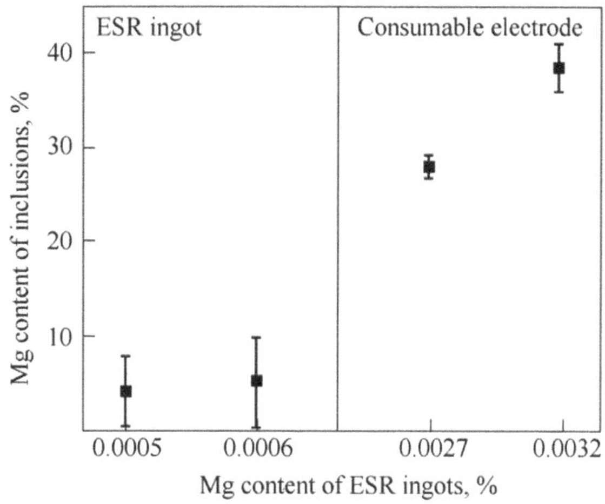

Fig. 5.19 Relation between the Mg content of ESR ingots and the Mg content of inclusion

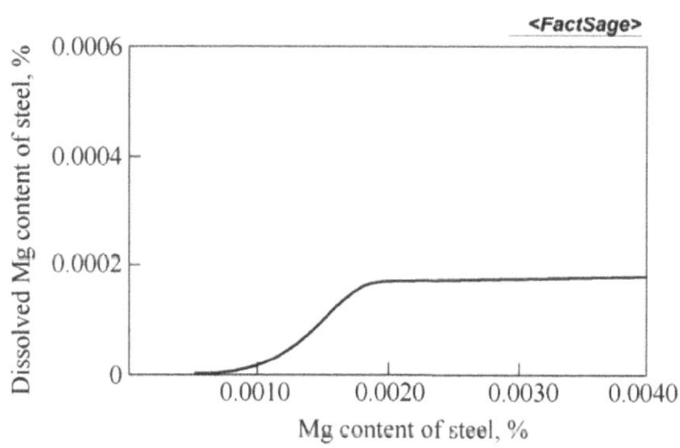

Fig. 5.20 Relation between total Mg content of steel and dissolved Mg content

is also the cause of the decreasing of Mg content in steel. The variation of Mg, Al and O content of $MgO{\cdot}Al_2O_3$ inclusions with the increasing of Mg content in steel at 1873 K is calculated by FactSage7.0, the result is shown in Fig. 5.21.

As can be seen in Fig. 5.21, the Mg content of inclusions is around 10% when the Mg content of steel is 0.0005% in the equilibrium state. The avaerage Mg content of inclusions in ESR ingots are merely 5%, it can be inferred that reaction (5.12) occured in the liquid metal pool during ESR process, which induced the burning loss of Mg content of $MgO{\cdot}Al_2O_3$ inclusions:

$$3MgAl_2O_4 + 2[Al] = 4Al_2O_3 + 3[Mg] \tag{5.12}$$

The reaction is easy to proceed as the Mg content of liquid metal pool is relative low, which cause the Mg content of $MgO{\cdot}Al_2O_3$ inclusions is lower than equilibrium state, and resulting the gasification of dissolved Mg and entering the atmosphere.

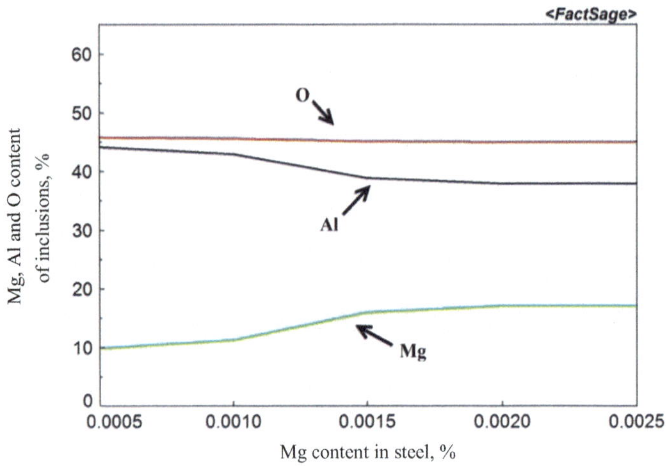

Fig. 5.21 Variation of Mg, Al and O content of $MgO{\cdot}Al_2O_3$ inclusions with the increasing of Mg content (1873 K)

5.2 Analysis on the Effect of Mg-Containing Inclusion on the Carbide

5.2.1 Effect of Mg Addition on Carbides in H13 Steel

Table 5.3 shows the chemical compositions of H13 steel. The Mg content of remelted ingots are 0, 0.0014 and 0.0018%, respectively.

5.2.1.1 Thermodynamic Calculation of Effect of Mg on Precipitates in H13 Steel

The equilibrium phase precipitation in H13 steel with various Mg contents was calculated using Thermo-Calc.

It can be seen from Fig. 5.22 that the types of carbides in steel are not changed, but the precipitation temperatures of precipitated phase are changed by adding Mg. The effect of Mg content on precipitates formation temperature is shown in Table 5.4.

Table 5.3 Chemical composition of H13 steel (wt/%)

Sample	C	Si	Mn	Cr	Mo	V	N	O	S	Als	Mg
D1	0.40	0.72	0.40	4.82	1.30	0.78	0.0300	0.0045	0.0020	0.036	0
D2	0.40	0.90	0.39	5.05	1.43	0.92	0.0073	0.0012	0.0013	0.024	0.0014
D3	0.39	0.89	0.40	4.94	1.39	0.92	0.0046	0.0018	0.0009	0.040	0.0018

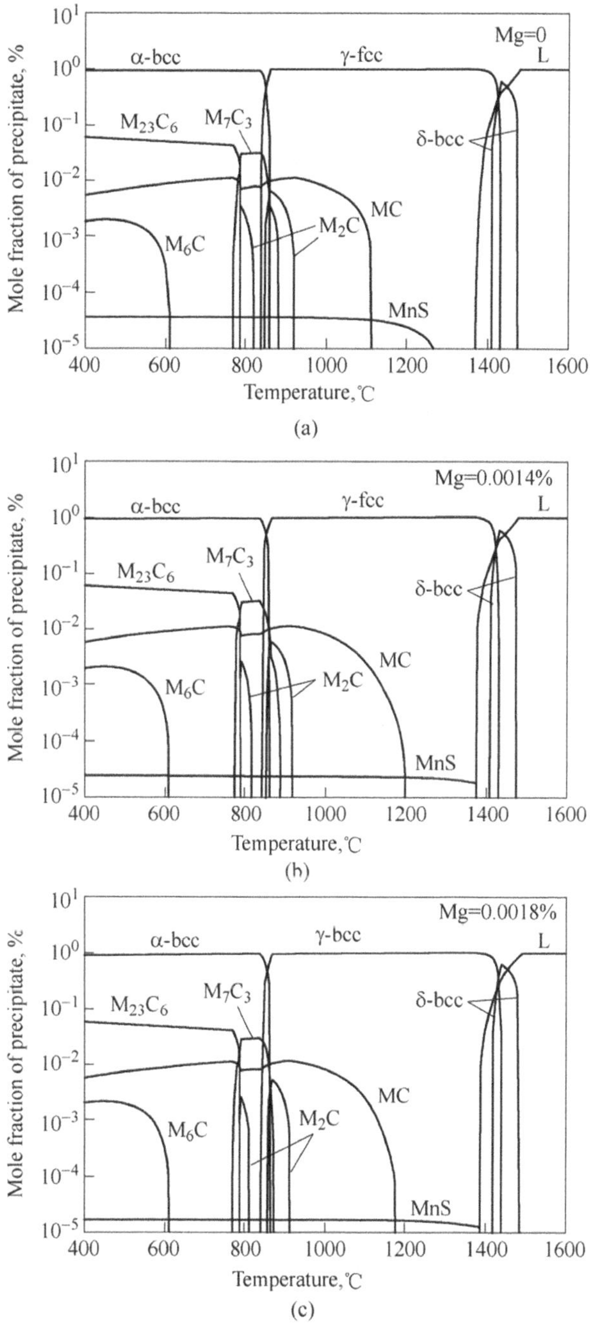

Fig. 5.22 Equilibrium phase precipitation in H13 steel with various Mg contents calculated using Thermo-Calc: **a** 0; **b** 0.0014%; **c** 0.0018%

Table 5.4 Thermodynamic calculation of precipitates formation temperature in Mg-containing H13 steel (°C)

Sample	MC	$M_{23}C_6$	M_6C	M_7C_3	M_2C	MnS
D1	1110	787	610	880	920	1290
D2	1200	792	613	891	920	1373
D3	1170	789	614	870	910	1380

Comparing Fig. 5.22a with Fig. 5.22b, it shows that the precipitation temperature of MC increases from 1110 to 1200 °C. Meanwhile, the precipitation temperature of $M_{23}C_6$ increases from 787 to 792 °C. With further increasing Mg addition, the precipitation temperature of MC decreases to 1170 °C. It indicates that MC is a highly stable phase. The precipitation temperature of MC is 1200 °C when Mg content is 0.0014% in steel, which is about 30 °C higher than that of no Mg steel. Consequently, the precipitation temperature of MC phase can be improved by adding Mg, which is able to refine austenite grains due to lots of undissolved carbides during high temperature heat treatment. It is noted that more Mg addition is less beneficial to phase precipitation. The precipitation temperature of MnS is increased from 1290 to 1380 °C by adding Mg.

5.2.1.2 Effect of Mg Content on the Morphology and Size of Carbides

In the final solidification of ESR ingots, a large amount of carbon and alloying elements are enriched, leading to the precipitation of a large number of carbides [15]. Figure 5.23 shows the morphology and size of carbides in ESR ingots.

The carbides in H13 cast ingot are V-rich and Mo-rich. The morphology of V-rich carbides is lamellar, and their size is about 10 μm. The size of Mo-rich carbide is about 3 μm, which is smaller than V-rich carbide. The carbides distribute along the grain boundary, as shown in Fig. 5.23b, c. The primary carbides mainly are V-rich carbides in cast ingot. The nucleation, precipitation and dissolution of carbides have a decisive effect on the performance of materials. The segregation of Mg in steel is equilibrium segregation. The segregation of Mg is the key to refine carbides. Mg can

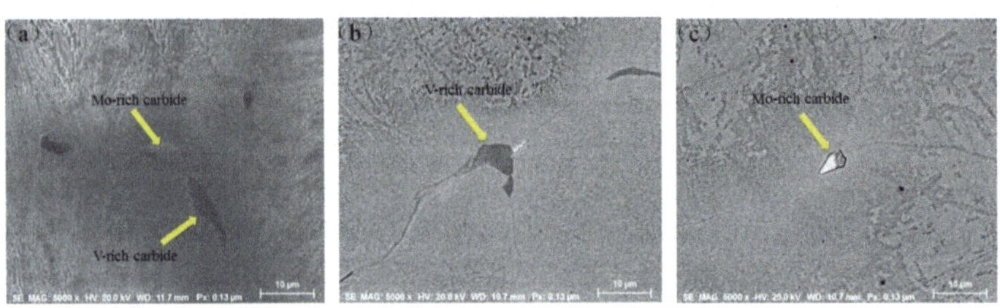

Fig. 5.23 Carbides precipitates along the grain boundary in ESR ingot

change energy of grain boundary, and dissolve into carbide [16]. The morphology, types, and size of carbides in steel after electrolysised were analyzed by SEM–EDS. As shown in Fig. 5.24, the carbides in Fig. 5.24a, c are V-rich while it is Mo-rich in Fig. 5.24b.

As can be seen in Fig. 5.24, the morphology of V-rich carbide in Mg-free steel is bone and square, and the morphology of Mo-rich carbide is short fish-bone. It can be obiously seen that the size of V-rich carbide is about 10 μm while the Mo-rich carbide is 10 μm.

Figure 5.25 shows the carbides in the steel with Mg addition after electrolysised.

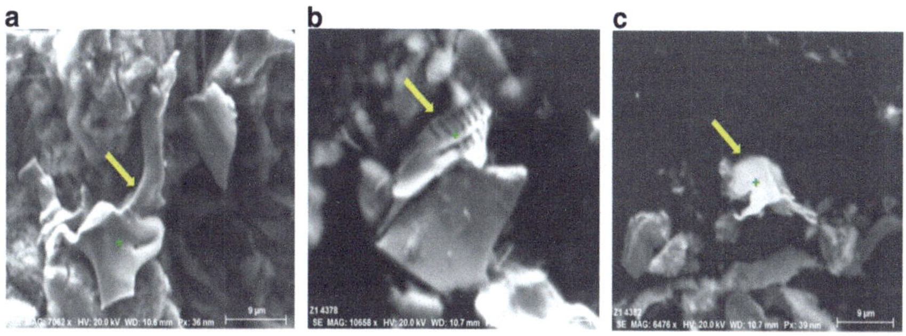

Fig. 5.24 Morphology of carbides in No–Mg containing H13 steel: **a**, **c** V-rich carbide; **b** Mo-rich carbide

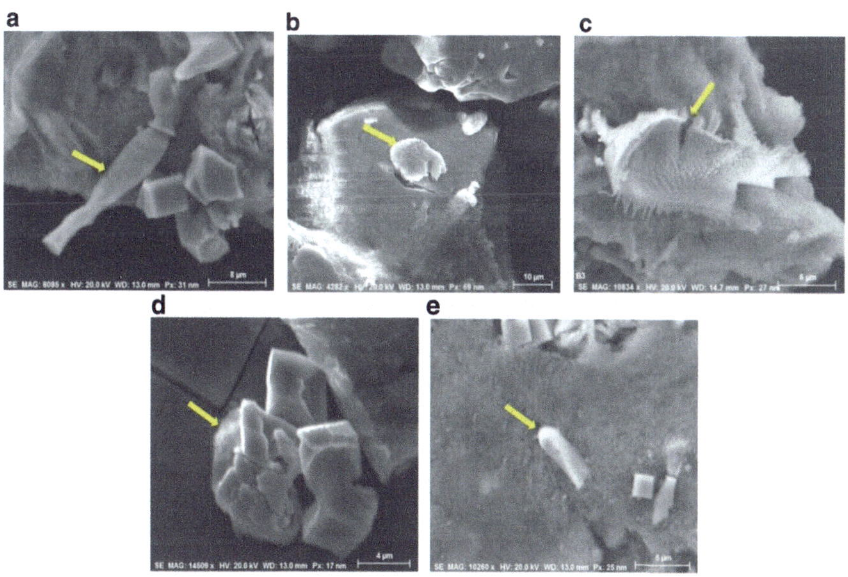

Fig. 5.25 Morphology of carbides in Mg containing H13 steel: **a**, **b**, **c** V-rich carbides; **d**, **e** Mo-rich carbides

As can be seen in Fig. 5.25, the morphology of V-rich carbide is long bar and granular, the size of most carbides is around 5 μm. The morphology of Mo-rich carbide is brain-like and short rodlike, the size is around 4 μm The size of carbide in the steel with high Mg addition is smaller than that without Mg addition.

It is noted that the morphology and size of carbide are changed by adding Mg. Table 5.5 shows the basic parameters of carbides.

As shown in Table 5.5, with increasing Mg content, the number and average diameter of carbides decreases. The total area of carbides also decreases. The size and number of carbides are shown in Fig. 5.26.

As shown in Fig. 5.26, the size of carbide decrease with the increasing of Mg content. The size of carbide is ranging approximately from 2 to 5 μm. The size of carbide ranging from 1 to 2 μm is increasing with Mg addition. At the same time, the large size of carbide ranging from 7 to 9 μm is decreasing with Mg content increasing. It is expected that the carbide in steel is greatly influenced by Mg.

The shape factor (roundness) is calculated using Eq. (5.13) [17]. The calculated results are shown in Table 5.6.

Table 5.5 Basic parameters of carbides

Sample	Number	A, μm^2	$\overline{D}$, μm	$\overline{A}$, μm^2	S, mm^2
D1	89.00	1986.48	5.02	22.32	4.80
D2	72.00	1474.73	4.84	20.48	4.80
D3	56.00	887.42	4.21	15.85	4.80

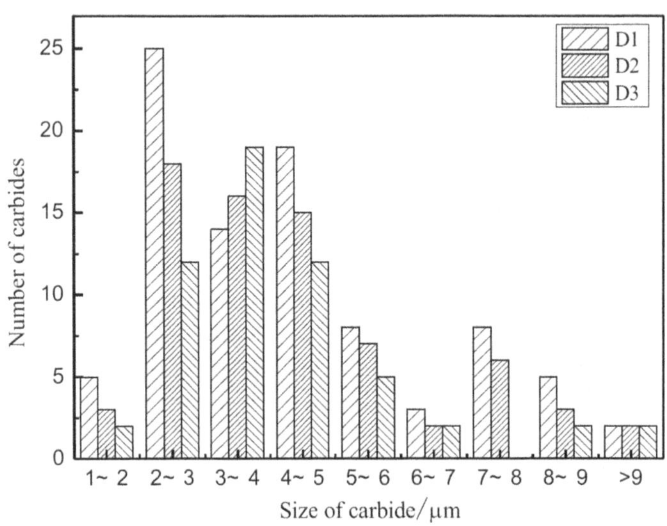

Fig. 5.26 Effect of Mg content on the size of carbides

Table 5.6 Effect of Mg content on carbide parameters

Number	$\overline{D}_{max}$, μm	Roundness
D1 (Mg: 0)	10.05	0.30
D2 (Mg: 0.0014%)	9.27	0.37
D3 (Mg: 0.0018%)	8.58	0.36

$$\text{Roundness} = 4 \times \frac{[\text{Area}]}{\pi[\text{Major} - \text{axis}]^2} \quad (5.13)$$

Where area is the area of carbides; major-axis is length of the long axis. The value of roundness increases with increasing of circularity and becomes 1 for a perfect circle. Table 5.6 demonstrates that the roundness of precipitate carbides in grain boundaries improved with increasing Mg content. It is expected that the grain boundary cohesive bond and vacancy formation energy are increased by Mg content, which results in the change in morphology of carbide. Table 5.6 shows the largest size of carbide was also decreased from 10.05 to 8.58 mm.

As shown in Table 5.6, the maximum diameter of carbide decreases and roundness increases with the increasing of Mg content. The morphology of carbides changed, because Mg can be dissolved in carbides and act as the core of carbide formation. Some of the elongated carbides change to spherical as can be seen in Figs. 5.24 and 5.25. Therefore, the size of the primary carbide decreases after Mg treatment, and the carbides are tend to be spherical.

From the analysis above, it can be conclude that Mg is mainly concentrated near the grain boundary or phase boundary. Due to the great difference in atomic radius between Mg atoms and Fe atoms, additional lattice distortion and elastic distortion energy will inevitably occur in the single cell and parent phase of carbide. According to the theory of Prines [18], the elastic distortion energy generated by replacement of solid dissolved atoms in the matrix can be calculated by Eqs. (5.14) and (5.15):

$$Q = \frac{24\pi K G r_1^3 \varepsilon^2}{3K + 4G} N_A \quad (5.14)$$

$$\varepsilon = \frac{r_1 - r_0}{r_1} \quad (5.15)$$

K: Shear modulus of elasticity;
G: Matrix shear modulus of elasticity;
r_l: radius of solute;
r_0: replaced atomic radius.

Table 5.7 shows the parameters used in the calculation.

The elastic distortion energy of Mg dissolved in the cementation body after replacing Fe can be calculated according to Eqs. (5.14) and (5.15). When the cementite was replaced with Fe at 1473 K, the elastic distortion at the beginning of precipitation of MC was 42,082.7, 50,713.7 J/mol for $M_{23}C_6$ and 54,536.2 J/mol for M_6C.

Table 5.7 Parameters used in the calculation

Element	Shear modulus of elasticity, MPa	Atom radius, nm	Dependent variableε	Matrix shear modulus of elasticity, MPa
Mg	4.1×10^4	0.160	0	–
Fe	77970–7.021 *T*	0.127	0.2025	98637–44.45 *T*

It can be seen from the calculation results that Mg produces a large elastic distortion energy in the cements. Due to the large atomic radius of Mg on the phase boundary, it tends to converge and enter the two-phase lattice, which can increase the additional elastic deformation energy obtained. When Mg enters the carbide phase, it will produce great elastic distortion, which makes the lattice on both sides of the interface mismatch and increases the elastic energy. Mg has the largest elastic distortion energy on M_6C, so it has a significant influence on spheroidization of M_6C carbides. M_6C is mainly Mo-rich carbide according to the thermodynamic calculation, it can be seen in Figs. 5.24 and 5.25 that the morphology of Mo-rich carbide are changed significantly after Mg addition.

The facilitation of heterogeneous nucleation of carbide is achieved by existence of $MgO{\cdot}Al_2O_3$ and TiN in molten steel until the end of solidification, as shown in Fig. 5.27a. It is expected that $MgO{\cdot}Al_2O_3$ inclusion promotes the crystallization of TiN. What's more, the crystallized TiN facilitates the heterogeneous nucleation of VC. This is considered as a chain of heterogeneous nucleation. It is found that the size of inclusions was refined by Mg addition [19–21]. The optimal size of inclusions which promoted heterogeneous nucleation was smaller than 1 mm. Figure 5.27b shows that Mg is dissolved into carbides. It is stated that the morphology of carbide is transformed by dissolved Mg. From the results of Fig. 5.27, it is proved that the transformation in morphology and size of carbides is attributed to the heterogeneous nucleation of carbide by $MgO{\cdot}Al_2O_3$ and TiN. And the increasing roundness of carbide as shown in Table 5.6 is attributed to the dissolved Mg in carbides.

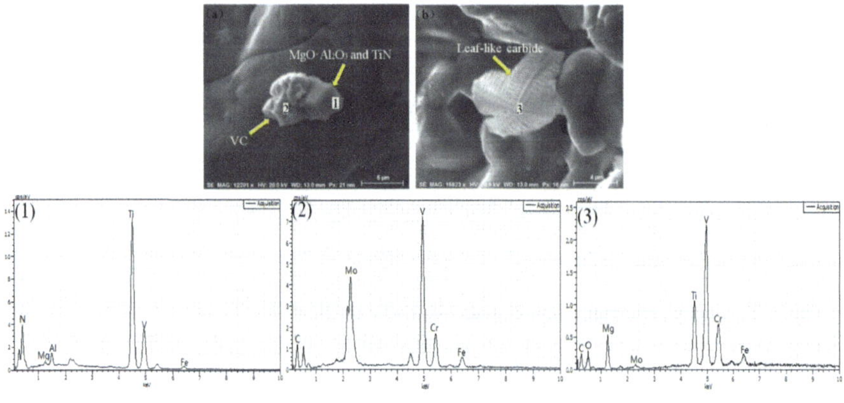

Fig. 5.27 The SEM–EDS results of carbides heterogeneous

5.2.2 Effect of Mg Content on Segregation of Alloy Elements

Figure 5.28 shows the microstructure evolution of as-cast steel with different Mg addition.

It can be observed in Fig. 5.28 that the as cast microstructure of H13 steel is composed of martensite, retained austenite and primary carbides. It can be seen that the segregation of network is reduced with the addition of Mg. The amount of the retained austenite decreases with increasing Mg content, while the ratio of martensite increases. With the addition of Mg, the as-cast microstructure of steel changes from finely meshed distribution to long chain and isolated distribution gradually. It can be concluded that the amount of carbides decreases and the segregation of cast ingot microstructure is relieved by Mg.

Alloy elements such as Mg, V, Cr and Mo were analyzed by electron probe micro-analysis (EPMA). The results of lining scanning are shown in Fig. 5.29.

As shown in Fig. 5.29, the intensity of Mg in matrix with Mg addition is higher than that without Mg addition. The intensity of Mg in the steel withoutMg addition is keeping stable as a line, while that is changing acutely with high Mg content. According to Fig. 5.29a, b, it is expected that the location of high intensity of Mg is the area of segregation of carbides. It means that Mg can distribute to the area of segregation of carbides. It can be obviously seen that the carbides distribute along grain boundaries. Mg also distributes along grain boundaries, which is in agreement with the fingdings reported by Bor et al. [22] and Ge et al. [23] The alloy elements in the formation of carbide are V, Cr and Mo. The effect of Mg on the formation of carbides' elements is shown in Fig. 5.29d–f. It is noted that the segregation of V and Mo with Mg addition is lower than that without Mg addition. There is not obviously influence on Cr. The results above show that the Mg in steel distributes along grain boundaries, and the segregation of V and Mo is inhibited by adding Mg, resulting in decreasing carbides.

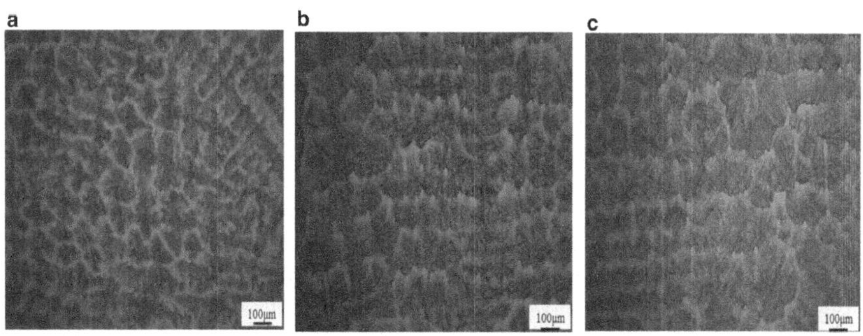

Fig. 5.28 Microstructure of ESR ingot with different Mg content: **a** 0; **b** 0.0014%; **c** 0.0018%

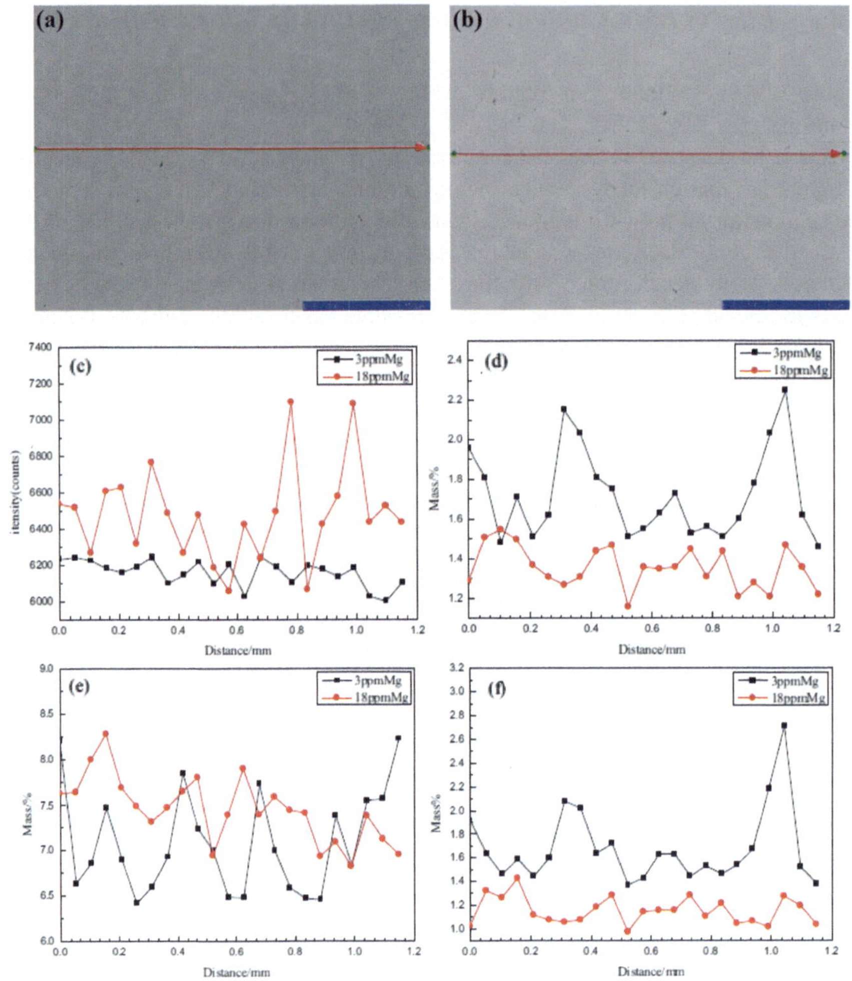

Fig. 5.29 Alloy elements segregation in steel with different Mg content: **a** Mg free; **b** 0.0018% Mg; **c** Mg; **d** V; **e** Cr; **f** Mo

Figure 5.30 shows the results of electron probe microanalysis (EPMA) map scanning of the segregation of carbides. It can be seen that Mg distributes in carbide, the dark area of the figure means a lower alloy elements content.

It means that Mg is dissolved into carbide. According to previous study [24], the carbide is more easily spherized with Mg addition. Combined with the results shown in Fig. 5.30, the reason probably is that Mg is dissolved into carbides.

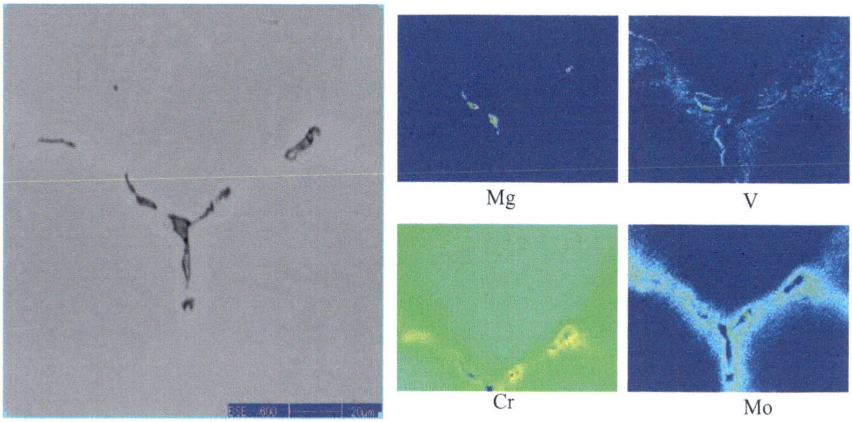

Fig. 5.30 Element mappings of typical carbide in the steel with Mg addition

5.2.3 *Mechanism of Refining and Spheroidizing Carbides by Mg*

The banded segregation in Mg bearing steel is improved and the average size and roundness of carbides are decreased compared with Mg free H13 die steel. On the basis of statistical thermodynamics, Mclean [25] assumes that the crystal defect area is an ideal solid solution without rules, and the interaction between the partial polymerization elements is ignored. When the concentration of solute elements in the crystal is far less than 1%, the thermodynamic formula of equilibrium partial polymerization at the defect is derived as follows:

$$C_g = C_o \exp\left(\frac{\Delta U}{RT}\right) \rightarrow \beta = \frac{C_g}{C_o} = \exp\left(\frac{\Delta U}{RT}\right) \tag{5.16}$$

C_o: concentration at defect;
ΔU: partial polymerization energy of 1 mol of solute (function of temperature);
R: molar gas constant;
T: temperature;
β: enrichment coefficient.

According to the theory put forward by Mott and Nabarro [26], ΔU can be expressed as:

$$\Delta U \leq \frac{1}{3\pi} \cdot \frac{1+v}{1-v} \cdot G \cdot |\Delta V| \cdot N_A \tag{5.17}$$

Table 5.8 Parameters of Mg and matrix [27]

Atom	Atom diameter with the coordination number is 12, nm	Atom volum, nm^3	Atom volume deviation ΔV, nm^3	Matrix shear modulus, MPa	Temperature range, K
α-Fe	0.25537	0.008720	0.008589	89,334–29.688 T	293–1184
γ-Fe	0.25787	0.008978	0.008331	98,637–44.45 T	1184–1665
Mg	0.32094	0.017309	–		

v: poisson ratio;
G: Matrix shear modulus of elasticity;
ΔV: atom volume deviation;
N_A: Avogadro constant.

The effect of solute on the elastic modulus of α-Fe in H13 steel is directly proportional to the mole fraction of the solution element. G can be increased by 0.35% per 1% mole fraction Cr, while G can be reduced by 0.12, 0.28 and 1.08%, respectively by per 1% mole volume Si, Mn and Ni. Solute elements also have similar effects on poisson ratio.

Assume that the effect of solute element on G in H13 steel is offset, and poisson ratio is 0.30. Put the parameters of Mg atom and matrix (as shown in Table 5.8) into Eq. (5.17) to get the expression of ΔU (take the maximum value). Put ΔU into Eq. (5.16) to get the relationship between the enrichment coefficient β and temperature T (5.18):

$$\beta = \begin{cases} \exp\left(\dfrac{11295.41}{T} - 3.76\right) & (293 - 1184\ \text{K}) \\ \exp\left(\dfrac{11691.28}{T} - 5.27\right) & (1184 - 1665\ \text{K}) \end{cases} \tag{5.18}$$

As can be derived from Eq. (5.18), the equilibrium partial polymerization constant of Mg is 8.69 at 1573 K. The equilibrium partial polymerization constants of Mg at annealing temperature (1133 K) and tempering temperature (853 K) were 497.4 and 13, 121.2 respectively. Mg has high degree of partial polymerization, and Mg atoms can diffuse to the vacancy and defect, reducing the growth of nets-chain-typed carbides [28]. In addition, Mg can also concentrate on the selective growth interface of carbides and hinder the growth of MC carbides [29, 30]. So the size of carbide in magnesium H13 die steel is small.

Mg has spheroidization effect on carbides. The formula of specific interfacial energy of coherent interface is as follows [27]:

$$\sigma = \frac{2}{3} G d \delta^2 \tag{5.19}$$

G: shear modulus of less rigid crystals on both sides of the interface;
δ: the elastic lattice distortion caused by the difference of atomic spacing between the two sides of the interface, that is, disregistry;
d: diameter of carbides.

Assume that the shear modulus of Fe remains constant at a certain temperature. The atom radius of Mg (0.16 nm) is larger than that of Fe (0.1276 nm). For the same annealing temperature and same carbides, additional lattice elastic distortion energy will be generated due to the large atom radius Mg in the phase interface and into the two-phase lattice, which will destroy the coherent relation and make the carbide and the matrix become semi-coherent or non-coherent during the growth process, resulting in the increase of mismatch degree (δ) and the increase of specific interface energy. In order to reduce the energy of the system, the phase interface area will be reduced as much as possible, and the carbides enclosed by Fe matrix grain will tend to spheroidize. The driving force of spherification increases with the increasing of specific interface energy. Therefore, carbide spheroidization is favorable for H13 steel after Mg treated.

VC is precipitated around the fine $MgAl_2O_4$ inclusion in Mg treated H13 die steel, and the carbide size is relatively small, as shown in Fig. 5.31.

According to Bramfitt's theory, in the process of heterogeneous nucleation, $\delta <$ 6% is a strong effective nucleation, $\delta = 6\% - 12\%$ is a normal effective nucleation, and $\delta > 12\%$ is an inefficient nucleation.

The equation proposed by Bramfitt for calculating lattice disregistry is expressed as follow [31]:

$$\delta_{(hkl)n}^{(hkl)s} = \sum_{i=1}^{3} \frac{\frac{\left| d_{[uvw]_s^i} \cos\theta - d_{[uvw]_n^i} \right|}{d_{[uvw]_n^i}}}{3} \times 100 \tag{5.20}$$

where s stands for the substrate (nucleating agent) of nucleating agent and n the nucleated solid, δ_n^s.

$(hkl)_s$ = a low-index plane of the substrate (nucleating agent),
$[uvw]_s$ = a low-index direction in $(hkl)_s$,
$(hkl)_n$ = a low-index plane in the nucleated solid,
$[uvw]_n$ = a low-index direction in $(hkl)_n$,
$d[uvw]_n$ = the interatomic spacing along $[uvw]_n$,
$d[uvw]_s$ = the interatomic spacing along $[uvw]_s$,
θ = the angle between the $[uvw]_s$ and $[uvw]_n$.

The adopted lattice constants of the related phases are presented in Table 5.9.

The effects of the main elements in H13 steel on lattice constants of γ-Fe were considered. The lattice constants were taken as their values at 1400 °C, which shows the primary austenite forming at 1420 °C. The computation method was introduced

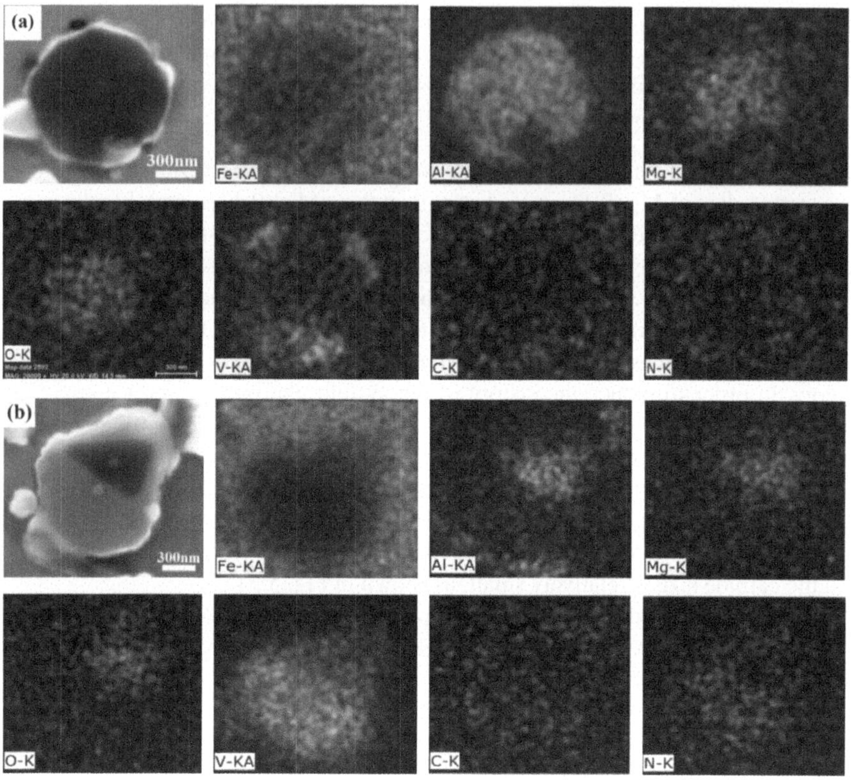

Fig. 5.31 Morphology of carbide precipitates around Mg bearing inclusion

Table 5.9 Lattice constants of the studied phases

Phase (crystal system)	Lattice constant at 1400 °C, nm	Coefficient of linear expansion (1/8C)
$MgAl_2O_4$ (cubic)	$a_1 = 0.8132$	7.6×10^{-6}
Al_2O_3 (hexagonal)	$a_2 = 0.4812$	7.5×10^{-6}
γ-Fe (cubic)	$a_3 = 0.3686$	23×10^{-6}
M(CN) (cubic)	$a_4 = 0.4209$	8.29×10^{-6}
M_6C (cubic)	$a_5 = 1.118$	10×10^{-6} (assumed)

in detail in Ref. [32] and the calculated values of disregistry between phases are listed in Table 5.10.

It can be seen from Table 5.10 that the disregistries between the $MgAl_2O_4$ and other phases mentioned are much smaller than those of Al_2O_3 so that the γ-Fe, M(CN) and M_6C can more easily form in the Mg-bearing steel.

Table 5.10 Disregistry values between related phases

	Dismatch value, %	Match plane	Modified lattice constant
$\delta_{\gamma-Fe}^{Al_2O_3}$	11.6	$(a100)_{\gamma\text{-}Fe}//(0001)_{Al_2O_3}$	$2\times\frac{\sqrt{2}}{2}a3 \leftrightarrow a2$
$\delta_{\gamma-Fe}^{MgAl_2O_4}$	4.0	$(100)_{\gamma\text{-}Fe}//(100)_{MgAl_2O_4}$	$3\times\frac{\sqrt{2}}{2}a3 \leftrightarrow a1$
$\delta_{M_6(CN)}^{Al_2O_3}$	13	$(100)_{M_6(CN)}//(100)_{Al_2O_3}$	$a5 \leftrightarrow 2a2$
$\delta_{M_6(CN)}^{MgAl_2O_4}$	2.8	$(100)_{M_6(CN)}//(100)_{MgAl_2O_4}$	$a5 \leftrightarrow 2\times\frac{\sqrt{2}}{2}a1$
$\delta_{M(CN)}^{Al_2O_3}$	11.3	$(110)_{M(CN)}//(0001)_{Al_2O_3}$	$2\times\frac{\sqrt{2}}{2}a4 \leftrightarrow a2$
$\delta_{M(CN)}^{MgAl_2O_4}$	3.0	$(100)_{M(CN)}//(100)_{MgAl_2O_4}$	$2a4 \leftrightarrow a1$
$\delta_{M_6(CN)}^{\gamma-Fe}$	1.1	$(100)_{M_6(CN)}//(100)_{\gamma\text{-}Fe}$	$a5 \leftrightarrow 3a3$
$\delta_{\gamma-Fe}^{M_6(CN)}$	1.1	$(100)_{\gamma\text{-}Fe}//(100)_{M_6(CN)}$	$3a3 \leftrightarrow a5$
$\delta_{M(CN)}^{\gamma-Fe}$	12.1	$(100)_{M(CN)}//(100)_{\gamma\text{-}Fe}$	$a4 \leftrightarrow a3$
$\delta_{\gamma-Fe}^{M(CN)}$	13.8	$(100)_{\gamma\text{-}Fe}//(100)_{M(CN)}$	$a3 \leftrightarrow a4$
$\delta_{M_6(CN)}^{M(CN)}$	6.5	$(100)_{M_6(CN)}//(100)_{M(CN)}$	$a5 \leftrightarrow 4\times\frac{\sqrt{2}}{2}a4$
$\delta_{M(CN)}^{M_6(CN)}$	6.1	$(100)_{M_6(CN)}//(100)_{M(CN)}$	$4\times\frac{\sqrt{2}}{2}a4 \leftrightarrow a5$

5.3 Effect of Heat Treatment Process on Carbides Type and Distribution of Mg Contained H13 Steel

5.3.1 Evolution of Carbides in H13 Steel in Heat Treatment Process

Carbide is important as a second phase in H13 steel. The type, quantity, size, morphology and distribution of carbide have important effect on the performance of steels. The compositions of H13 ingot are listed in Table 5.11. The heat treatment processes in detail are listed in Table 5.12.

5.3.1.1 Microstructure of H13 Steel

The microstructure of H13 steel after different heat treatment process are shown in Fig. 5.32.

Table 5.11 Chemical compositions of experimental H13 steel (wt%)

C	Si	Mn	Cr	Mo	V	P	S	Fe
0.41	0.99	0.29	5.01	1.22	0.93	0.023	0.006	–

Table 5.12 Heat treatment process

Forging process	Annealing process	Quenching process	Tempering process
Forged between 860 °C to 1150 °C and forged into Φ325 mm in diameter	Heated to 760 °C for 2 h → reheated to 860 °C for 8 h → furnace cooled to 500 °C → air cooled to room temperature	Heated to 850 °C for 1 h → reheated to 1050 °C for 100 min → oil quenched	Heated to 590 °C for 4 h → air cooled to room temperature

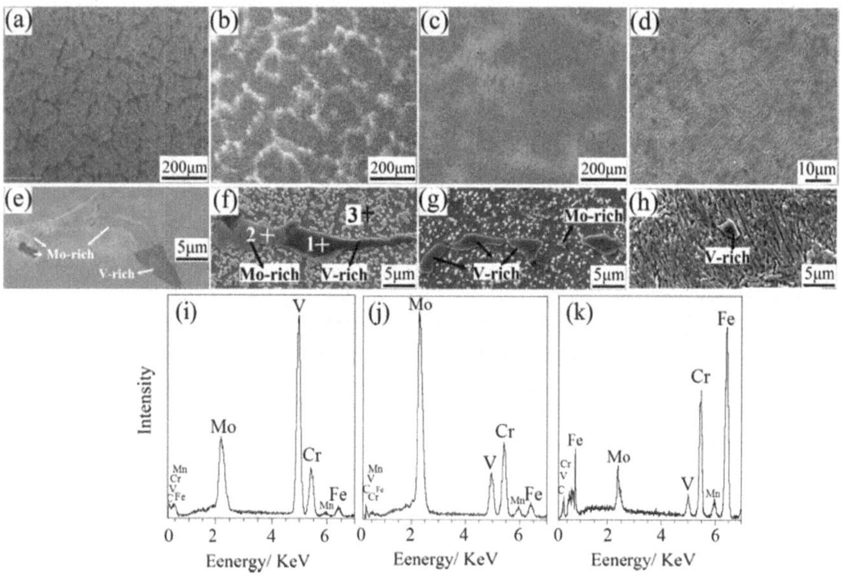

Fig. 5.32 Morphologies of different heat treated H13 steel: **a** and **e** ESR ingot; **b** and **f** Annealing; **c** and **g** Forging + annealing; **d** and **h** Quenching + tempering; EDS spectrums of point **i** 1; **j** 2 and **k** 3 in (**f**)

As shown in Fig. 5.32a, the dendritic segregation was widely distributed in H13 steel. The interdendritic space presents a net-shaped microstructure, where elements with portion coefficient less than 1 are concentrated and secondary phases precipitate out when solute elements are supersaturated. Figure 5.32e presents the coexisted primary V-rich and Mo-rich carbides in the interdendritic space. In fact, the amount of primary V-rich carbides is far more than that rich in Mo and there are no Cr-rich primary carbides in the ESR ingot. After annealing, a large amount of secondary carbides precipitate in the interdendritic space and the white net-shaped carbide segregation area is composed, as shown in Fig. 5.32b. Figure 5.32f shows the carbides in carbide segregation area where most of the massive secondary carbides with fine size are detected to be rich in Cr and the primary carbides are not decomposed. However, the carbide segregation was strongly decreased after the hot forging and re-annealing process, as shown in Fig. 5.32c. The carbide segregation is weakened sharply, most of Mo-rich and part of V-rich primary carbides are decomposed (Fig. 5.32g). Further, the carbide segregation was not detected after the quenching and tempering process and only a small fraction of primary V-rich carbides remained in steel, as shown in Fig. 5.32d. Moreover, secondary carbides get finer and more dispersive gradually by comparing Fig. 5.32f–h.

The carbide types in H13 steel after each stage of heat treatment process were determined by XRD. The XRD spectrums are documented in Fig. 5.33.

As can be seen in Fig. 5.33a, the carbides in ESR ingot are MC, M_6C and less M_7C_3. The observed results confirmed that the MC and M_6C are usually in the forms of primary carbides, and M_7C_3 as secondary carbide. Carbides in annealing sample and forging + annealing sample are M_7C_3, MC and less M_6C, which was in accordance with that a large amount of secondary M_7C_3 carbides formed during annealing process shown in Fig. 5.32b, e. Most of primary M_6C and minority of

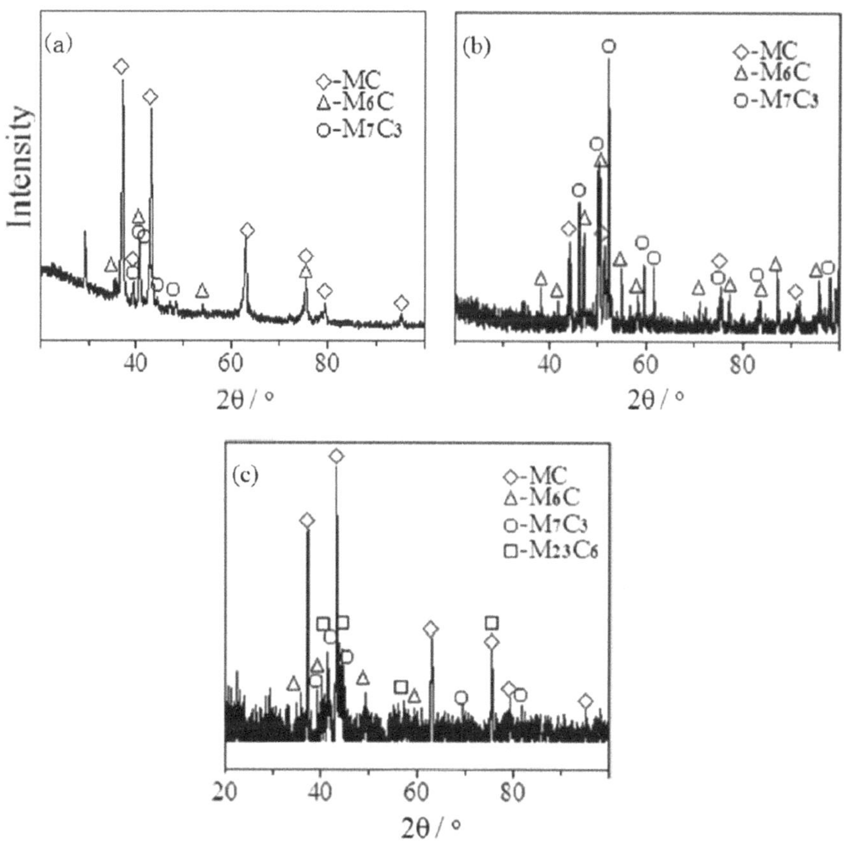

Fig. 5.33 XRD spectrums of carbides in different heat treated H13 steel: Carbides types in **a** ESR ingot; **b** Annealing; **c** Quenching + tempering

primary MC carbides are decomposed and may be replaced by a small part of precipitated secondary M_6C and MC in forging + annealing sample. Carbides in quenching + tempering sample are MC, $M_{23}C_6$, M_7C_3 and M6C, which reveals that the MC and M_6C re-precipitated out after carbides dissolution. $M_{23}C_6$ carbides can be formed by the in-situ transformation of M_7C_3, so $M_{23}C_6$ carbides may be formed as a new phase by the transformation of some M_7C_3 carbides during the tempering process.

5.3.1.2 Thermodynamic Calculation of Carbide Phases in Heat Treated H13 Steel

The relationship between carbides and temperature was calculated by Thermo-calc software as plotted in Fig. 5.34a. The related data is listed in Table 5.13.

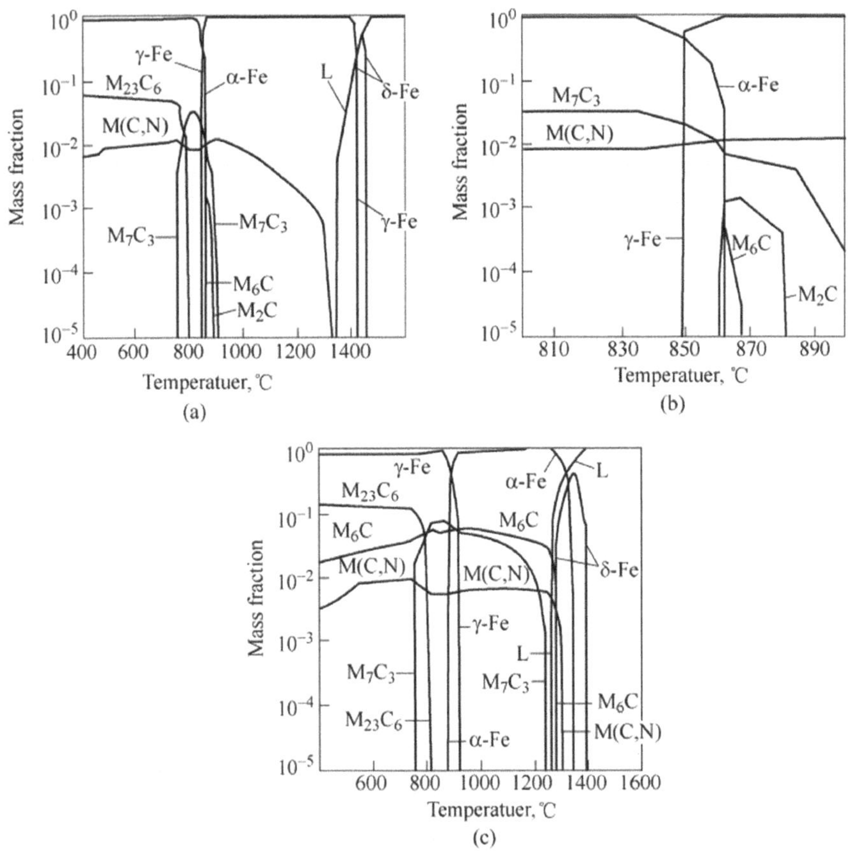

Fig. 5.34 Calculation results of carbide phases in the H13 steel: **a** 400–1600 °C; **b** 800–900 °C; **c** Simulated situation in segregation area

Table 5.13 Related data of carbides in H13 steel by Thermo-Calc

	MC	M_7C_3	$M_{23}C_6$	M_6C	M_2C
Dominated elements	V	Cr	Cr	Mo	Mo
Temperature intervals, °C	–1328	759–907	–795	861–867	862–882

Table 5.14 Contents of carbides in different heat treated steel (wt%)

	Ingot	Annealed	Forged + annealed	Quenched + tempered
Calculated values	6.5	2.2	2.2	6.0
Experimental values	3.2	4.5	4.1	2.9

Note the calculated values are the sum of carbide contents at differentheat treatment temperatures according to Fig. 5.34a

The carbides formed in the temperature range of 800–900 °C is shown in Fig. 5.34b. It is considered that the segregation of solute elements promote the formation of primary carbides during the solidification process, so the mass fraction of non-metallic elements was assumed to enlarge two times and the metallic elements enlarge three times in the segregation area except for Cr (enlarging two time as well) due to its better diffusion ability. The assumed compositions were calculated by Thermo-calc software to simulate the carbide formation process in the segregation area. The result is shown in Fig. 5.34c.

Based on equilibrium thermodynamic calculation, carbides in H13 steel are secondary MC, M_7C_3, M_2C, M_6C and $M_{23}C_6$ according to Fig. 5.34a, whereas M_7C_3, M_6C and M_2C exist only in a narrow temperature range. However, in non-equilibrium state, the temperature range for M_7C_3, M_6C are largely extended due to the element segregation, as shown in Fig. 5.34c. The solidi cation process of H13 steel can be summarized as follows according to Fig. 5.34:

(1) MC began to precipitate at 1328 °C after the liquid phase of H13 steel disappeared (1346 °C), the austenite formed completely, and then existed in the steel all through the heat treatment process;
(2) M_7C_3 began to precipitate when the temperature is reduced to 907 °C. M_2C began to precipitate at 882 °C afterwards, and hence the precipitation rate of M_7C_3 decreased;
(3) M_2C is unstable and easily decomposed as M_6C and MC. The mass percent of M_6C reached a maximum 0.11% at 862 °C and decomposed rapidly with the decreasing of temperature, the precipitation of M_7C_3 began to accelerate till the content of M_7C_3 reach the maximum;
(4) M_7C_3 transformed to $M_{23}C_6$ at 795 °C and then reach the flat peak.

Both thermodynamic and kinetic conditions are needed to be met for the precipitation and decomposing of carbides. Therefore, the formation of carbides in H13 steel at different heat treatment stage can be interpreted by combining the thermodynamic condition, kinetic condition and the heat treatment process [33].

It indicates that $M_{23}C_6$ carbides exist in the equilibrium state of H13 ESR ingot, as shown in Fig. 5.34a, c. But the XRD result (Fig. 5.33a.) is contradictory to the calculated results by Thermo-Calc, since the rapid cooling rate of ESR process and lead to the insufficient of kinetic conditions for the formation of $M_{23}C_6$.[34] Owing to both thermodynamic condition and kinetic condition were met in the liquid phase, MC, M_6C and M_7C_3 were detected in ESR ingot. M_7C_3, MC and few M_6C were found in the annealed H13 steel, which is accordance with the carbides in H13 ESR ingot. There is a large difference in the content of each type of carbide, this is because a large amount of secondary M_7C_3 precipitated, M_6C dissolved and MC remained unchanged under the sufficient thermodynamic and kinetic conditions during annealing process. M_2C is difficult to generate during annealing, because the conditions of both thermodynamics and kinetics of $M_6C + MC \rightarrow M_2C$ are poor. The carbides in quenched-tempered H13 steel are MC, $M_{23}C_6$ and a small amount of M_7C_3 and M_6C. The evolution of carbides in H13 steel can be speculated according to the generating curves of three types of carbides in Fig. 5.34a, c: primary carbide MC cannot be dissolved completely in H13 steel. The reason is that the temperature range of forging and annealing thermodynamically ensure the inevitability of its existence. But the harmful effect of primary carbide MC can be reduced by forging and high temperature insulation for a long time to change its morphology in the dynamic equilibrium [35]. With the improvement of composition segregation, the temperature range of M_7C_3 and M_6C precipitation in the segregation zone narrowed and carbides remained partially during the quenching and tempering process. $M_{23}C_6$ precipitated easily during tempering process with a long time.

5.3.1.3 Quantitative Chemical Analyses of Carbides in Different Heat Treated H13 Steel

Table 5.14 shows the contents of carbides in H13 steel at different heat treatment stage.

As can be seen in Table 5.14, a large amount of primary carbides precipitated after the ESR process, and secondary carbides precipitated little or no. The content of carbides in annealed sample is the maximum while ingot sample is the minimum, this is because the severe segregation in the ESR ingot, a large amount of secondary carbides precipitated and primary carbides cannot be eliminated after annealing. A large amount of primary carbides were broken and dissolved partially into the matrix after forging and annealing. The segregation is improved. But a large amount of secondary carbides M_7C_3 still precipitated, so the content of carbides in forged + annealed sample is next to that of annealed sample. The segregation was further eliminated after high-temperature quenched. Since the precipitated amount of secondary carbides was few, and which size distribution was tiny and diffused, the content of carbides in quenched + tempered sample was the least. The working temperature of H13 die steel is high, which is equivalent to multiple tempering treatments. As the carbides precipitated and grew continuously, the content of carbides was increased gradually. The precipitation and transformation of carbides during the heat treatment require an incubation period [36], similarly, the growth of carbides takes time. It indicates that each phase of carbides could not reach equilibrium state both on temperature and time during the heat treatment process, resulting in the large difference between the calculated values of carbides contents and the measured results.

The size of carbides in H13 die steel obviously decreased after heat treatment. The alloying elements diffuses rapidly at a high temperature in annealing treatment, but the equilibrium of the phases in the carbide is hard to be reached. The quantitatively analyzed compositions of carbides in annealed H13 die steel are listed in Table 5.15, in which W is a residual element. MC and M_6C are designated as M(C, N) and M_6(C, N) respectively because they contain trace amount of nitrogen.

The Cr-rich M_7C_3 is the dominant carbide which occupies a mass fraction of 64.82% (=2.856/4.406) of total carbides according to Table 5.15. It is believed that the precipitated M_7C_3 is formed by the transformation of M_3C [37], which is usually replaced by M_7C_3 when the temperature increases. One or more types of transitional carbides may form during annealing before the stability is achieved [38].

The mass percentage of M_6(C, N) in annealed H13 die steel is smaller than that of M_7C_3 and M(C, N) as shown in Table 5.15. This is due to the narrow thermodynamic stability region of M6(C, N) in this steel (see Fig. 5.34a) and the lower affinity of Mo with C in comparison with V. The secondary carbides M(C, N) precipitated during annealing of H13 die steel have high stability, small size and dispersive distribution, which present as a precipitation strengthening in steel [39].

Thermo-Calc software is employed to calculate the change of the chemical compositions of these three types of carbides in H13 die steel with temperature, as shown in Fig. 5.35.

Based on the calculated results shown in Fig. 5.35, the mass ratios of main elements in the carbides in the steel annealed at 860 °C are calculated as listed in Table 5.16. The measured mass ratios of Cr/Fe, Mo/Fe and V/(Cr + Mo) in the corresponding carbides are 1.80, 2.13 and 3.51, respectively. The measured values are close to the calculated values at 860 °C as shown in Table 5.16, but there are still certain differences. The difference is attributed to the fact that the formation of the carbides can not reach equilibrium during annealing at 860 °C. The concentrations of alloying elements in the carbides will approximate the calculated value with the prolonging of holding time of heat treatment [40]. It has been demenstrated by Pigrova [41] that there are experimental illustrations that the dominant element contents in corresponding carbides increase with prolonged annealing time.

Table 5.15 Phase compositions of carbides in annealed H13 die steel

	Crystal type	Cr	Fe	Mn	Mo	W	V	C	N	Σ
M_7C_3	Complex hexagonal crystal	1.495	0.832	0.018	0.098		0.165	0.248		2.856
		$(Cr_{0.5961}Fe_{0.3088}Mn_{0.0068}Mo_{0.0211}V_{0.0672})_7C_3$								
M_6(C, N)	Complex cubic crystal	0.049	0.173	0.010	0.368	0.013	0.028	0.015	0.003	0.659
		$(Cr_{0.1091}Fe_{0.3572}Mn_{0.0205}Mo_{0.4427}W_{0.008}V_{0.0625})_6(C_{0.847}N_{0.153})$								
M(C, N)	Cubic crystal	0.084			0.076	0.004	0.562	0.137	0.028	0.891
		$(Cr_{0.118}Mo_{0.059}W_{0.002}V_{0.821})(C_{0.850}N_{0.150})$								
Σ		1.628	1.005	0.028	0.542	0.017	0.755	0.400	0.031	4.406

Note the carbon contents are calculated values

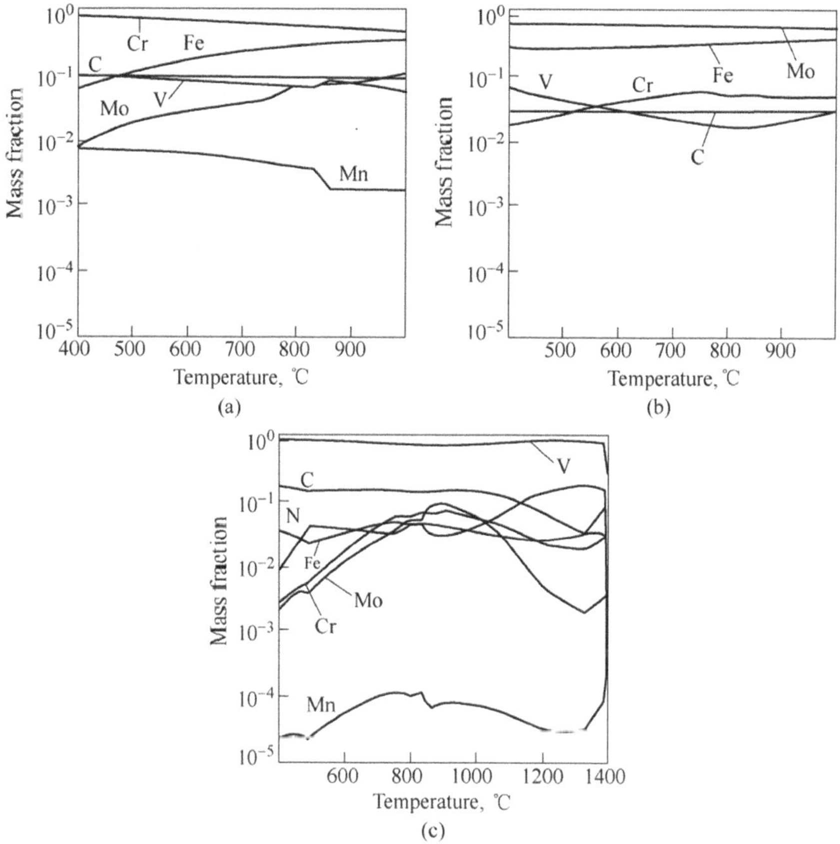

Fig. 5.35 Change of chemical compositions of carbides with temperature in annealed H13 steel: **a** M_7C_3, **b** M_6C, **c** M(C, N)

Table 5.16 Mass Ratios of main elements in carbides in annealed steel

	MC (V/(Cr + Mo + Fe))	M_6C (Mo/Fe)	M_7C_3 (Cr/Fe)
Calculated values	4.67	1.78	1.69
Experimental values	3.51	2.13	1.80

Table 5.17 Mean content and maximum content of alloying elements of matrix (wt%)

No	V	Mo	Cr	V_{max}	Mo_{max}	Cr_{max}
Mg free	1.84	1.79	8.81	8.41	2.34	10.15
Mg content 0.0018%	1.56	1.58	8.02	4.21	2.42	9.15

5.3.2 *Effects of Mg on Carbide Type and Distribution in H13 Die Steel After Annealing*

5.3.2.1 Carbides Type and Particle Size Distribution of H13 Steel After Annealing

The carbides of ESR ingots with Mg content 0, 0.0006, 0.0010, 0.0019 and 0.0032% after annealing were analyzed. The numbers were set as Z1, Z2, Z3, Z4 and Z5, respectively. The XRD results are shown in Fig. 5.36.

As shown in Fig. 5.36, the typical carbides of ESR ingots after annealing are Cr-rich M_7C_3, Mo-rich M_6C and VC, and the amount of M_7C_3 is dominated carbides of all. The maximum peak value of M_7C_3 and M_6C in Mg containing ESR ingots are higher than that of Mg free ESR ingot, but the peak value of MC type carbide is low. The size distribution of carbides in ESR ingots after annealing is shown in Fig. 5.37.

As can be seen in Fig. 5.37, the size of carbide in Mg free annealed H13 die steel is mainly distributed in 1.5–2.5 μm, and the amount of carbides in 0.4–0.6 μm is the most. The size of carbide decreases with the increase of Mg content of steel. The size of carbide is mainly distributed between 0.8 and 1.5 μm when Mg content of steel is 0.0032%, among which the amount of carbide between 0.3 and 0.5 μm is the most. It can be concluded that the growth and coarsening of carbide at high temperature is inhibited with the increasing of Mg content of steel effectively.

5.3.2.2 Morphology of Carbides of H13 Steel After Annealing

The structure of H13 die steel after annealing with different Mg content is shown in Fig. 5.38.

As shown in Fig. 5.38, the carbide distribution of annealed H13 die steel is very uneven and a lot of banded structures along grain boundary can be detected. The carbides in Mg free steel are in the shape of fish bone. However, the carbides in the Mg containing steel are relatively less and dispersed. The large size carbides which were precipitated along the grain boundary of steel, are shown in Figs. 5.39 and 5.40.

The large bulk Mo-rich M_6C and MC carbides can be detected in carbides segregation zone of Mg free H13 steel, which sizes are over 10 μm. VC type carbides are the main large size carbides in Mg containing H13 steel, which size is 5–10 μm. The morphology and quantity of carbide in carbides unsegregation zone are shown in Fig. 5.41.

As can be seen in Fig. 5.41, the size of carbides in carbides unsegregation zone decreases with the increasing of Mg content in steel. The carbide roundness in steel is counted, and the results are shown in Fig. 5.42. As can be seen from Fig. 5.42, the roundness of carbides in H13 steel decreases gradually with the increasing of Mg content of steel. It indicates that Mg has a spheroidization effect on carbides.

After Mg atom is segregated to the grain boundary, the solute atoms with less mismatch are expelled from the grain boundary to the inner lattice or grain boundary.

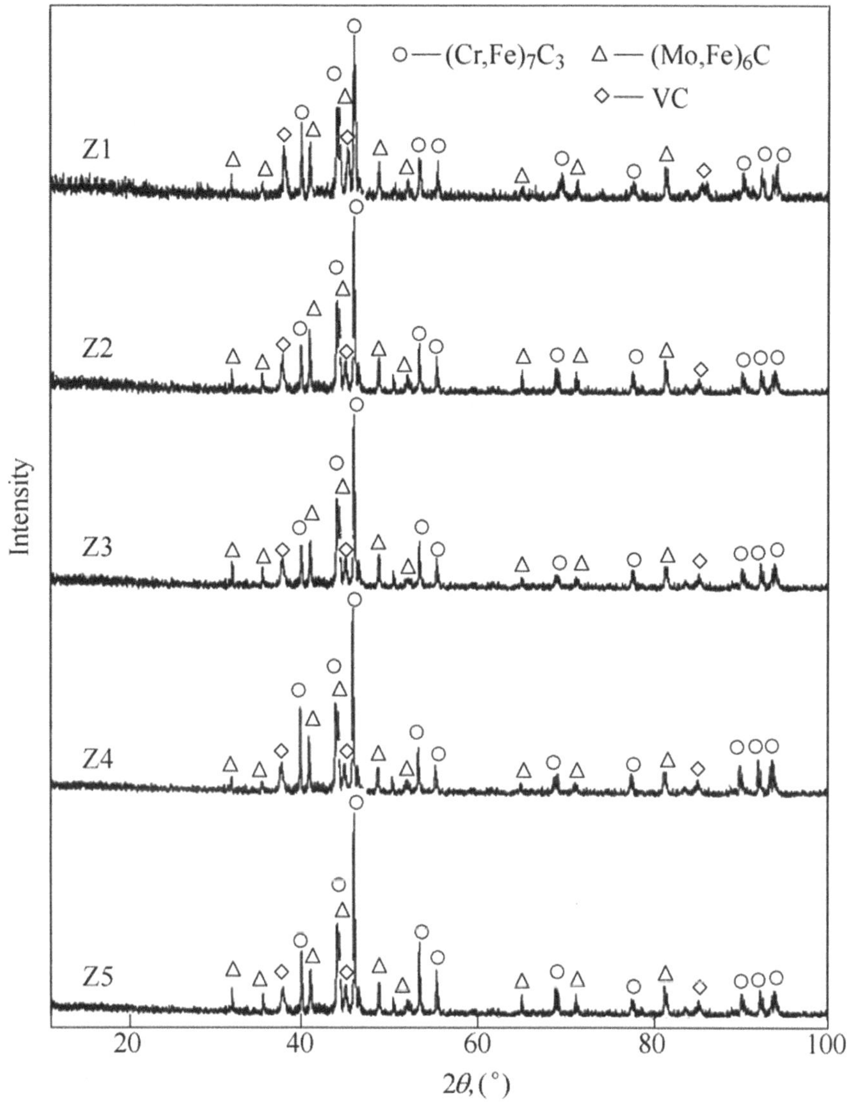

Fig. 5.36 XRD analysis of carbides in ESR ingots after annealing

Mg can also enhance the interstitial atoms to grain boundary deviation. Mg atom enters into the grain boundary phase unit cell, promotes the grain boundary phase spheroidization and reduces the stability. The area micrograph of carbides in ESR ingot with Mg content 0.0018% after annealing was analyzed by EPMA, as shown in Fig. 5.43.

As shown in Fig. 5.43 the primary and secondary carbides are mostly spherical and diffusedly distributed. The primary carbide is shown in Fig. 5.43a with black

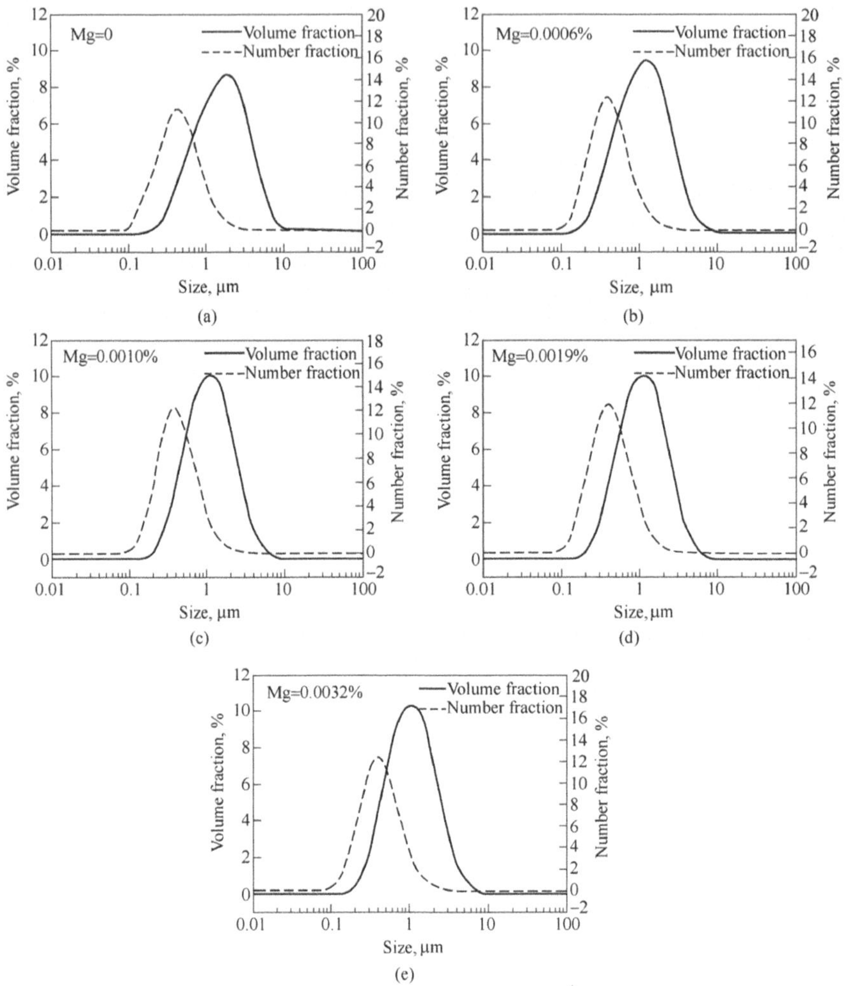

Fig. 5.37 Size distribution of carbides in H13 die steel after annealing

blocks in the center, while the secondary carbide is shown in Fig. 5.43a with smaller white bright particles. $MgO{\cdot}Al_2O_3$ is formed after Mg treatment in H13 steel, which is beneficial to promote heterogeneous nucleation of carbide. The morphology and size of carbide depend on nucleation and growth. The nucleation entropy of carbide is increased after Mg dissolved into carbides, and the microstructure of alloy system with higher nucleation entropy is finer. Mg can dissolve into carbide in the form of inclusions and change the morphology and size of carbide by influencing the nucleation growth process of carbides [42].

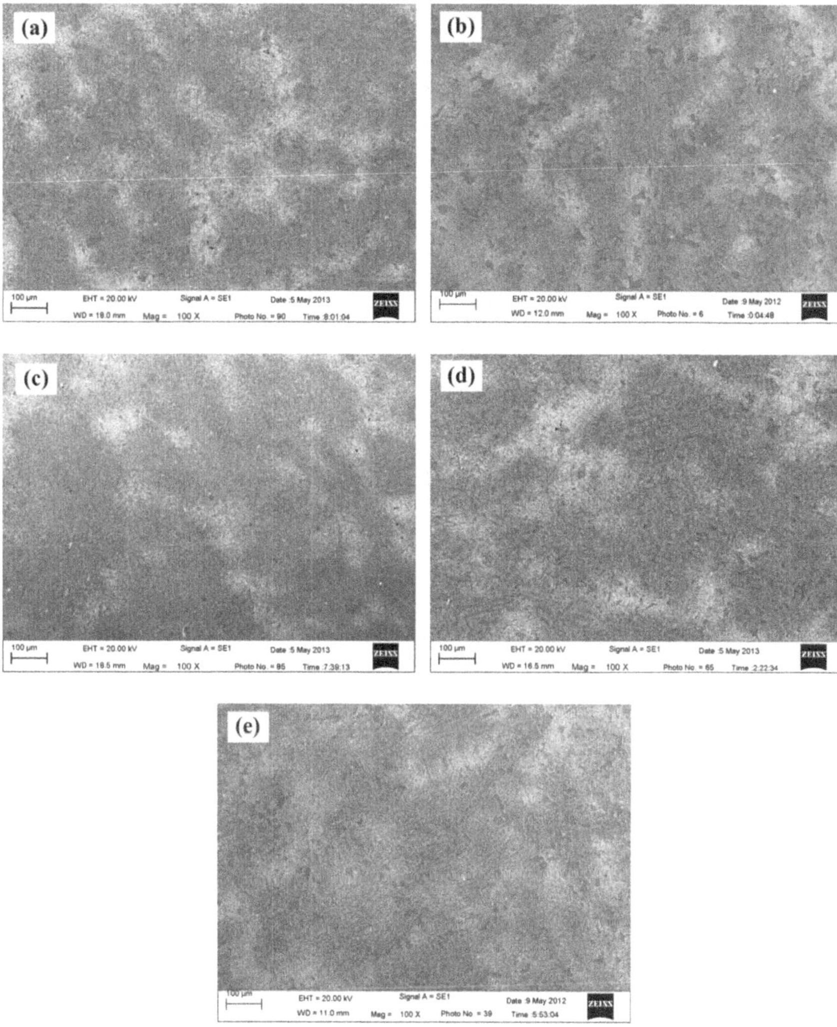

Fig. 5.38 H13 die steel structure after annealing treatment: **a** Mg content; **b** Mg content 0.0006%; **c** Mg content 0.0010%; **d** Mg content 0.0019%; **e** Mg content 0.0032%

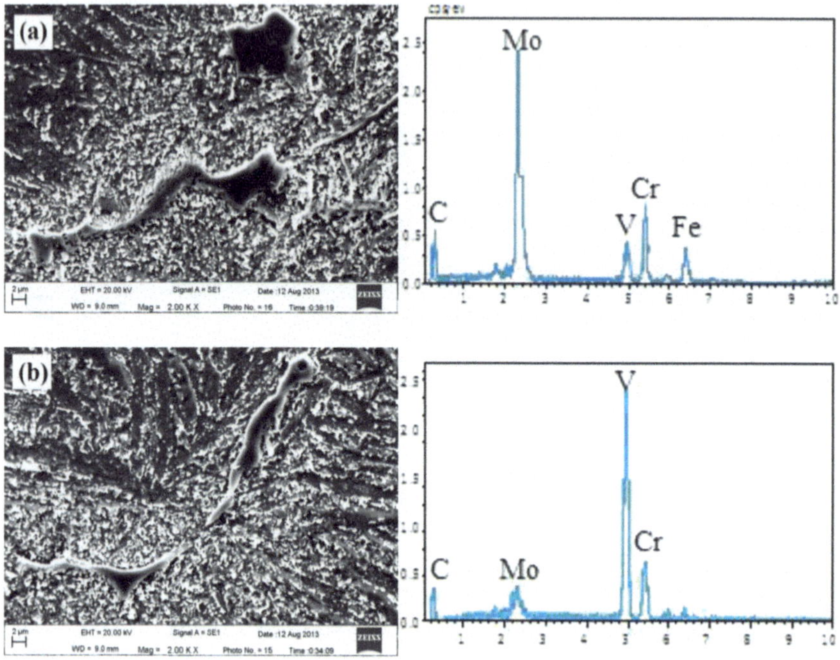

Fig. 5.39 Large size carbides of Mg free H13 steel

5.3.3 Effect of Mg on Carbides Type and Distribution in H13 Die Steel After Quenching and Tempering

5.3.3.1 Carbides Type and Size Distribution in H13 Steel After Quenching and Tempering

The H13 steel with Mg content 0, 0.0006, 0.0010, 0.0019 and 0.0032% (number H1, H2, H3, H4 and H5, respectively) were quenched and tempered, the XRD analyzation results are shown in Fig. 5.44.

As can be seen in Fig. 5.44, VC and $M_{23}C_6$ are the mainly carbides of H13 steel after quenching and tempering. The type of carbides remain unchanged after Mg treatment in H13 steel, while the peak value of $M_{23}C_6$ type carbides increase slightly. The size distribution of carbides is shown in Fig. 5.45.

As can be seen in Fig. 5.45, the size distribution of carbides in Mg free H13 die steel after tempering is in the range of 120–300 nm, the carbides amount in range of 80–160 nm is the most. The size carbides in Mg containing quenched and tempered H13 steel is mainly distributed between 80 and 200 nm. The carbide quantity curve gradually moves to the left with the increasing of Mg content of H13 steel. The size between 50 and 100 nm is the maximum amount of carbides in the Mg content 0.0032% H13 steel.

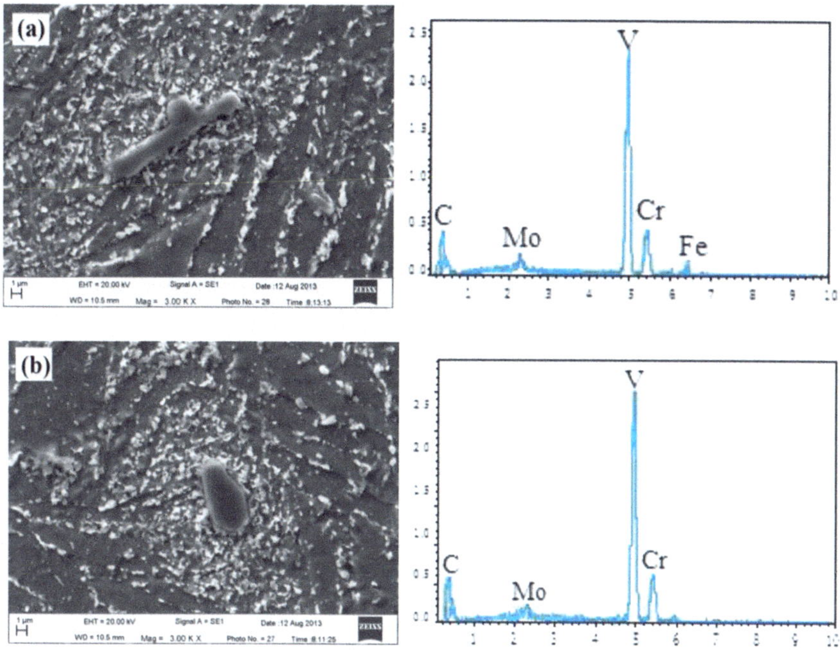

Fig. 5.40 Large size carbides of Mg content 0.0010% H13 steel

5.3.3.2 Morphology of Carbides of H13 Steel After Quenching and Tempering

The carbides of H13 steel after quenching and tempering are tiny, the distributions of carbides were observed by TEM, as shown in Fig. 5.46.

As can be seen in Fig. 5.46, the size of carbides in Mg free H13 steel is large, and the size of carbides in Mg containing H13 steel is relatively small. As shown in Fig. 5.47, the large carbides precipitated in Mg free H13 steel are V-rich carbides, which size are in the range of 10–20 μm with angular in shape and large size. The hardness of this type of carbides is large and hard to dissolve in austenite, which induced the reducing of toughness of H13 steel [43].

The V-rich carbides in size of smaller than 10 μm are precipitated on the periphery of Mg bearing inclusions in Mg containing H13 steel, and the shape is changed from angular irregular shape to spherality, as shown in Fig. 5.48.

Quantitative analysis of Mg, V, Cr and Mo in matrix after quenching and tempering are shown in Fig. 5.49. The Mg free H13 steel is shown in D1, and the Mg 0.0018% content H13 steel is shown in D3, respectively.

According to Fig. 5.49b–d and the mass fraction of V, Cr and Mo precipitation in the matrix during tempering process, the maximum V content of Mg content 0018% H13 steel is 4.0%, while the maximum V content of Mg free H13 steel is 8.0%. The maximum Cr content of Mg content 0018% H13 steel is 9.15%, while the

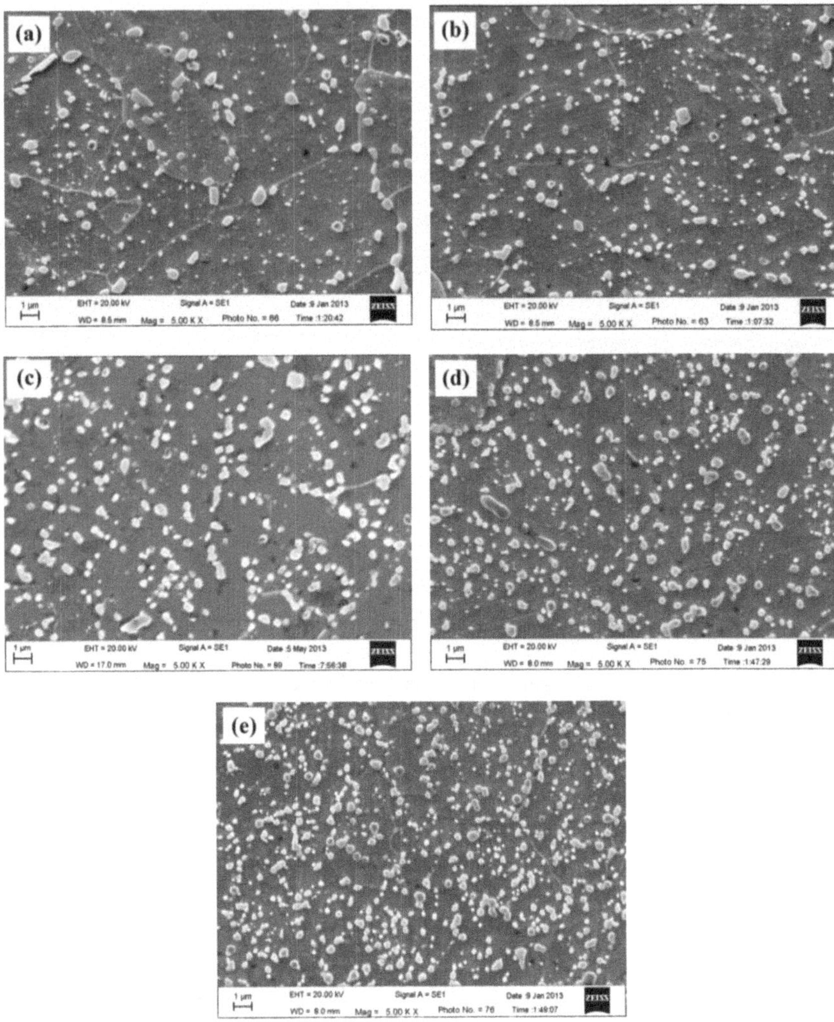

Fig. 5.41 Morphology of carbides in carbides unsegregation zone:**a** Mg content 0; **b** Mg content 0.0006%; **c** Mg content 0.0010%; **d** Mg content 0.0019%; **e** Mg content 0.0032%

maximum Cr content of Mg free H13 steel is 10.15%. It can be concluded that the precipitation of carbides in H13 steel matrix decreases by means of inhibiting the segregation of alloying elements (V, Mo, Cr) in the steel matrix by Mg, meanwhile the mechanical property of H13 steel is improved. The mean content and maximum content of alloying elements of matrix are listed in Table 5.17. It can be seen that the segregation of alloying elements in H13 steel after quenching and tempering treatment is significantly smaller than that of anneal H13 steel, and the segregation

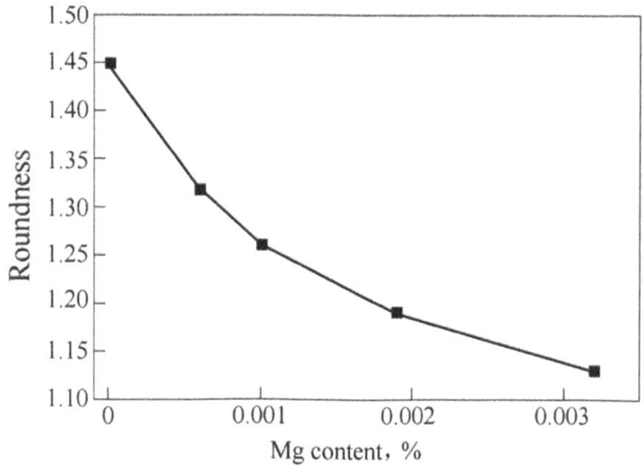

Fig. 5.42 Effect of Mg on carbides roundness in steel

of alloying elements in Mg content 0.0018% steel is significantly smaller than that of the Mg free H13 steel.

5.4 Effect of Magnesium on Mechanical Properties of H13 Die Steel

5.4.1 Effect of Magnesium on Phase Transformation of H13 Die Steel

The CCT curve of H13 die steel after annealing treatment was determined, and its composition is shown in Table 5.18.

The microstructure of H13 steel without magnesium at different cooling rates is shown in Fig. 5.50.

Figure 5.50 shows that when the cooling rate is 100°C/h, the microstructure of the steel is ferrite (F) and pearlite (P). When the cooling rate is 200 and 500 °C/h, the microstructure of the steel is bainite (B) and martensite (M). When the cooling rate is greater than 0.5 °C/s, the microstructure of the steel is martensite. Combined with the expansion curve analysis results, the CCT curve of H13 steel without magnesium is obtained, as shown in Fig. 5.51.

The microstructure of H13 die steel with 0.0032% magnesium at different cooling rates is shown in Fig. 5.52.

Figure 5.52 shows that when the cooling rate is 100 °C/h, the microstructure of the steel is ferrite and pearlite. Compared with H13 steel without magnesium, the ratio of pearlite structure increases. When the cooling rate is 200 °C/h and 500 °C/h, the microstructure of steel is bainite and martensite. Compared with H13 steel without

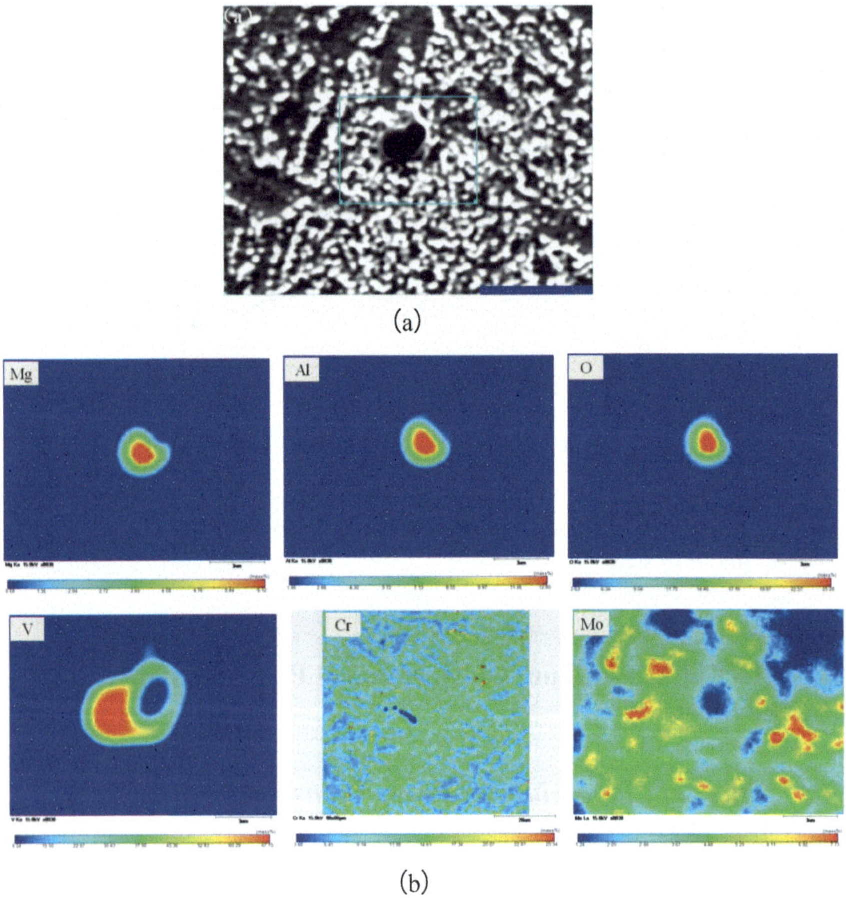

Fig. 5.43 Area scanning of composite growth surface of $MgO{\cdot}Al_2O_3$ and carbide

magnesium, the ratio of bainite reduces. When the cooling rate is greater than 0.5 °C/s, the microstructure of steel is martensite.

Combined with the expansion curve analysis results, the CCT curve of H13 die steel with 0.0032% magnesium is obtained, as shown in Fig. 5.53.

Figures 5.51 and 5.53 show that the CCT curve features of the two steels are slightly different. The effect of magnesium on pearlite and bainite transformation is different. The pearlite transformation zone of H13 steel containing magnesium moves to the left and the incubation period becomes shorter. The reason is that magnesia-alumina spinel inclusions are the effective nucleation cores of ferrite and promote the nucleation and transformation of pearlite, which causes the left moved of C curve and improve the hardenability of steel. While the influence of magnesium on bainite transformation is opposite. The bainite transformation zone of magnesium containing steel becomes smaller and the incubation period becomes longer. The reason is

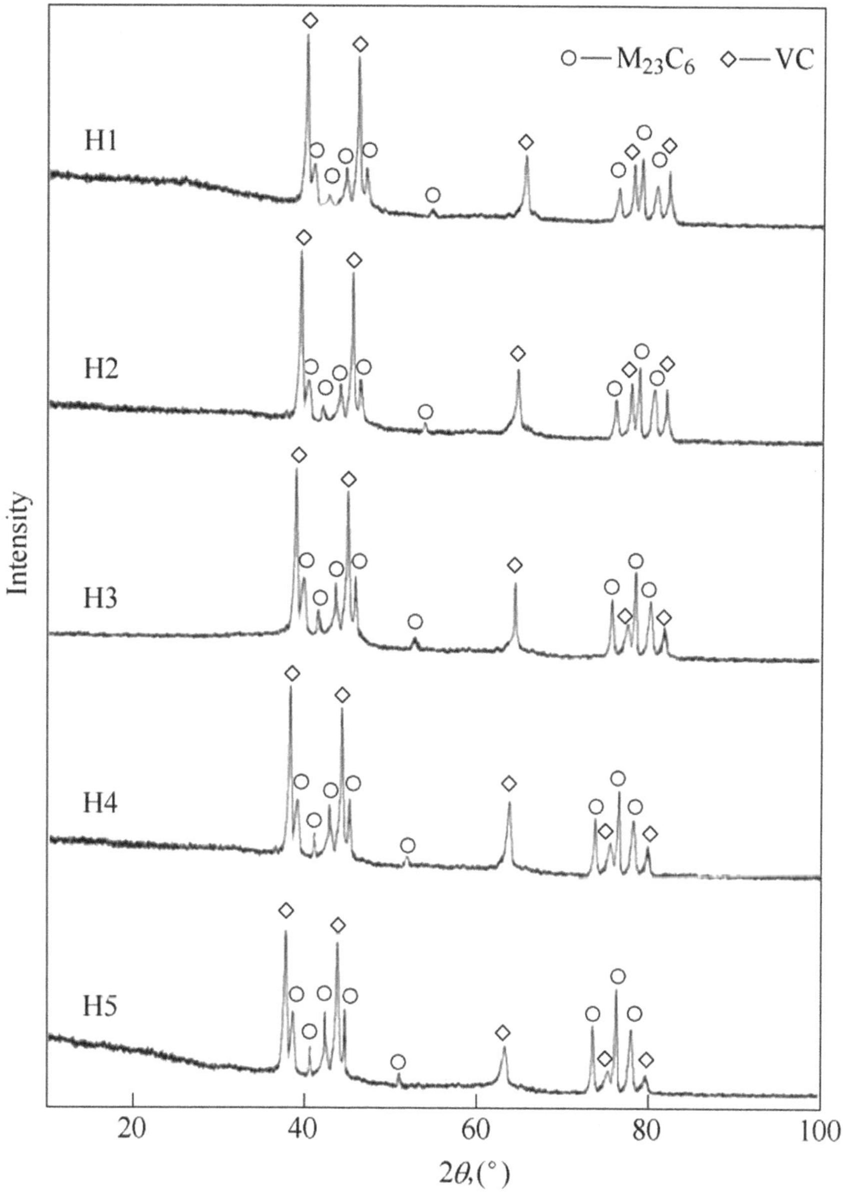

Fig. 5.44 XRD analyzation results of carbides in quenched and tempered steel

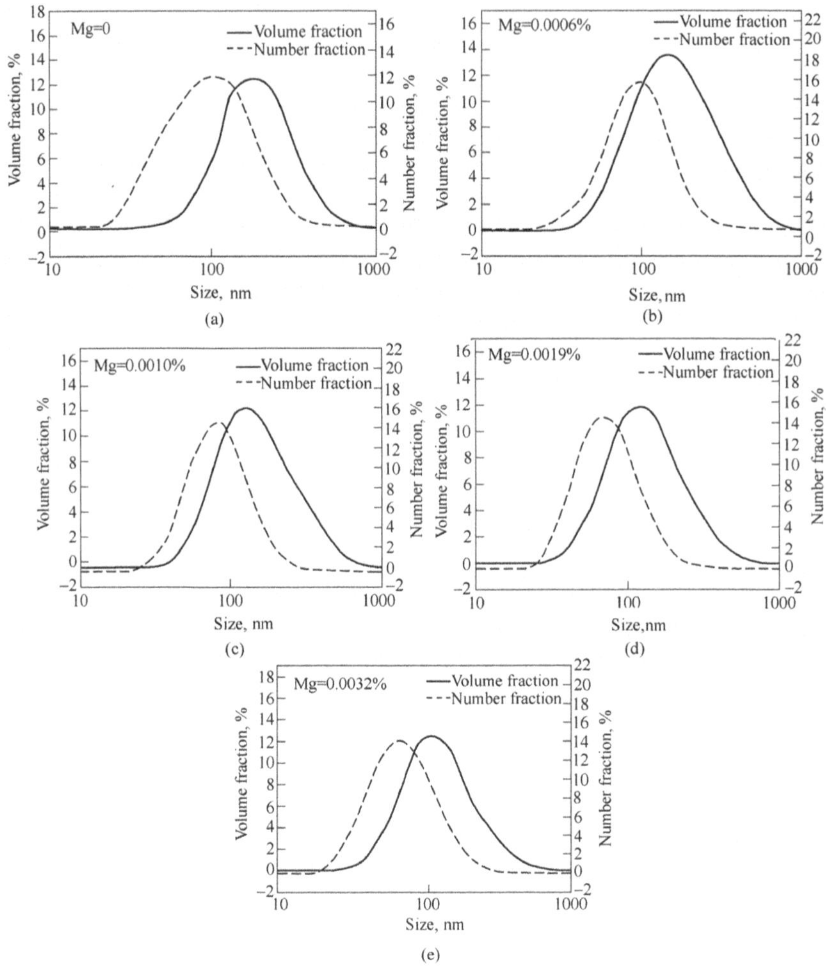

Fig. 5.45 Size distribution of carbides in tempered H13 die steel

that magnesium tends to segregate at austenite grain boundary in low temperature, which reduces the energy of grain boundary and the favorable nucleation position of cementite at grain boundary.

5.4.2 *Effect of Magnesium on Thermal Stability of H13 Die Steel*

The thermal stability of steel is the ability to maintain its internal structure and mechanical properties during heating at a certain temperature. Therefore, thermal

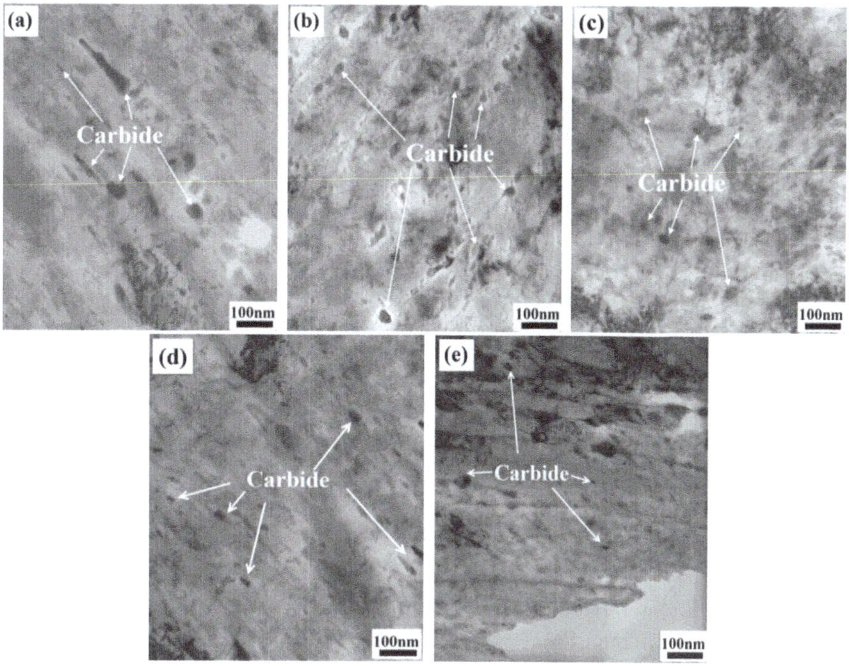

Fig. 5.46 Morphology of carbides by TEM: **a** Mg content 0; **b** Mg content 0.0006%; **c** Mg content 0.0010%; **d** Mg content 0.0019%; **e** Mg content 0.0032%

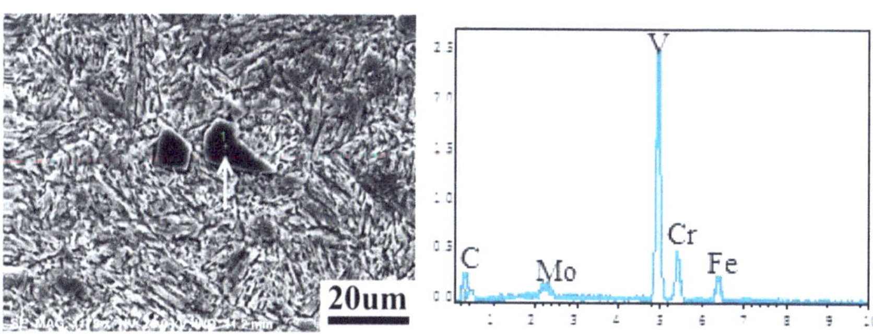

Fig. 5.47 Large size V-rich carbides in Mg free H13 steel

stability is one of the most important properties of hot work die steel. The thermal stability of hot work die steel depends on the decomposition degree of tempered solid solution, precipitation amount of carbide and the aggregation and growth degree of alloy compounds during high temperature holding [44].

The heat stability experiment of steel is to heat and hold the quenched and tempered steel after reaching the predetermined hardness, and measure the hardness under different holding time to investigate the thermal stability of hot working

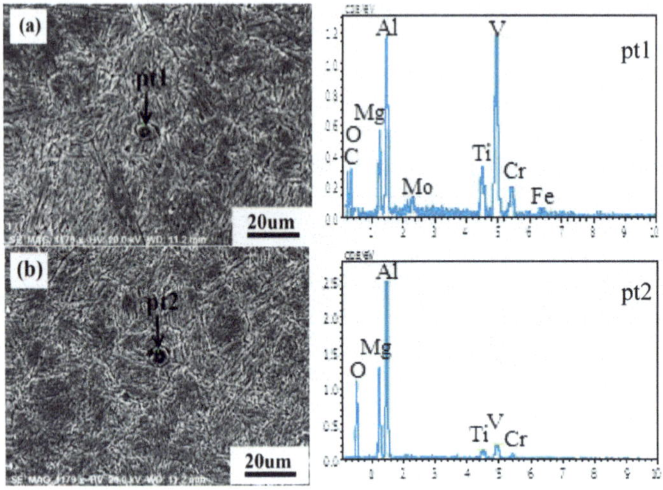

Fig. 5.48 Large size carbides in Mg containing H13 steel

die steel under high temperature for a long time. The H13 die steel without magnesium and 0.0032% magnesium content is held at 580 °C, and the hardness change with time is shown in Fig. 5.54.

Figure 5.54 shows that the hardness of H13 die steel with 0.0032% magnesium content is always higher than that of H13 steel without magnesium when holding at 580 °C. After holding at 580 °C for 20 h, the hardness value of H13 die steel without magnesium has been reduced to HRC 33.7 after long time high temperature tempering, and the hardness value of H13 die steel with 0.0032% magnesium content can still reach HRC 37.6.

The TEM images of the steel before and after thermal stability experiment are shown in Fig. 5.55.

Figure 5.55a, c show that the average width of martensitic lath in steel with 0.0032% magnesium content after quenching is smaller than that of steel without magnesium. The matrix structure of steel with 0.0032% magnesium content after quenching is also smaller, and the hardness is slightly larger than that of steel without magnesium. Figure 5.55b, d show that the strip morphology of martensitic plate in the two steels basically disappears after holding at 580 °C for 20 h. The high-density dislocations almost disappear in the magnesium free steel, while there are a large number of high-density dislocations and large elliptical carbides in the steel with 0.0032% Mg content. The energy spectrum analysis shows that these carbides are vanadium-rich and chromium-depleted carbides with the size larger than 200 nm. The carbides in the two steels coarsen, while the size of carbides in the steel with 0.0032% magnesium content is smaller than that of steel without magnesium, which leads to good tempering softening resistance. This was due to the high segregation tendency of magnesium at tempering temperature. Magnesium segregates to the preferred growth interface of carbides, which hinders the growth of carbides.

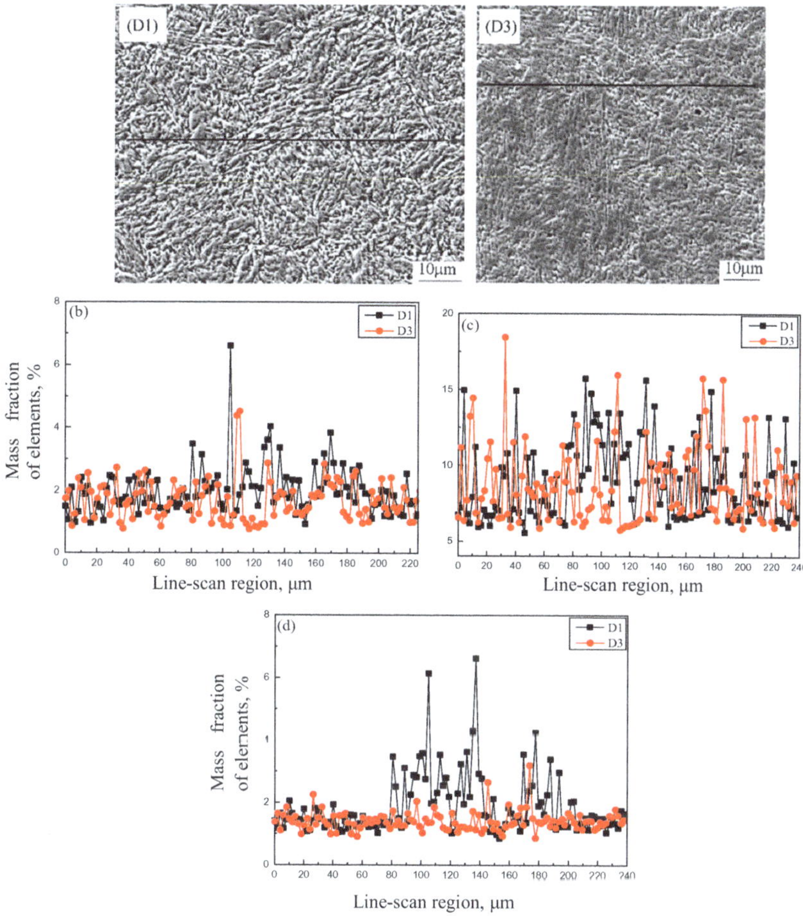

Fig. 5.49 Quantitative analysis of alloy elements in H13 steel after quenching and tempering **b** V; **c** Cr; **d** Mo

Table 5.18 Materials composition (wt%)

No	C	Si	Mn	Cr	Mo	V	Ni	P	S	Mg
1#	0.41	0.99	0.29	5.01	1.22	0.93	0.13	0.023	0.006	0
2#	0.42	0.98	0.28	5.02	1.22	0.92	0.14	0.026	0.007	32×10^{-4}

The hardness of steel with or without magnesium content after holding at different temperatures for 4 h is shown in Table 5.19.

The microstructure transformation of steel is caused by the reduction of solid solution amount of alloy elements such as carbon, chromium, molybdenum and vanadium, as well as the precipitation of carbides, the aggregation and growth of

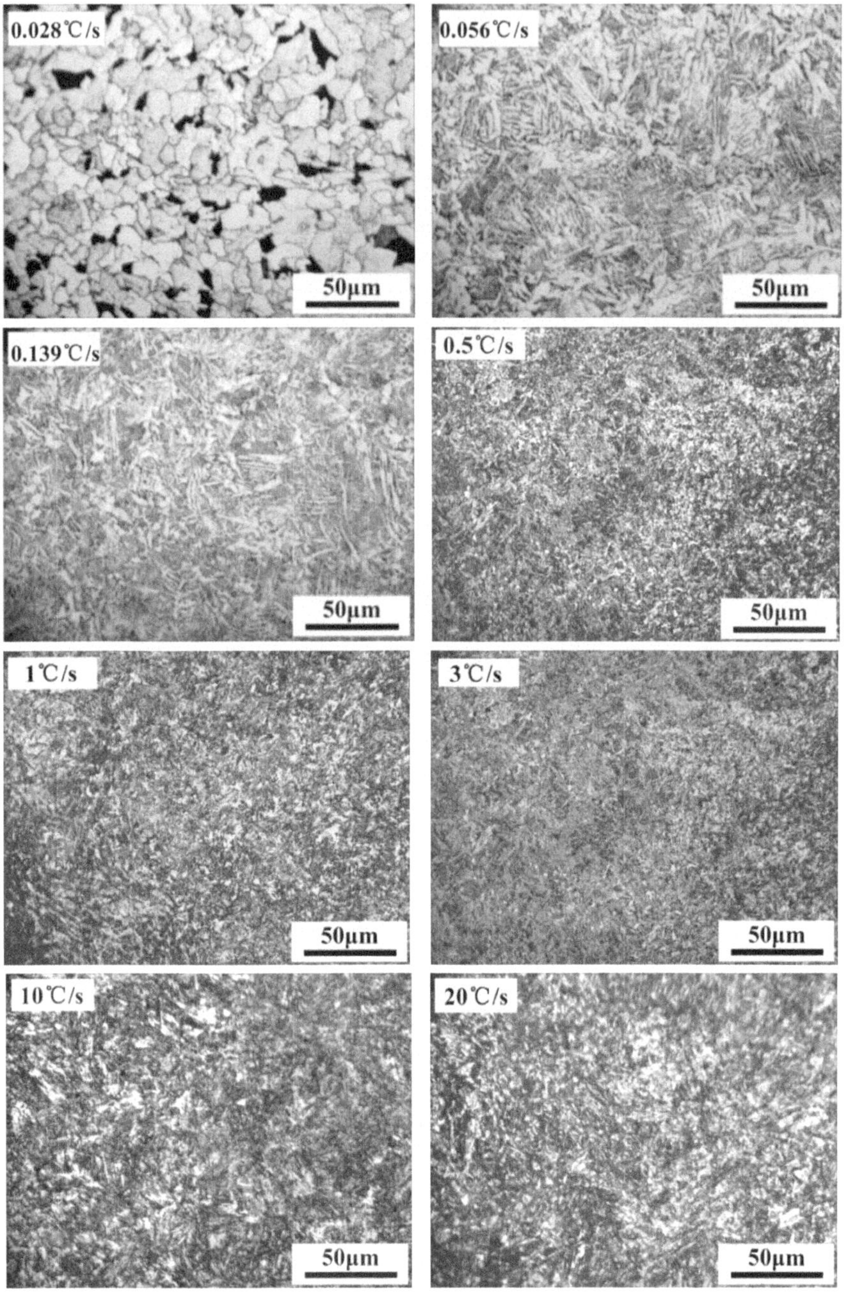

Fig. 5.50 Microstructure of H13 steel without magnesium at different cooling rates

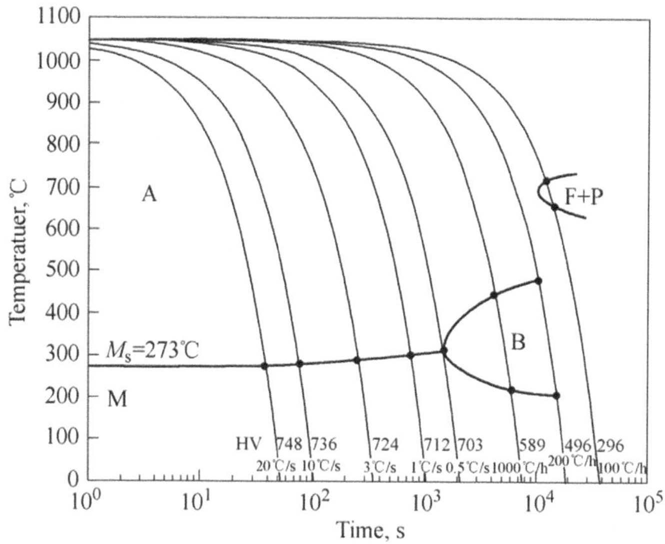

Fig. 5.51 CCT curve of steel without magnesium

dispersed carbides [45]. The process of tissue transformation with diffusion mechanism should satisfy Arrhenius equation model. Li et al. [46] deduced from alloy thermodynamics that the relationship between hardness change and holding time is as follows:

$$\frac{(\Delta HRC)^2}{t} = A \times \exp\left(-\frac{Q}{RT}\right) \tag{5.21}$$

where t is holding time, s; Q is diffusion activation energy, kJ/mol; R is gas constant 8.315 J/(mol·K); T is heating temperature, K.

From Eq. (5.21), it can be concluded that:

$$2\ln(\Delta HRC) = (\ln A + \ln t) - \frac{Q}{RT} \tag{5.22}$$

According to Eq. (5.22), the curve of 2ln(ΔHRC)-Q/RT was obtained, and linear fitting was performed, as shown in Fig. 5.56.

The equation of the straight line is as follows:

$$y_1 = 27.31 - 19352.68x \tag{5.23}$$

$$y_2 = 26.82 - 19495.82x \tag{5.24}$$

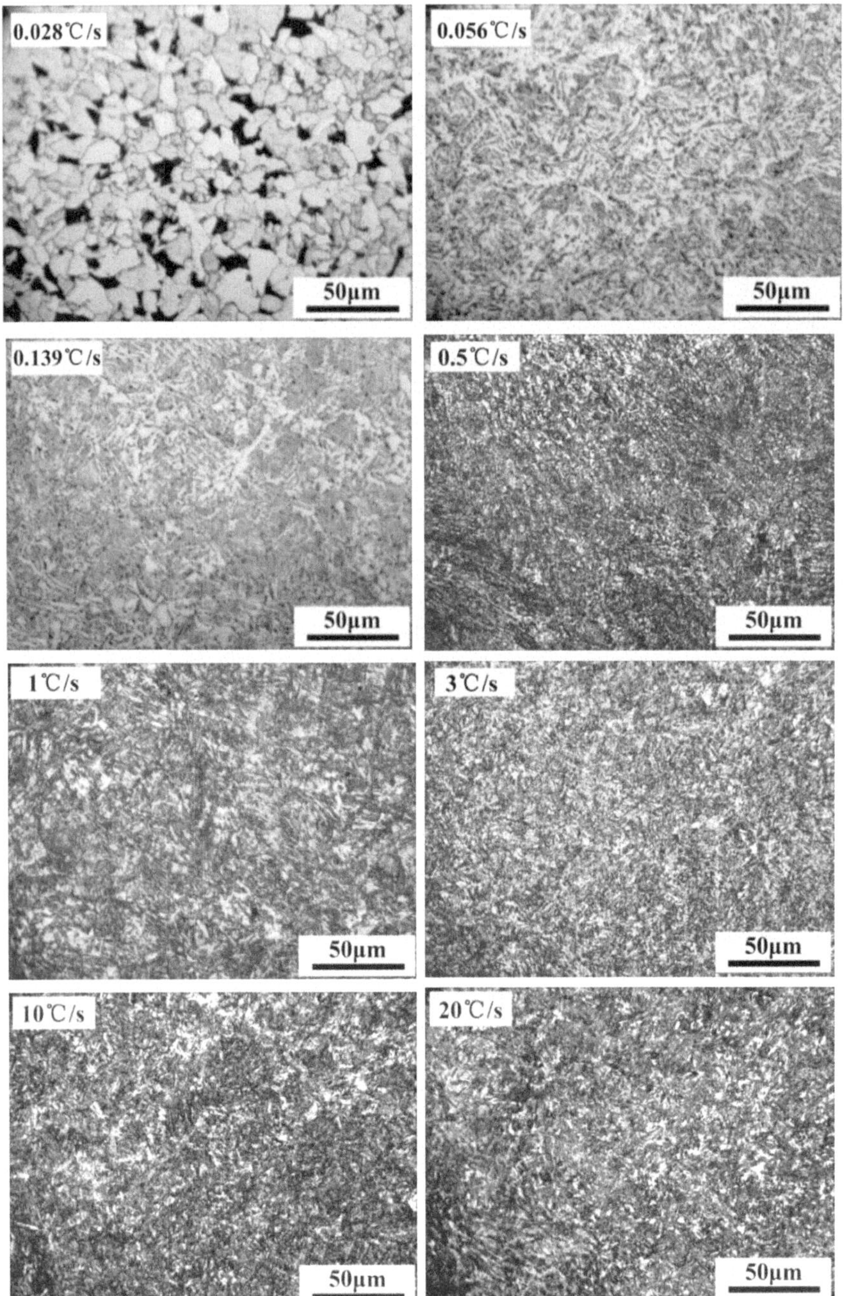

Fig. 5.52 Microstructure of steel with 0.0032% magnesium at different cooling rates

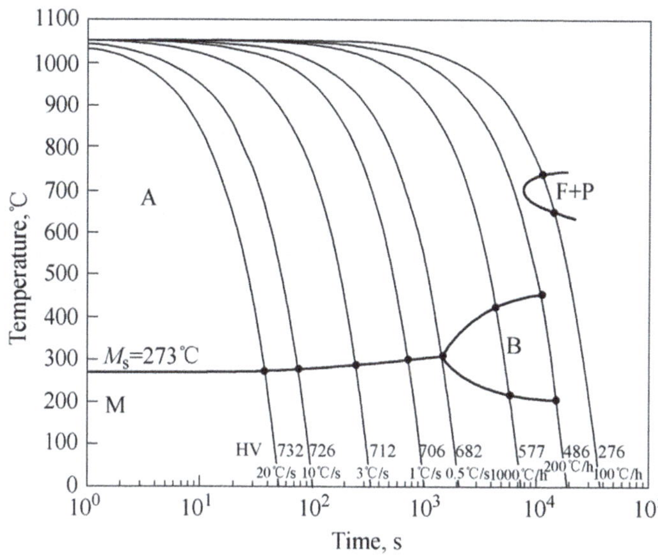

Fig. 5.53 CCT curve of steel with 0.0032% magnesium

Fig. 5.54 Thermal stability curve of H13 steel at 580 °C

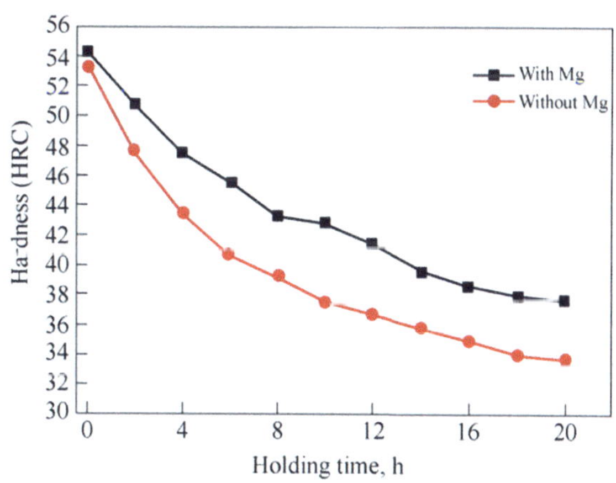

By comparing Eqs. (5.22), (5.23) and (5.24), $A_1 = 50{,}522{,}862$, $A_2 = 30{,}909{,}883$, $Q_1/R = 19{,}352.68$, $Q_2/R = 19{,}495.82$ are gotten. The relationship among hardness change, holding time and heating temperature of the two steels is as follows:

$$\frac{(\Delta HRC_1)^2}{t} = 50522862 \times \exp\left(-\frac{19352.68}{T}\right) \tag{5.25}$$

$$\frac{(\Delta HRC_2)^2}{t} = 30909883 \times \exp\left(-\frac{19495.82}{T}\right) \tag{5.26}$$

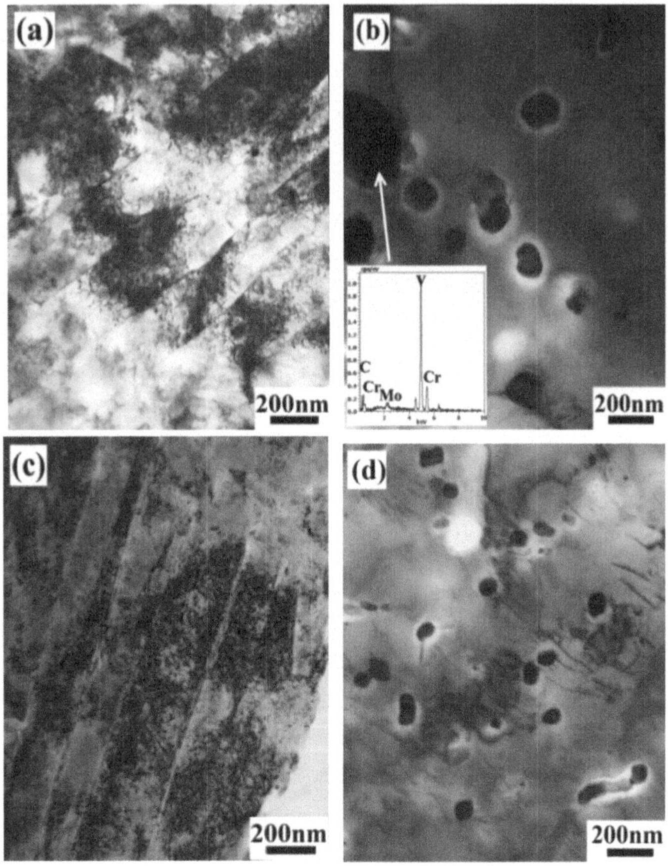

Fig. 5.55 TEM images of steel before and after thermal stability experiment: **a** Microstructure of steel without magnesium after quenching, **b** Morphology of carbide after thermal stability experiment of steel without magnesium, **c** Microstructure of steel with 0.0032% magnesium content after quenching; **d** Morphology of carbide after thermal stability experiment of steel with 0.0032% magnesium content

Table 5.19 Hardness of steel after holding at different temperatures for 4 h

Temperature	As-quenched	580 °C	600 °C	620 °C	640 °C	660 °C
Steel without magnesium	53.2	43.5	39.8	35.8	32.3	29.7
Steel with 0.0032% magnesium content	54.4	47.5	44.6	41.7	38.9	35.8

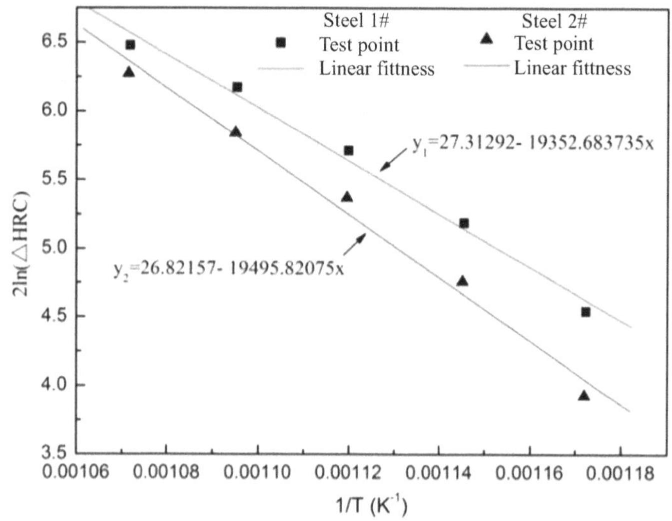

Fig. 5.56 Curve of 2ln(ΔHRC)-1/*T*

According to Eq. (5.26), it can be calculated that the maximum heating temperature when the hardness of steel without magnesium decreases to the failure hardness (HRC35) of hot working die steel is 656.7 and 626.7 °C respectively after holding for 2 and 4 h. The corresponding maximum heating temperature for the hardness of steel with 0.0032% magnesium content is 692.2 and 660.2 °C respectively. Under the same holding time, the maximum heating temperature of magnesium containing steel is 33.6 and 33.5 °C higher than that of steel without magnesium, respectively. The "maximum heating temperature" can represent the maximum working temperature that H13 die steel can adapt to in normal working for 2 or 4 h. In the meantime, "maximum heating temperature" also intuitively presents the thermal stability capacity of each steel [46]. Therefore, the heat stability of H13 die steel containing magnesium is better than that of H13 steel without magnesium.

5.4.3 *Effect of Magnesium on Mechanical Properties of H13 Die Steel After Annealing*

The mechanical properties of H13 steel with different magnesium content after annealing are shown in Table 5.20. H13 steel with 0, 0.0006, 0.0010, 0.0019 and 0.0032% magnesium contents are Z1, Z2, Z3, Z4 and Z5 respectively.

Table 5.20 shows that the tensile strength and hardness of the die steel without magnesium are higher than those of the die steel containing magnesium, while the plasticity and impact toughness are lower than those of the die steel containing

Table 5.20 Effect of magnesium on mechanical properties of H13 steel after annealing

No	Elongation after fracture, %	Tensile strength, MPa	Hardness (HRC)	Impact energy, J
Z1	8.0	903.8	27.4	4.7
Z2	11.2	842.9	21.1	4.6
Z3	10.8	842.3	23.1	5.7
Z4	15.3	816.4	20.5	6.8
Z5	18.6	764.6	18.8	8.5

magnesium. The microstructure of H13 die steel after annealing treatment is shown in Fig. 5.57.

Figure 5.57 shows that the microstructure of the steel is a mixture of lamellar pearlite and granular pearlite. There are more granular pearlites in the magnesium treated ESR ingot. Granular pearlite is transformed from flake pearlite, and its surface energy is smaller than that of spherical pearlite [47]. The strength and hardness of granular pearlite are lower than that of flake pearlite, so the strength and hardness of H13 die steel without magnesium are higher, while the plasticity ratio is lower [48]. The tensile fracture morphology of H13 die steel after annealing is shown in Fig. 5.58.

Figure 5.58 shows that the fracture zone of magnesium free steel is large, accounting for about half of the whole fracture area. There are large carbides in the tearing area as the fracture source, and no dimples are found, which is the cleavage fracture morphology [49]. There is no large tearing zone in the fracture surface of the steel with 0.0006% magnesium content, which is quasi cleavage fracture. The fracture surface of crack propagation zone is not smooth, which is multi-source cracking. There are micro-pores and dimples in the tear edge. The fracture surface of the steel with 0.0010% magnesium content is relatively flat with small dimples distributed on the tear edge, which is quasi cleavage fracture. The fracture surface of the steel with 0.0019% magnesium content and 0.0032% magnesium content is cup-cone shaped, and the tear zone is dimple, which is ductile fracture. Compared with the steel with 0.0019% magnesium content, the dimple size of the steel with 0.0032% magnesium content is larger. The number and size of dimples reflect the toughness and brittleness of the material. The large size, depth and quantity of the dimples indicate that the crack has undergone large local plastic deformation and makes the macro plastic deformation of the material before fracture also higher, and the overall performance of the material is better in terms of plasticity and toughness [50]. It is found that there are clusters of VN carbide and Al_2O_3 inclusions in the fracture of magnesium free steel, with the size more than 10 μm, as shown in Fig. 5.59. These inclusions are harmful to the steel as the crack source. However, there are VC carbides and $MgO{\cdot}Al_2O_3$ inclusions in the fracture surface of magnesium bearing steel, which are smaller than 5 μm, as shown in Fig. 5.60. Therefore, with the increase of magnesium content in steel, the toughness of steel gradually increases, while the hardness gradually decreases.

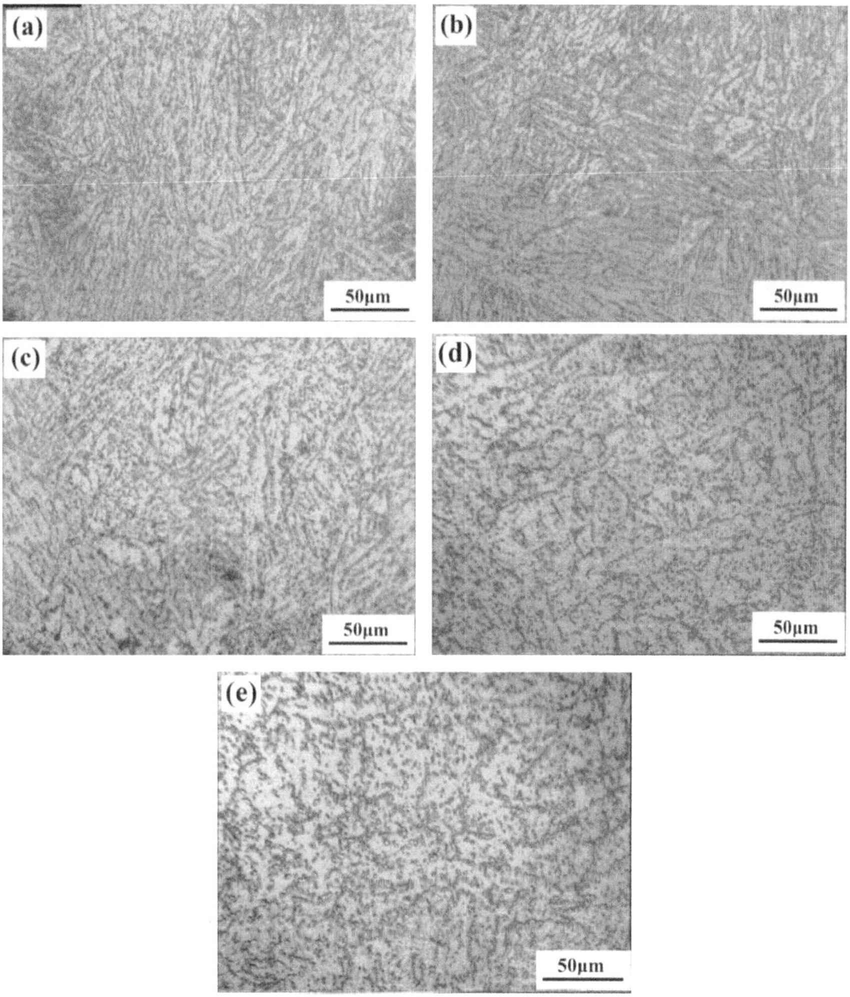

Fig. 5.57 Microstructure of ESR ingot with different magnesium content after annealing. **a** 0, **b** 0.0006%, **c** 0.0010%, **d** 0.0019%, **e** 0.0032%

5.4.4 Effect of Magnesium on Properties of H13 Die Steel After Quenching and Tempering

The mechanical properties of H13 steel (No. L1, L2, L3, L4 and L5) with magnesium content of 0, 0.0006, 0.0010, 0.0019 and 0.0032% are shown in Table 5.21.

Table 5.21 shows that the tensile strength and hardness of H13 die steel gradually increase with the increase of magnesium content in the steel. When the content of magnesium in steel is 0.0006 and 0.0010%, the elongation after fracture decreases. When the content of magnesium in steel is more than 0.0010%, the elongation after

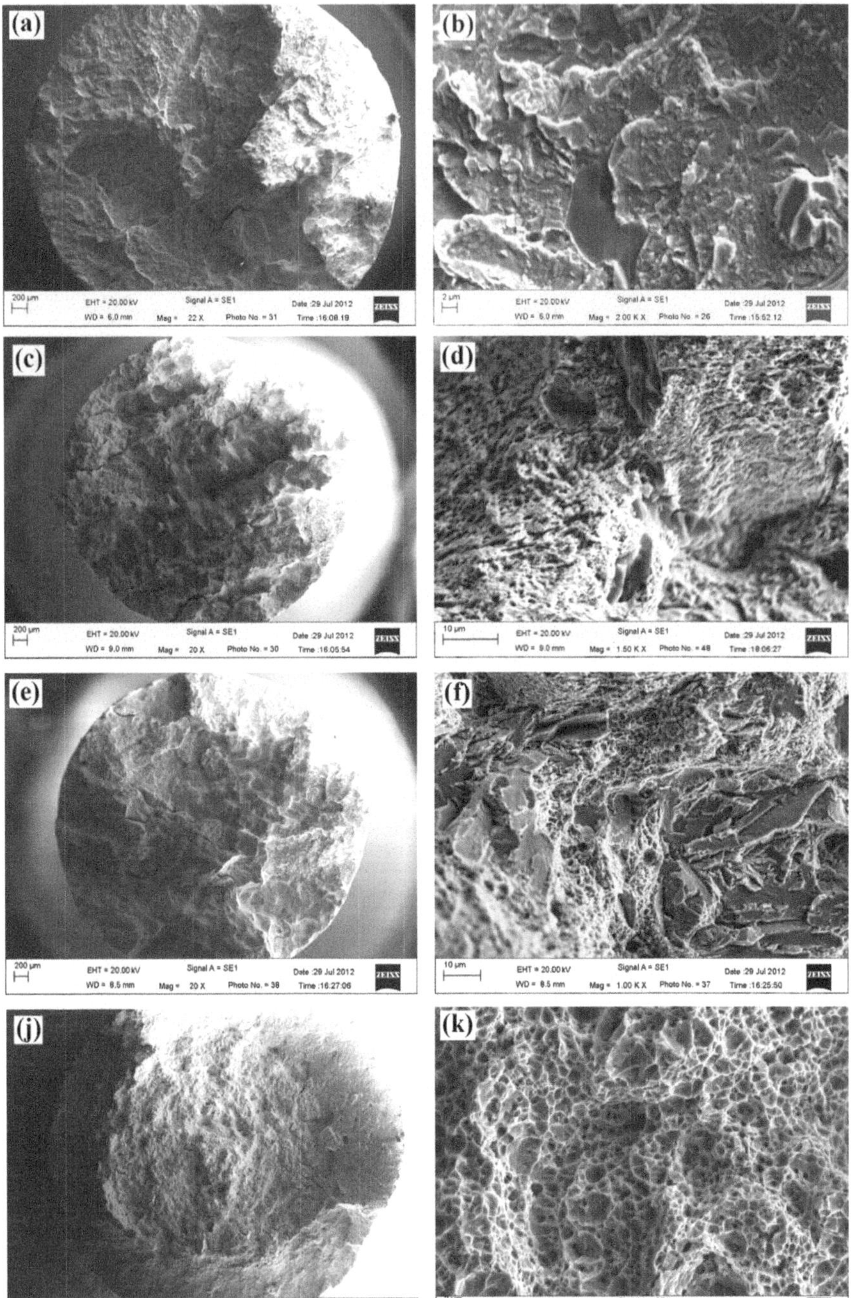

Fig. 5.58 Tensile fracture morphology of steel with different magnesium content. **a**, **c**, **e**, **j**, **m** are macro fracture morphology of steel with 0, 0.0006, 0.0010, 0.0019 and 0.0032% magnesium content, respectively. **b**, **d**, **f**, **k** and **n** are tear zone morphology of steel with 0, 0.0006, 0.0010, 0.0019 and 0.0032% magnesium content, respectively

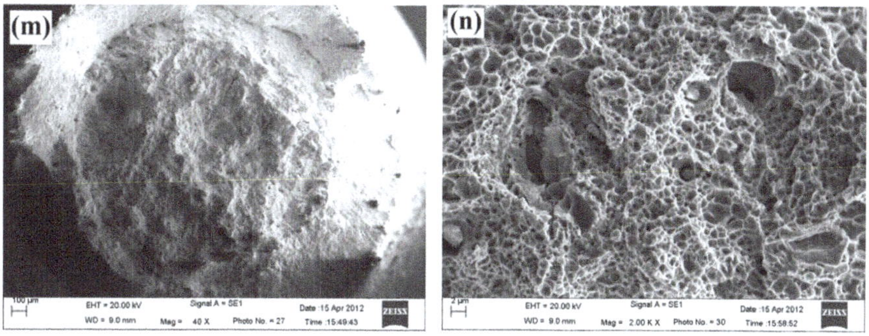

Fig. 5.58 (continued)

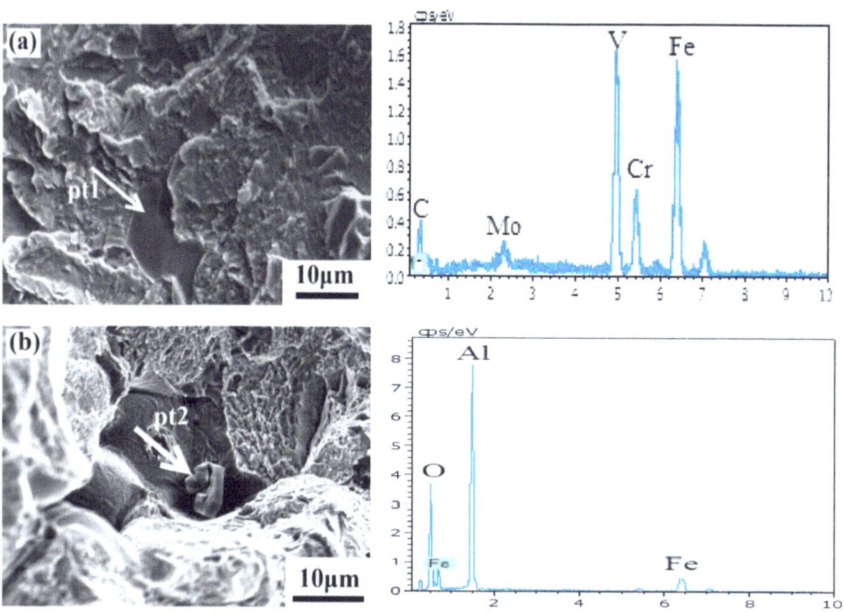

Fig. 5.59 Carbides and inclusions in fracture of the steel without magnesium

fracture increases with the increase of magnesium content. The impact energy fluctuation is similar to the elongation after fracture, which decreases first and then increases with the increase of magnesium content. The microstructure after quenching and tempering is shown in Fig. 5.61.

Figure 5.61 shows that the microstructure of the steel without magnesium is tempered sorbite structure, while the microstructure of magnesium containing steel is mixed structure of tempered sorbite and tempered troostite. During tempering, cementite dissolves into α-Fe again and forms a mixture of fine acicular ferrite and fine cementite. At this time, the microstructure is tempered troostite. With the

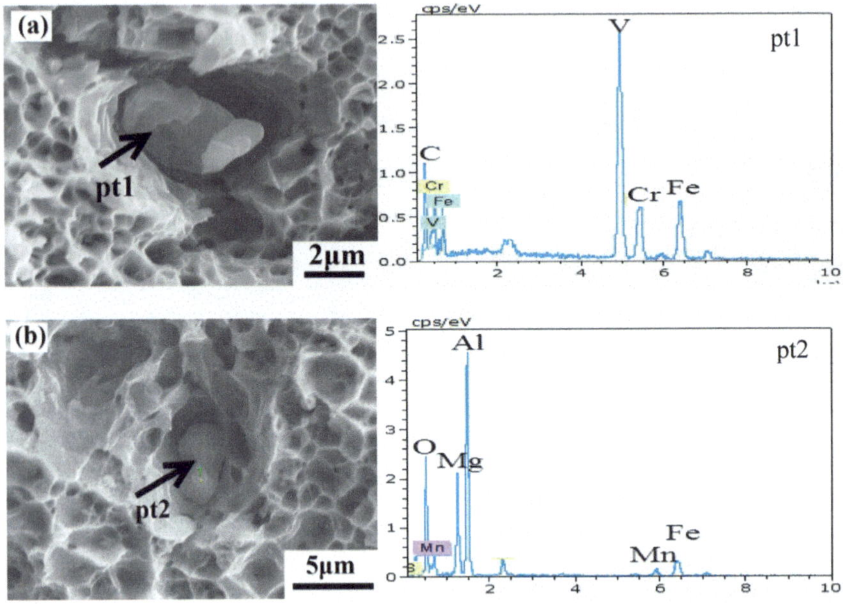

Fig. 5.60 Carbides and inclusions in fracture of the steel with 0.0032% magnesium content

Table 5.21 Effect of magnesium on mechanical properties after quenching and tempering

Number	Elongation after fracture, %	Tensile strength, MPa	Hardness (HRC)	Impact energy, J
L1	7.7	1617.6	45.7	11.9
L2	6.3	1690.2	47.3	9.5
L3	3.6	1702.9	47.8	9.1
L4	6.7	1824.4	49.6	10.7
L5	8.5	1967.3	51.4	13.4

increase of temperature, acicular ferrite recrystallizes into tempered sorbite with equiaxed ferrite and granular cementite. The strength and hardness of tempered troostite matrix are higher than that of tempered sorbite, so the strength and hardness of magnesium containing H13 die steel are higher than those of H13 die steel without magnesium [51].

The tensile fracture morphology of H13 steel after quenching and tempering is shown in Fig. 5.62.

Figure 5.62 shows that the tensile fracture of H13 die steel after tempering treatment is typical cup-cone shaped. Figure 5.62a shows that the area of the fracture zone of steel without magnesium accounts for about half of the whole fracture area. The tearing zone is transgranular cleavage fracture, with small shallow dimples on the tear edge, as shown in Fig. 5.62b. The fracture tearing area of the steel with 0.0006%

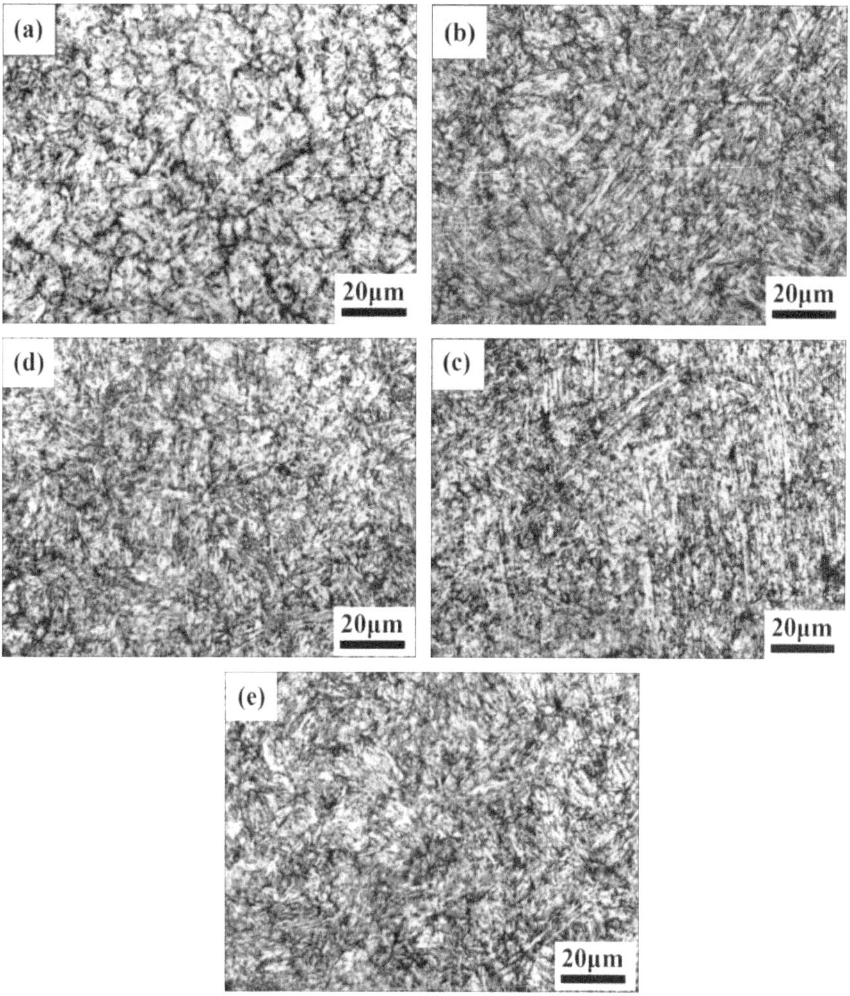

Fig. 5.61 Microstructure after quenching and tempering of the steel with different magnesium contents in mass fraction. **a** 0, **b** 0.0006%, **c** 0.0010%, **d** 0.0019%, **e** 0.0032%

magnesium content accounts for about three quarters of the whole fracture area with intergranular secondary cracks, as shown in Fig. 5.62c. There are obvious radial edges around the central fracture source, and the radiation area is obviously larger than that of steel without magnesium, as shown in Fig. 5.62d, and therefore the tensile strength of the steel increases and the toughness decreases. The fracture surface of the steel with 0.0010% magnesium content is smooth without large tearing ridge or pit, indicating that the fracture surface is brittle fracture. The crack propagation zone accounts for about 3/5 of the whole area, and the shear lip area is the smallest, as shown in Fig. 5.62e, and therefore the plasticity and toughness of steel with 0.0010%

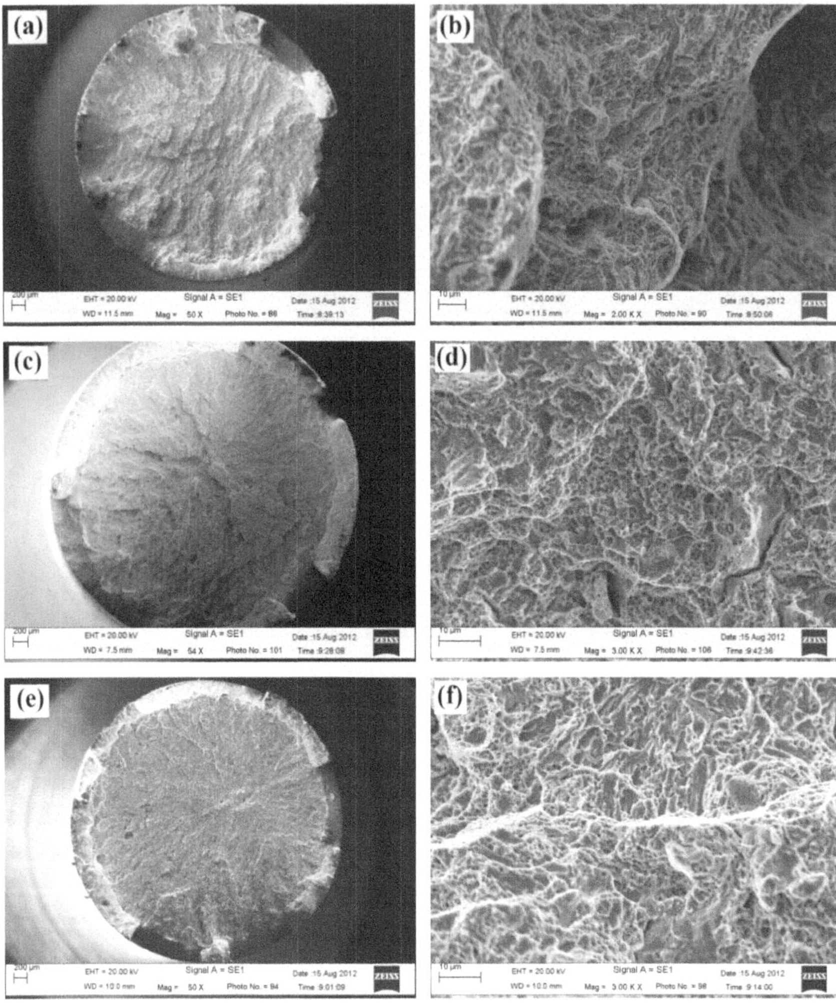

Fig. 5.62 Tensile fracture morphology of steel with different magnesium contents in mass fraction. **a**, **c**, **e**, **j**, **m** are macro fracture morphology of steel with 0, 0.0006, 0.0010, 0.0019 and 0.0032% magnesium content respectively. **b**, **d**, **f**, **k** and **n** are tear zone morphology of steel with 0, 0.0006, 0.0010, 0.0019 and 0.0032% magnesium content respectively

magnesium content is the worst. The results show that the crack propagation zone of the steel with 0.0019% magnesium content accounts for about 4/5 of the whole area, and the strength increases. The shear lip area is similar to that of the steel with 0.0006% magnesium content. The whole fracture surface of the steel with 0.0032% magnesium content is very smooth, and the central instantaneous fracture area is also quite flat. There are few tear edges on the fracture surface while deep dimples appear around the tear edges, and the shear lip area is widened, as shown in Fig. 5.62n.

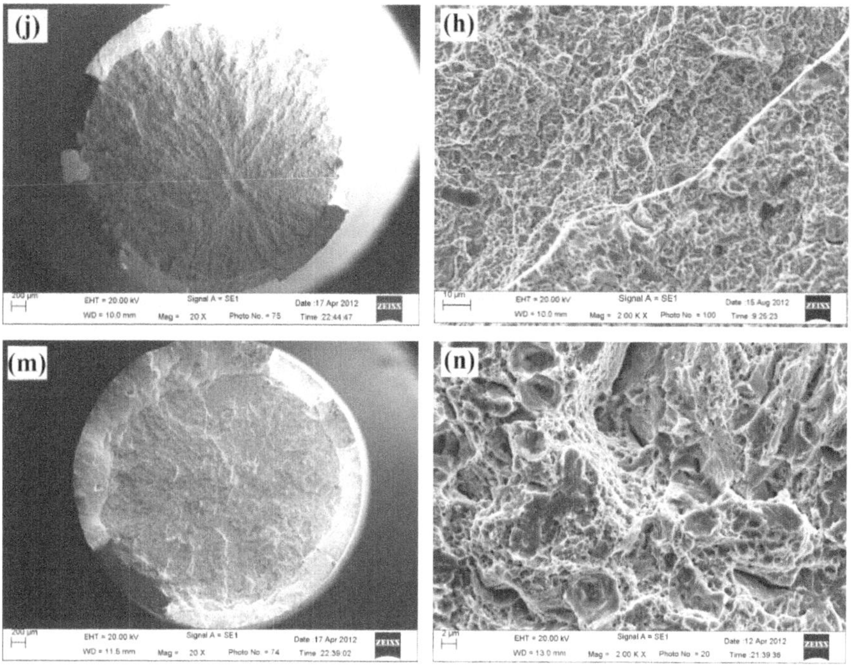

Fig. 5.62 (continued)

Therefore, the strength and toughness of the steel with 0.0032% magnesium content is the best.

The SEM results show that there are mainly irregular Al_2O_3–SiO_2 composite inclusions in the fracture of the steel without magnesium. As shown in Fig. 5.63a, the rhombic AlN is about 10 μm in size. These two inclusions mostly exist around the tearing ridge and the tear zone near the surface, which are prone to cause cracks in the steel. The size of subglobose Al_2O_3 is about 5 μm, as shown in Fig. 5.63c.

There are few inclusions in the fracture surface of magnesium bearing steel, which are mainly $MgO{\cdot}Al_2O_3$ inclusions (shown in Fig. 5.64a) and MgO–Al_2O_3–SiO_2–CaO inclusions (shown in Fig. 5.64a, c). The size of inclusions is less than 5 μm and the shape is spherical or subglobose.

There are many massive carbides embedded in the tearing ridge of the fracture surface near the tearing zone. The morphology is shown in Fig. 5.65.

Figure 5.65 shows that the prismatic carbides on the fracture tearing ridge are mainly composed of high vanadium low chromium and molybdenum carbides. The melting point and hardness of these carbides are high, and these carbides distribute near the grain boundary, which reduces the strength and adhesion of grain boundary, promotes the generation of secondary cracks at grain boundary and finally becomes crack sources. The carbides in the fracture surface of the steel without magnesium is clustered with size in the range of 10–20 μm, as shown in Fig. 5.65a. However,

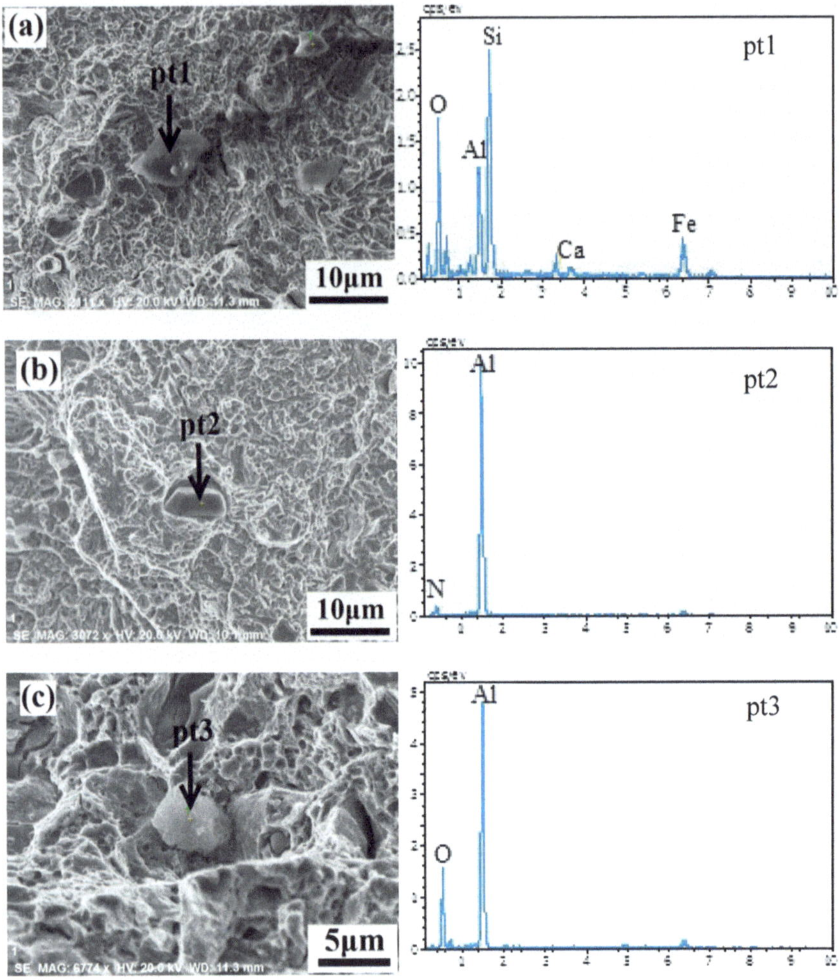

Fig. 5.63 Inclusions in fracture of H13 die steel without magnesium

the distribution of carbides in the steel containing magnesium is relatively uniform and most of the carbides are spherical or subglobose with size less than 10 μm.

The influence of carbide morphology and size on fracture behavior can be explained by McClintock model [52]. The shear fracture diagram is shown in Fig. 5.66.

According to McClintock fracture theory, when the following conditions are satisfied, the material will fracture in time.

$$\frac{1}{\sigma}\left|\frac{d\sigma}{d\varepsilon}\right| < \sqrt{\frac{3}{8}}F_b^2\left(\frac{2b}{l_b}\right)^2\sqrt{\left(\frac{b}{a}\right)^2+1} \tag{5.27}$$

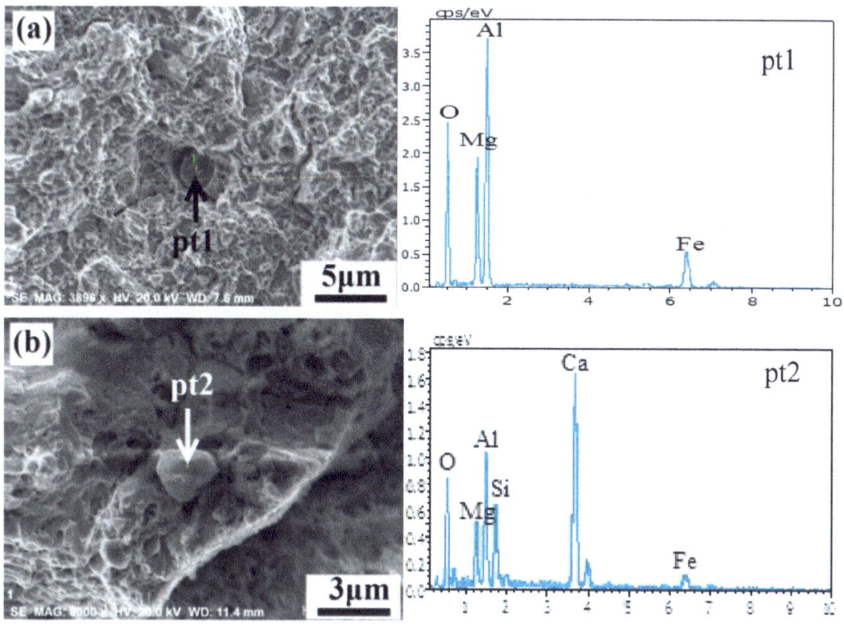

Fig. 5.64 Inclusions in fracture of H13 die steel containing magnesium

where σ is the true stress, ε is the true strain, a and b are the long and short half axis of the elliptical hole respectively, l_b is the spacing of the hole on the b axis; F_b is the growth factor of the hole in the b direction.

The hole spacing l_b in b direction can be expressed as follows.

$$l_b = \frac{4b}{3f}(1 - f) \tag{5.28}$$

where f is the volume fraction of carbide.

The b/a is approximately equal to the morphology ratio of carbides $1/w$. Equation (5.29) can be obtained by substituting l_{b}.

$$\frac{1}{\sigma}\left|\frac{d\sigma}{d\varepsilon}\right| < kF_b^2\left(\frac{2f}{1-f}\right)^2\sqrt{\left(\frac{1}{w}\right)^2 + 1} \tag{5.29}$$

where k is a constant, d_σ/d_ε is considered to be independent of material conditions, and f is a constant under certain alloy composition.

Equation (5.29) shows that the decrease of morphology ratio leads to the decrease of the right side of the equation, which makes the tearing propagation of shear band more difficult and increases the impact fracture energy at the same volume fraction.

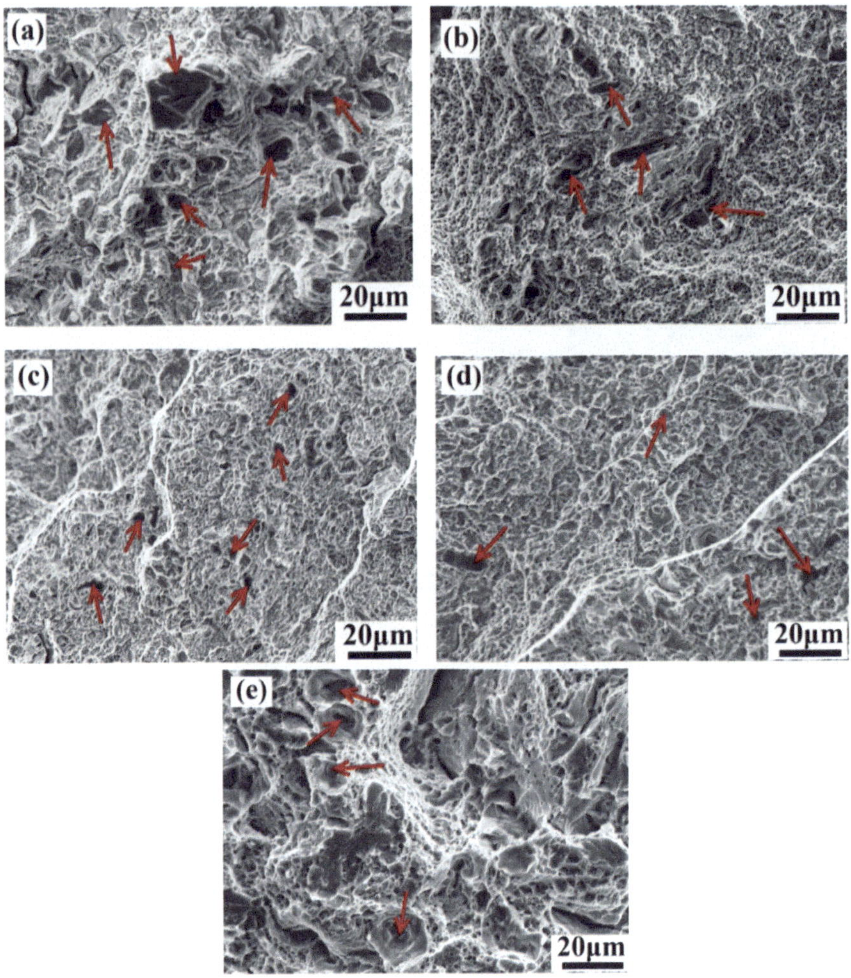

Fig. 5.65 Fracture carbides of the steel with different magnesium contents in mass fraction. **a** 0, **b** 0.0006%, **c** 0.0010%, **d** 0.0019%, **e** 0.0032%

The carbide in magnesium containing steel is relatively small and presents spherical or subglobose, which improves the impact toughness of the steel.

5.4.5 Effects of Magnesium on Wear Resistance of H13 Die Steel

Wear performance is measured by wear rate, which is the ratio of wear volume to wear distance, in mm^3/mm. The lower the wear rate corresponds to a better the wear

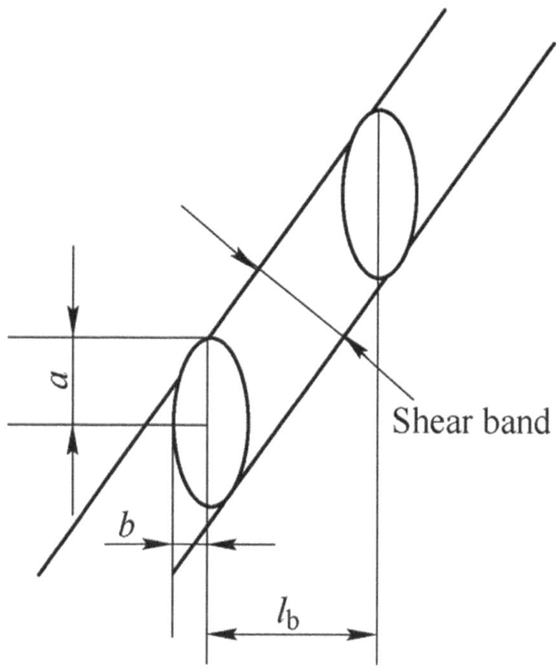

Fig. 5.66 Shear fracture diagram of elliptical cavity

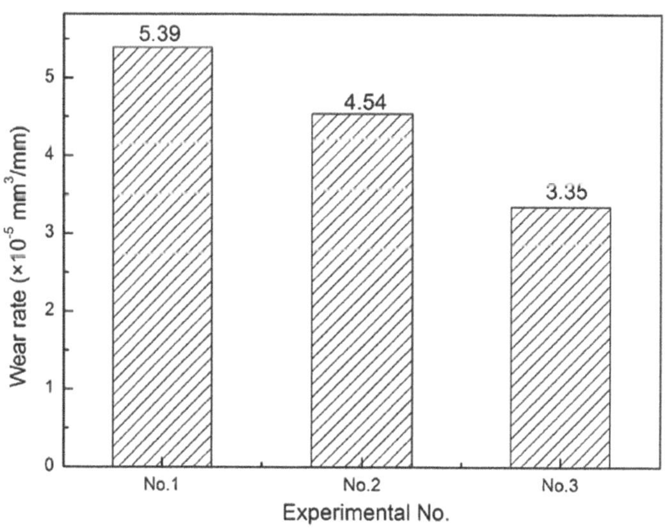

Fig. 5.67 Variation of wear rate of tempered H13 die steel with the different Mg contents

resistance of the steel. Figure 5.67. shows the variation of wear rate of the tempered H13 die steel with different Mg contents.

It can be seen from Fig. 5.67 that the wear rate gradually decreased with increasing magnesium. The wear rate is 5.39×10^{-5} mm^3/mm for the specimen without magnesium, but is 3.35×10^{-5} mm^3/mm for the specimen with 10 ppm magnesium. SEM morphologies of worn surface of H13 steel with the different magnesium content are shown in Fig. 5.68.

It is clear that the worn surfaces were mostly covered with oxide particles and adhesive trace was found on worn surfaces (Fig. 5.68a). The adhesive trace resulted

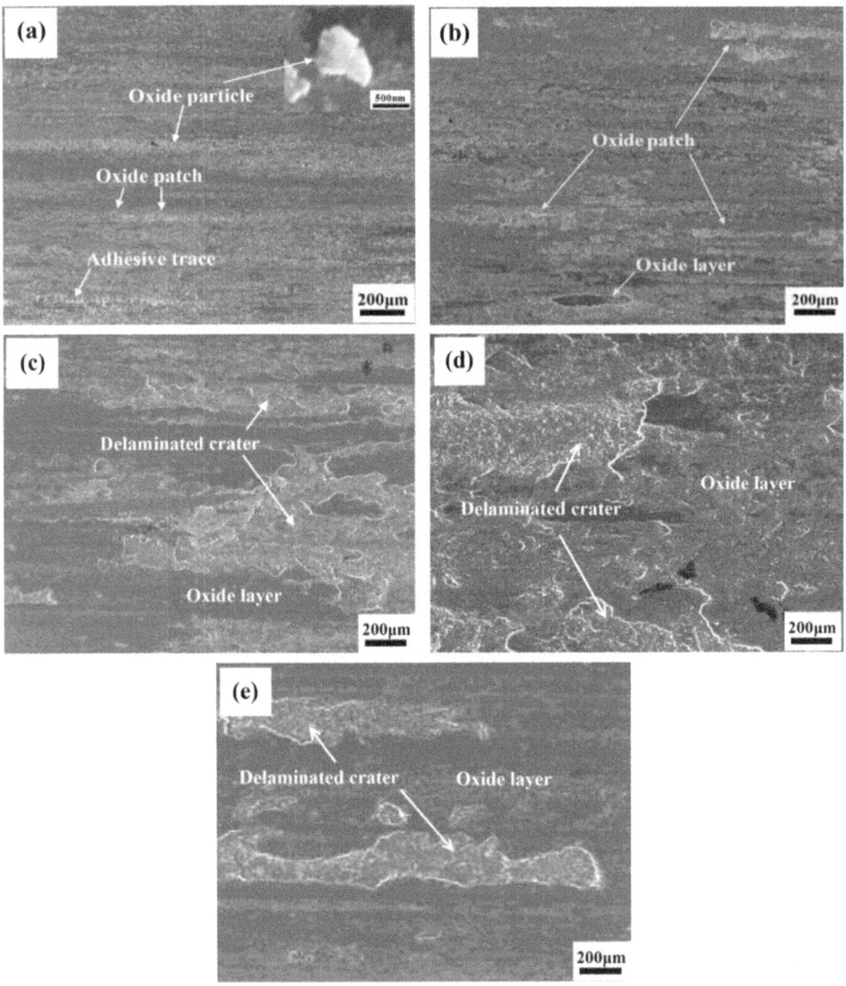

Fig. 5.68 SEM morphologies of worn surfaces of the specimens, Mg content: **a** 0; **b** 0.0006%; **c** 0.0010%; **d** 0.0019%; **e** 0.0032%

from the shear and rupture of adhesive junctions between asperity tips on contacting metal surfaces due to high local contact pressure during sliding, which was the typical characteristic of adhesive wear. In addition, there were a very small amount of oxide patches. For the specimen with 6 ppm magnesium, there were a mass of oxide patches and chain-like oxide patches were existed (Fig. 5.68b), which formed from the agglomeration of oxide particles at the adhesive trace. In addition, a smooth glaze oxide layer cover the partial worn surface just like the oxide island. In this case, the adhesive wear mechanisms dominated over other mechanisms. In this regime, the oxide patches can reduce metal-to-metal contact and causing a considerable reduction in wear. Moreover, the adherence of the oxidized wear particles to the sliding surfaces yields compacted, protective oxide films, which also reduce the wear rate [53]. Thus the specimen with 6 ppm magnesium possessed the better wear resistance than that without magnesium. With further increase in magnesium content, the degree of oxide layer coverage increased greatly, which result in the wear rate dropped further. In addition, some delaminated craters produced on the worn surface as shown in Fig. 5.68c. This is the typical oxidative wear [54, 55]. Figure 5.69 shows the friction coefficient as a function of the sliding distance for H13 steel with different magnesium content.

The friction coefficient of all specimens becomes stable in sliding distance between 20 and 30 m. This gradual increase in friction coefficient during the initial stage of the test may be due to in situ surface tempering [38]. For the specimen without magnesium, the friction coefficient exhibits the highest value (about 0.8). With increase of magnesium content, the friction coefficient decreased gradually due to an increase of the degree of oxide patches [38].

According to the Archard's equation [52]:

$$W_{\mathrm{v}} = k\frac{LS}{3H}. \tag{5.30}$$

Where W_{v} is the volumetric wear loss, k is the wear coefficient, L is the normal load, S is the sliding distance, H is the hardness of the metal.

According to Eq. (5.30), the wear loss of material is proportional to friction coefficient, is inversely proportional to hardness of the material. According to the results shown in Fig. 5.69 and Table 5.21, it is found that the friction coefficient of tempered H13 die steel decreases with the increase in the Mg content of the steel, whereas the hardness increases. Therefore, the wear loss of the steel decreases with increasing the Mg content of the steel. The large-sized carbides induce cracking and increase in the wear loss of the steel [56]. Mg-containing H13 die steel exhibits a higher wear resistance because the carbides in the steel are finer and more uniform [16, 42] in comparison with the H13 die steel without Mg addition.

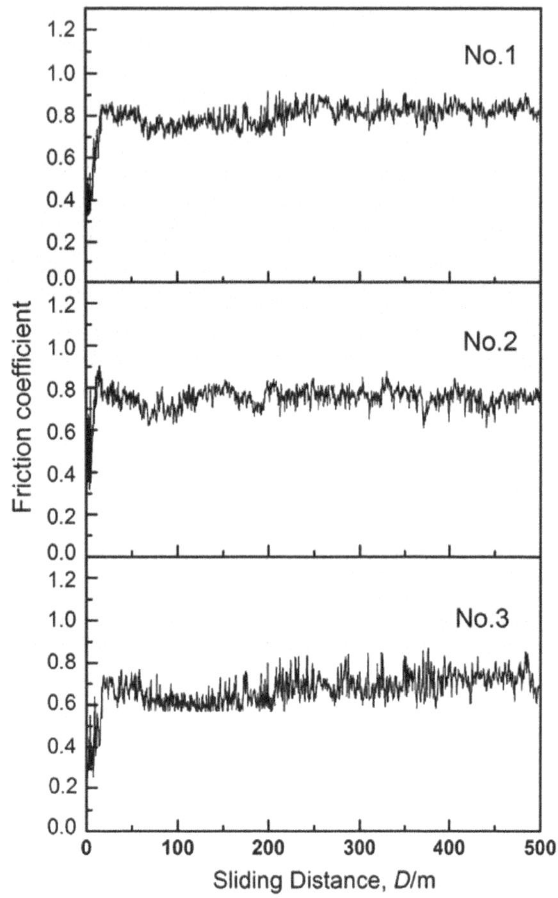

Fig. 5.69 Relationship of friction coefficient with sliding distance for specimens

References

1. Chen B, Jiang M, Wang CG et al (2007) Theoretical research of applying Mg to ultra clean steel production. Iron and Steel 42(7):30–33
2. Zhou DG, Fu J, Li J et al (2002) Study on control and effect of magnesium in bearing steel. Iron and Steel 37(7):23–25
3. Wang H, Li J, Wang LL (2014) Inclusions modification of magnesium treatment in H13 die steel. China Sci Paper 9(2):175–177
4. Wang LL, Li J, Zhou W et al (2012) (2012) A study of the effects of Mg on the inclusion in cutlery steels. South Metals 3:1–3
5. Li J, Wang FM, Zhang XY et al (2006) Effects of Mg contained deoxidizer on cleanliness of liquid steel and mechanical properties of sheet. International conference of Thin Slab Casting and Rolling, Guang Zhou
6. Park JH, Todoroki H (2010) Control of MgO·Al2O3 spinel inclusions in stainless steels. ISIJ Int 50(10):1333–1346
7. Sun WS, Ding GR, Fu J et al (1997) Effect of Mg in 35CrNi3MoV steel. Ordnance Mater Sci Eng 20(4):3–8

8. Liu J, Lu QL, Li Z et al (2011) Study on the mechanism of trace magnesium improving carbide in bearing steel. J Iron Steel Res 23(5):39–44
9. Li GY, Li TQ (2010) The influence of inclusions containing Mg on solid transformation of one pipeline steel. Iron and Steel 45(7):76–80
10. Guo LF, Li H, Wang Y (2012) Floating critical diameter and removal ration of Al2O3 particles in interface between steel and slag. Iron and Steel 47(4):3–27
11. Zhang LF, Thomas BG (2002) Alumina inclusion behavior during steel deoxidation[C]. In: 7th European electric steelmaking conference, Venice, Italy
12. Wikström J, Nakajima K, Shibata H et al (2008) In situ studies of the agglomeration phenomena for calcium–alumina inclusions at liquid steel–liquid slag interface and in the slag. Mater Sci Eng A 495(1–2):316–319
13. Du G, Li J, Wang Z et al (2017) Effect of magnesium addition on behavior of collision and agglomeration between solid inclusion particles on H13 steel melts. Steel Res Int 88(3):1600185
14. Wang H, Li J, Shi CB et al (2016) Evolution of Al2O3 inclusions by magnesium treatment in H13 hot work die steel. Ironmak Steelmak 44(2):128–133
15. Zhou J, Ma DS, Liu BS et al (2012) Research of band segregation evolution of H13 steel. J Iron Steel Res 24(4):47–52
16. Li J, Li J, Shi CB et al (2016) Effect of trace magnesium on carbide improvement in H13 steel. Can Metall Q 55(3):321–327
17. Mirzadeh H, Niroumand B (2009) Effects of rheocasting parameters on the microstructure of rheo-centrifuged cast Al–7.1 wt% Si alloy. J Alloys Compd 474(1–2):257–263
18. He B, Li J, Shi C et al (2018) Effect of Mg addition on carbides in H13 steel during electroslag remelting process. Metall Res Technol 115(501):501
19. Wu Z, Li J, Shi C et al (2014) Effect of magnesium addition on inclusions in H13 die steel. Int J Miner Metall Mater 21(11):1062–1067
20. Jiang ZH, Wang C, Gong W et al (2016) Evolution of inclusions and change of as-cast microstructure with Mg addition in high carbon and high chromium die steel. Ironmak Steelmak 42(9):669–674
21. Zou X, Zhao D, Sun J et al (2018) An integrated study on the evolution of inclusions in EH36 shipbuilding steel with mg addition: from casting to welding. Metall and Mater Trans B 49(2):481–489
22. Bor HY, Chao CG, Ma CY (1997) The influence of magnesium on carbide characteristics and creep behavior of the Mar-M247 superalloy. Scripta Mater 2(38):329–335
23. Ge HL, Youdelis WV, Chen GL (1989) Effect of interfacial segregation of magnesium on high carbon (18% Cr) cast steel. Mater Sci Technol 5(12):1207–1211
24. Sun XB (2003) Study of distribution of trace Mg in GH99 Alloy. Sichuan Metall 6:16–19
25. Mclean D (1957) Grain boundaries in metals[M]. Oxford University Press, London
26. Speight MV (1986) Growth kinetics of grain-boundary precipitates. Acta Metall 16(1):133–135
27. Yong QL (2006) Secondary phases in steels[M]. Metallurgical Industry Press, Beijing
28. Zhou J, Ma DS, Zhang CM (2012) Influence of different annealing process on microstructure and mechanical properties of H13 steel. Heat Treat Met 37(5):53–58
29. Bor HY, Chao CG, Ma CY (1999) The effects of Mg microaddition on the mechanical behavior and fracture mechanism of MAR-M247 superalloy at elevated temperatures. Metall and Mater Trans a 30(3):551–561
30. Bor HY, Ma CY, Chao CG (2000) The influence of Mg on creep properties and fracture behaviors of Mar-M247 superalloy under 1255 K/200 MPa. Metall and Mater Trans A 5(31):1365–1373
31. Bramfitt LB (1970) The effect of carbide and nitride additions on the heterogeneous nucleation behavior of liquid iron. Metall Trans 1(7):1987–1995
32. Yin HB, Shibata H, Emi T et al (1997) "In-situ" observation of collision, agglomeration and cluster formation of alumina inclusion particles on steel melts. ISIJ Int 37(10):936–945
33. Wang H, Li J, Shi C et al (2017) Evolution of carbides in H13 steel in heat treatment process. Mater Trans 58(2):152–156

34. He B, Li J, Shi CB et al (2016) Effect of cooling intensity on carbides in Mg-containing H13 steel during the electroslag remelting process. Chinese J Eng 38(12):1720–1727
35. Du YQ, Sun JL, Zheng YX (2015) Research on microstructure and performance of die steel H13 produced by gas-protective ESR furnace. Forg Stamp Technol 40(11):71–76
36. Ning A, Guo H, Chen X et al (2015) Precipitation behaviors and strengthening of carbides in H13 steel during annealing. Mater Trans 56(4):581–586
37. Janovec J, Vyrostková A, Svoboda M (2004) Evolution of secondary phases in Cr-V low-alloy steels during aging. Metall Mater Trans A 35(3):751–759
38. Bahrami A, Anijdan SHM, Golozar MA et al (2005) Effects of conventional heat treatment on wear resistance of AISI H13 tool steel. Wear 258(5):846–851
39. Honeycombe: Steels (1980) Microstructure and properties. Edward Arnold, London, pp 140–165
40. Pigrova GD (1996) Kinetics of carbide reactions in Cr–Mo–V steel. Met Sci Heat Treat 38(7–8):321–323
41. Pigrova GD (1996) Carbide diagrams and precipitation of alloying elements during aging of low-alloy steels. Metall and Mater Trans A 27(2):498–502
42. Li J, Li J, Wang L et al (2016) Study of the effect of trace Mg additions on carbides in die steel H13. Met Sci Heat Treat 58(5–6):330–334
43. Medvedeva A, Bergström J, Gunnarsson S et al (2009) High-temperature properties and microstructural stability of hot-work tool steels. Mater Sci Eng, a 523(1):39–46
44. Elias CN, Viana CSDC (1992) Effect of ausforming on microstructure and hardness of AISI H-13 tool steel modified with niobium. Mater Sci Technol 8(9):785–790
45. Ning AG, Guo HJ, Chen XC et al (2014) Precipitation behaviors of carbides in H13 steel during ESR, forging and tempering. J Univ Sci Technol Beijing 36(7):895–902
46. Li PG, Gao J (1997) Study on heat stability of hot die steels. Heat Treat Met 12:10–12
47. Wen JB (2011) Metal materials[M]. China Machine Press, Beijing
48. Li YY, Wang LL, Ning B (2013) Effect of magnesium on mechanical properties of H13 die steel after anneal. South Metals 191(2):48–53
49. Zhao MH, Zhang J, Feng D (2006) Atlas of fracture analysis of superalloy [M]. Metallurgical Industry Press, Beijing
50. Zhang WQ, Guo J (2012) Flow characteristics of 42Cr Mo steel on unidirectional tension at different temperatures. Heat Treat Met 37(7):94–97
51. Wang LL, Li J, Li YY et al (2013) Effect of Mg on structure and mechanical properties of hot-working-die steel H13. Spec Steel 34(5):68–70
52. Archard JF (1953) Contact and rubbing of flat surfaces. J Appl Phys 24(8):981–988
53. Stott FH, Glascott J, Wood GC (1985) Models for the generation of oxides during sliding wear. Proc R Soc A 402:167–186
54. Quinn TFJ (1983) Review of oxidational wear part I: the origins of oxidational wear. Tribol Int 16(5):257–271
55. Quinn TFJ (1983) Review of oxidational wear part II: recent developments and future trends in oxidational wear research. Tribol Int 16(6):305–315
56. Oh H, Yeon K, Yun Kim H (1999) The influence of atmospheric humidity on the friction and wear of carbon steels. J Mater Process Tech 95(1):10–16

Chapter 6
Effect of Rare Earth on the Carbide in Steel

Abstract Rare earth elements can affect solidification structure, inclusions and carbides of steel, reduce dendritic segregation of alloying elements and grain size. It is found that the amount of inclusions in 8Cr13MoV steel decreases obviously with the addition of Ce after electroslag remelting (ESR), and the Al_2O_3 inclusions in the steel are modified to $CeAlO_3$ inclusions first and then spherical Ce_2O_2S, leading to plasticity of inclusions. With the decrease in atomic ratio of Ce/S in Ce-containing inclusions in steel, the morphology of inclusions tends to be spherical. The modified sulfide inclusions are rare-earth sulfide and rare-earth oxide-sulfide. Rare earth Y can reduce grains of austenite hot work die steel, increase the number of austenite intragranular twins, increase the area ratio of twin grain boundaries, and further refine austenite grains. Interlaced twins are used to restrain dislocation m

Keywords Rare earth element · Inclusions · Carbides · Heat treatment · Mechanical property

The rare earth inclusions have a small lattice mismatch with the Fe, which can be used as heterogeneous nucleation site in the solidification process of liquid steel and increase the distribution coefficient of alloy elements. Therefore, rare earth can affect the steel solidification microstructure, reduce the alloy elements segregation in dendrites and refine the grain size. Various defects in the crystal, such as grain boundary, phase boundary, dislocation, hole and so on, have a obvious role on promoting the nucleation of new phase. Because of the high energy around the defects in crystal, the nucleation of new phase in these locations is relatively easy, promoting the formation of new phase. Rare earth is enriched on the grain boundary and can fill in the defects such as grain boundary and holes in austenite, reducing the amount of defects and the interface energy of grain boundary, which increases the difficulty of carbide to nucleate at the grain boundary. Yttrium (Y) is a strong carbide forming element due to the unsaturated 5d layer in the outer layer electron arrangement. Y can be combined with carbon to form carbides with high-melting-temperature, such as YC, Y_2C_3 and YC_2. These carbides can disperse in the crystal at the first crystallization and become the nucleation core, reducing the number of carbides. In addition, the filling of defects such as holes and grain boundary by rare earth limits the diffusion of atoms through the grain boundary and the growth of

J. Li and C. Shi, *Carbide in Special Steel*, Engineering Materials,
https://doi.org/10.1007/978-981-16-1456-9_6

carbides along grain boundary [1–6]. Moreover, rare earth can change the shape, size, quantity and distribution of carbides in steel. When rare earth enters the cementite, it can change the composition and structure of the cementite. Rare earth can break the eutectic carbide network and change the three-dimensional structure of primary carbides in high alloy die steel.

The addition of rare earth to high manganese austenitic steel can refine the grains and make the carbides distributed uniformly, improving the strength and plastic toughness of high manganese steel [7]. Adding rare earth in tool steel D2 can refining the as cast microstructure and provide heterogeneous nucleation core for the formation of eutectic carbides M_7C_3 [8], which can significantly reduce the size and arear ratio of eutectic carbides and make the eutectic carbides disperse uniformly. The addition of rare earth in Fe85Cr4Mo8V2C1 obviously reduce the area ratio coarse eutectic carbides and decrease the lamellar spacing of eutectic carbides, making the eutectic carbides more compact and spheroidized [9].

Due to a large radium of atom, the solution of rare earth element causes lattice expansion near the grain boundary, increasing the grain boundary energy and promoting the nucleation of carbides. Moreover, the grain refinement increases the area of grain boundary, and the carbide precipitated is finer and more dispersed [10]. In high manganese steel, rare earth can be enriched on the surface of newly precipitated carbides as surface active elements, limiting the growth rate of carbides [11]. When rare earth elements are added to stainless steel and heat-resistant steel, after solution and sensitization treatment, the carbide on the grain boundary become granular or the precipitation on the grain boundary is reduced. Rare earth can prevent carbide precipitation along the grain boundary and broke carbide. The carbides precipitated on various types of grain boundaries have obviously different morphology, which is also affected by the energy of different types of grain boundaries. The energy of $\sum 3$ grain boundary is the lowest, the carbides precipitated is finest. And the energy of $\sum 27$ grain boundary is the highest, the carbides precipitated is largest. Adding rare earth to Inconel 600 alloy steel is beneficial to the formation of a large number of $\sum 3$ grain boundaries, contributing to control the precipitation of carbides along the grain boundaries [12]. Based on the effects of rare earth as described above, the effect of rare earth element on carbide in tool die steel, especially the effect of rare earth inclusions on carbide, was studied.

6.1 Effect of Rare Earth on Inclusion Behavior in Steel Before and After ESR

6.1.1 Effect of Rare Earth on the Quantity and Morphology of Inclusions

The chemical composition of 8Cr13MoV is shown in Table 6.1, the steel without adding Ce, containing 0.015% and containing 0.024 Ce are marked sample A, B and C respectively.

Table 6.1 Chemical composition of 8Cr13MoV smelted in induction furnace (wt%)

Samples	C	Si	Mn	S	Cr	Ni	Mo	V	Als	Ce
A	0.72	0.33	0.46	0.0041	14.02	0.25	0.20	0.14	0.004	0
B	0.79	0.47	0.50	0.0011	14.66	0.25	0.20	0.18	0.012	0.015
C	0.78	0.46	0.52	0.0012	14.78	0.22	0.22	0.20	0.017	0.024

After electroslag remelting (ESR) process, the composition of 8Cr13MoV is shown in Table 6.2. The sample A1, B1 and C1 after ESR are corresponded to the sample A, B and C before ESR respectively.

According to the thermodynamic calculation method of multiphase and multicomponent equilibrium, when Al element existed in steel, Ce can react with Al and O firstly and $CeAlO_3$ inclusion was formed. The reaction formula is shown in Eq. (6.1) [13, 14].

$$[Ce] + [Al] + 3[O] = CeAlO_3(s) \qquad \Delta G^{\theta} = -1366460 + 364.3T\,(J/mol) \tag{6.1}$$

Moreover, if Al_2O_3 exists in the molten steel, Ce can react with Al_2O_3 inclusion, the reaction formula is shown in Eq. (6.2) [14].

$$[Ce] + Al_2O_3(s) = CeAlO_3(s) + [Al] \qquad \Delta G^{\theta} = 423900 - 247.7T\,(J/mol) \tag{6.2}$$

At the steel-making temperature (1600 °C), the ΔG of Eq. (6.2) is smaller than zero, indicating that reaction can take place. The angular Al_2O_3 inclusion with high hardness can be transformed into granular rare earth aluminate inclusion with low hardness.

When the content of Ce is high, the reaction as shown in Eq. (6.3) can take place [14].

$$CeAlO_3(s) + [Ce] + [S] = [O] + [Al] + Ce_2O_2S(s) \qquad \Delta G^{\theta} = -288550 + 18.1T\,(J/mol) \tag{6.3}$$

As shown in Table 6.1, for the samples containing 0.015% Ce and 0.024% Ce, the content of Als is high, which is considered the reaction as shown in Eq. (6.2) take place. The reduction of S content in sample B and C can be explained as two

Table 6.2 Chemical composition of 8Cr13MoV after ESR process (wt%)

Samples	C	Si	Mn	S	Cr	Ni	Mo	V	Als	Ce
A1	0.72	0.30	0.41	0.0037	14.02	0.25	0.20	0.14	0.004	0
B1	0.79	0.45	0.50	0.0011	14.66	0.24	0.21	0.18	0.043	52×10^{-4}%
C1	0.78	0.43	0.47	0.0013	14.76	0.23	0.22	0.21	0.048	82×10^{-4}%

aspects. On the one hand, the rare earth plays a role on desulfurization. On the other hand, the Eq. (6.3) takes place and the S content was decreased.

Whether the Als content in steel is high or low, the amount of inclusions is influenced, resulting in Al_2O_3 inclusion retained in steel [15]. The addition of Ce into steel is beneficial to reduce the amount of inclusions.

6.1.2 Effect of Rare Earth on the Morphology and Composition of Inclusions

The morphologies of typical inclusions containing Ce in 8Cr13MoV steel are shown in Fig. 6.1.

As shown in Fig. 6.1, the shape of Ce-contained inclusions is irregular, such as spherical and long stripe. The composition of Ce-contained inclusions is shown in Table 6.3, it can be seen that the Ce content in these inclusions is in the range of 7.29–15.13%.

Combining the analysis of Fig. 6.1 and Table 6.3, the Ce-contained inclusions in steel can be divided into three types. One is spherical and the atomic ratio of Ce/S is close to 1:1, as shown in Fig. 6.1a and b. The second type of Ce-contained inclusion is irregular and almost like elliptical, and the atomic ratio of Ce/5 is between 2 and 5, as shown in Fig. 6.1c and d. The others is like shown in Fig. 6.1e and f, the Ce-contained inclusions is long strip and the atomic ratio of Ce/S is more than 10. It can be seen that, with the decrease of the atomic ratio of Ce/S, Ce combines with more

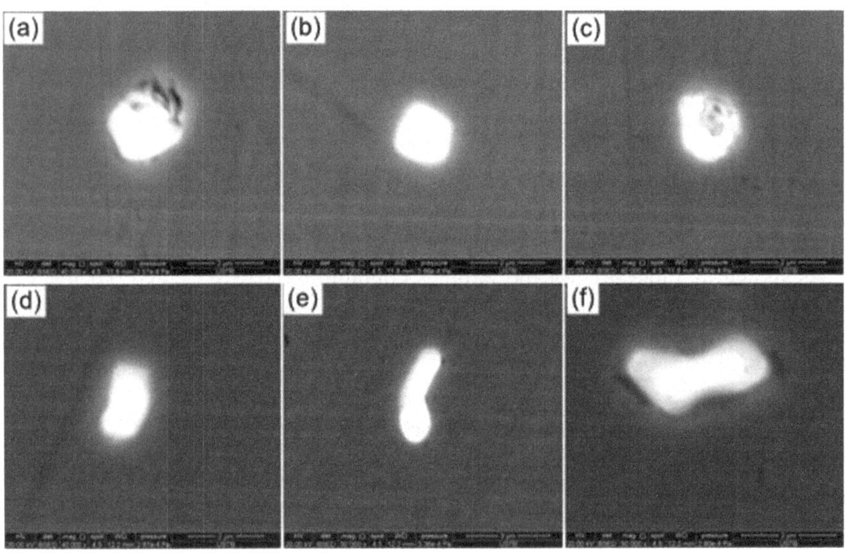

Fig. 6.1 Morphologies of typical inclusions containing Ce in 8Cr13MoV steel

Table 6.3 Composition of Ce-contained inclusions in 8Cr13MoV steel (at.%)

No	Fe	Cr	C	O	S	Ce
1	24.33	9.26	30.12	17.91	7.99	7.29
2	25.34	8.34	30.97	17.63	9.12	8.59
3	18.53	7.16	29.70	18.53	4.32	9.05
4	18.47	8.40	24.09	27.77	2.77	14.59
5	15.08	10.95	12.02	36.31	0.51	15.13
6	19.93	11.15	17.93	32.11	0.31	15.06

S atoms and the Ce-contained inlucions tends to be spherical. MnS inclusion can be modified by rare earth spherical and ellipsoidal rare earth sulfides and oxysulfides [16]. Spherical rare earth sulfides can improve the transverse toughness, weldability and fatigue property of steel during rolling [17].

The element mappings of Ce-contained complex inclusion in steel are shown in Fig. 6.2.

As shown in Fig. 6.2, the complex inclusion contains Ce, which can be considered the reaction production $CeAlO_3$ is the core of complex inclusion. Previous research indicated that the complex inclusion, a core of CeO_2 surrounded by $CeAlO_3$, can be formed after adding Ce [18]. For simplicity, the core is considered $CeAlO_3$ inclusion.

After adding rare earth into steel, the angular Al_2O_3 inclusion with high hardness can be modified to $CeAlO_3$, modified into spherical Ce_2O_2S inclusion further, which realized the plastic control of inclusions.

6.1.3 *Effect of Rare Earth on the Microstructure*

Using two-dimensional lattice disregistry to characterize the effectiveness of rare earth inclusions as the nucleation core, the lattice mismatch between Ce-contained inclusions and α-Fe at 1185 K was calculated in some researches [19, 20]. The results demonstrated that, the misfit values between Ce-contained inclusions and α-Fe were small. For example, the misfit between $CeAlO_3$ and α-Fe is 7%, indicating $CeAlO_3$ can promote the formation of intragranular acicular ferrite and refine microstructure.

The microstructures of steel containing different Ce contents are shown in Fig. 6.3.

As shown in Fig. 6.3, compared to the austenite grains of Ce-undoped steel, the size of austenite grains of Ce-doped steel was decreased. After adding rare earth, dissolved rare earth is enriched in the grain boundary through diffusion mechanism, reducing the segregation of impurity elements in the grain boundary and refining the grains [21]. At the meantime, rare earth inclusions can be the heterogeneous nucleation sites of intragranular acicular ferrite [22] and δ-Fe [23, 24]. Rare earth

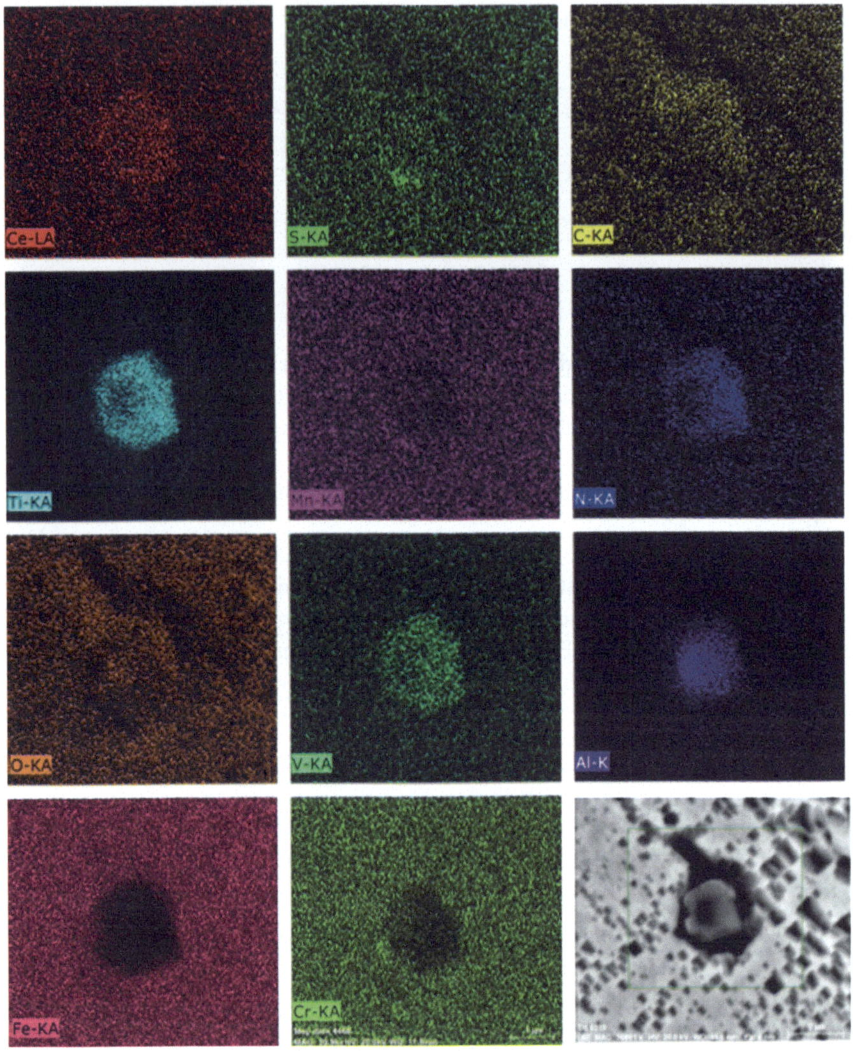

Fig. 6.2 Element mappings of Ce-contained complex inclusion

inclusions with high-melting-temperature in the molten steel can be the heterogeneous nucleation sites, decreasing supercooling degree of molten steel, refining the solidification microstructure, reducing segregation and control the microstructure during solidification process [13]. The content of dissolved rare earth is very low and about 95% of rare earth exists in inclusions [25]. Therefore, rare earth inclusion plays an important effect on refining microstructure.

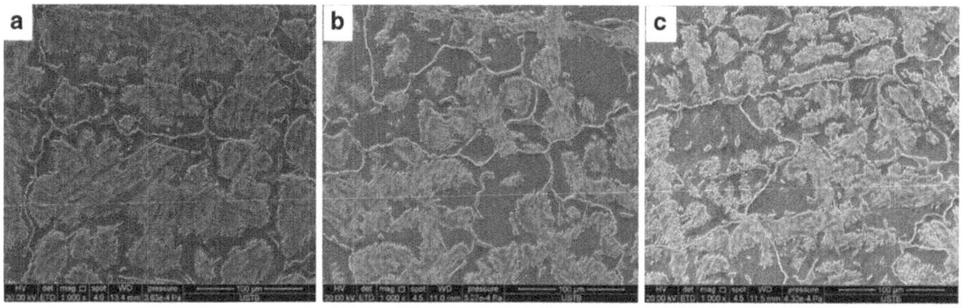

Fig. 6.3 Microstructures of steel containing different Ce content

6.2 Effect of Rare Earth Inclusion on the Carbide

6.2.1 *Effect on Rare Earth Inclusion on the Carbide of Austenitic Hot-Work Die Steel*

The chemical compositions of the four ESR ingots with different rare earth additions are listed in Table 6.4. They were marked as Y0, Y1, Y2 and Y3, respectively.

Figure 6.4 shows the microstructure in cross section in the ESR ingots with different addition of yttrium remelted through ESR process.

The dendritic grains distribute dense and display fine when the residual content of yttrium is 0.0060% in weight percent (i.e. in ingot Y2). The morphologies of dendritic microstructure gradually become refining with increasing residual content of yttrium in ESR ingot, while those turn into coarsening dendritic microstructure when the excess yttrium is added (i.e. in ingot Y3). The secondary dendrite arm spacing (SDAS) is an important parameter to characterize the quality of the ESR ingots, and quantitatively analysis of SDAS is shown in Fig. 6.5.

The result indicates that SDAS decreased from 48.3 μm to 39.1 μm with increasing residual content of yttrium from 0 to 0.0060%, but it would reversely increase to 50.2 μm when the residual content of yttrium in ingot Y3 reaches to 0.0086%. Combination observation in Fig. 6.4 and statistical result of SDAS in Fig. 6.5, it reveals that rare earth yttrium has an important effect on dendritic morphologies and SDAS. In addition, the appropriate residual content of yttrium to improve microstructure in experimental ESR ingot is 0.0060% in weight percent.

Table 6.4 Chemical compositions of ESR ingots (wt%)

	C	Si	Mn	Cr	Mo	V	Al	P	Y	S	O
Y0	0.70	0.54	14.90	3.53	1.55	1.73	0.012	0.013	0	0.0020	0.0017
Y1	0.71	0.59	15.38	3.28	1.91	1.78	0.010	0.012	0.0005	0.0021	0.0015
Y2	0.68	0.50	14.90	3.36	1.56	1.65	0.011	0.013	0.0060	0.0018	0.0016
Y3	0.71	0.54	14.50	3.37	1.60	1.65	0.013	0.011	0.0086	0.0016	0.0015

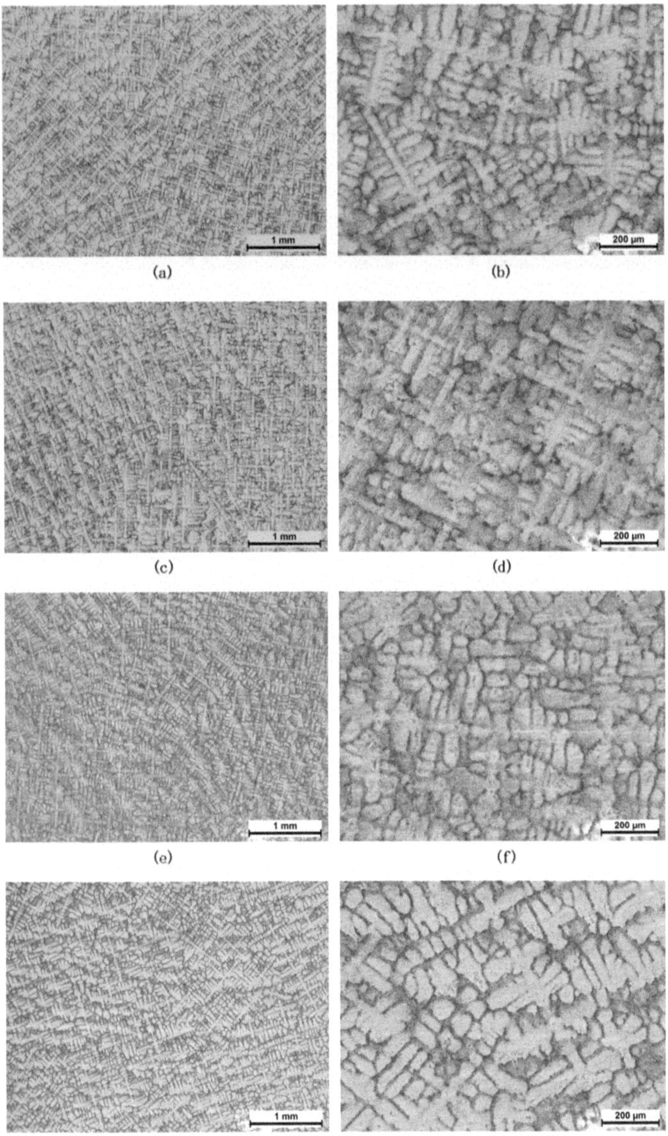

Fig. 6.4 OM micrograph of dendritic microstructure in ESR ingots: **a**, **b** 0RE; **c**, **d** 0.0005%RE; **e**, **f** 0.0060%RE; **g**, **h** 0.0086%RE

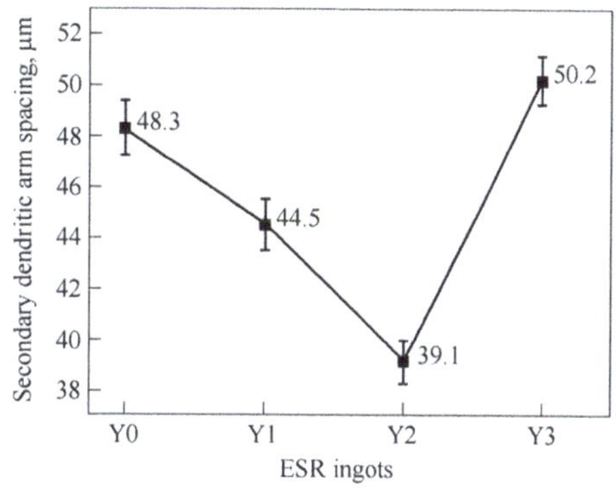

Fig. 6.5 Secondary dendritic arm spacing of ingots as-casted with different addition content of yttrium

The segregation ratio δ of alloying elements was applied to evaluate the extent of segregation of alloying elements ($\delta_i = c_{i\,\max}/c_{i\,\min}$, where $c_{i\,\max}$ is the maximum concentration of i element at interdendritic area and $c_{i\,\min}$ is the minimum concentration of i element at dendritic area). Electron probe microanalysis (EPMA) was used to detect the micro-components at dendritic and interdendritic areas. Five points measured for each sample and means calculated, as shown in Fig. 6.6.

The result indicates that segregation degree of each element almost alleviated with increasing residual content of yttrium by 0.0060%, while that would convert to deteriorate when the residual content of yttrium in ingot Y3 reaches to 0.0086%. This result is in good agree with the tendency of the influence of yttrium on dendritic microstructure and SDAS.

Differential thermal analysis (DTA) was used to detect the heat fluctuation during carbide decomposition, in order to analyze the effect of rare earth on the decomposition temperature of primary carbide in the ESR ingot. Figure 6.7 shows the decomposition transition curve of primary carbide in the ESR ingots with different rare earth contents. In the DTA experiments, the samples were heated to 1300 °C at the rate of 20 °C/s. After holding for 5 min, the samples were cooled to 600 °C at the rate of 20 °C/s. In the process, argon protection was always maintained.

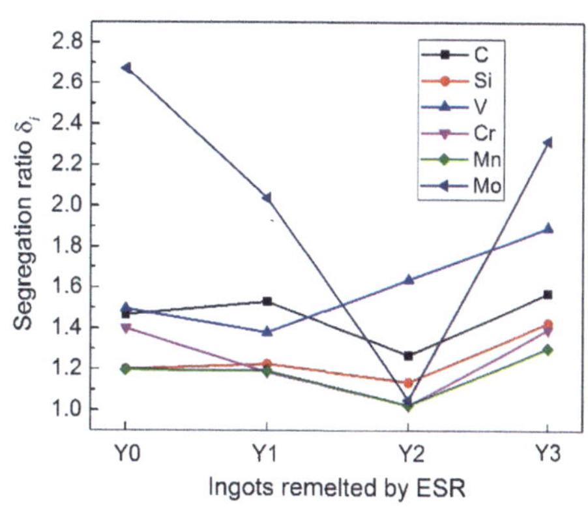

Fig. 6.6 Segregation degree of each element in ESR ingots with different content of yttrium

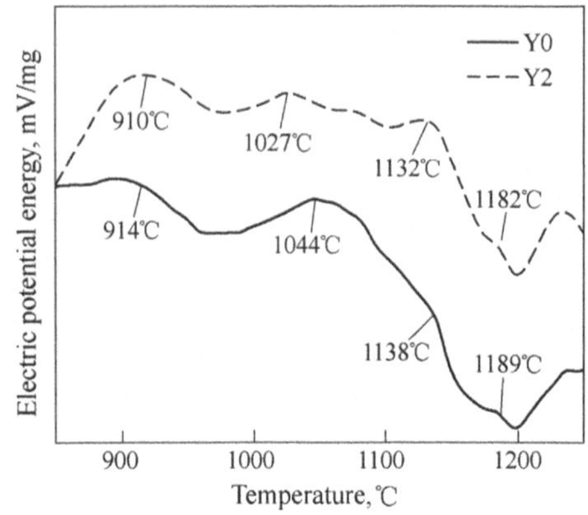

Fig. 6.7 Decomposition transition curve of primary carbide

From Fig. 6.7, it can be seen that the addition of rare earth obviously decrease the decomposition temperature of primary carbides, resulting in the reduction of precipitation temperature and amount of primary carbides during solidification process. Wang et al. [26] investigated the effect of rare earth on primary carbides in rolling steel, the results indicated that rare earth can obviously decrease the volume fraction of primary carbides. SEM micrographs in backscattered electron (BSE) mode in Fig. 6.8 show the effect of yttrium addition on the morphology and distribution of precipitated carbides in ESR ingots.

As can be seen in Fig. 6.8, the sizes and number density of carbides decreased to the limit values as well as dispersed distribution of carbides when the addition of yttrium reached to 0.0060%, but these turn to deterioration when excessive addition. The basic parameters and characteristic parameters of carbides are shown in Table 6.5.

As can be seen in Table 6.5, the volume fraction, average size and area fraction of primary carbide can be significantly reduced by appropriate rare earth content. However, the number of primary carbide increases with increasing rare earth content, indicating that the primary carbide is smaller. Therefore, the amount, size and distribution of primary carbide can be effectively controlled by adding appropriate rare earth.

Rare earth inclusions can be heterogeneous nucleation sites for the formation of primary carbides, refining carbides and improving the distribution. At the meantime, rare earth affects the driving force of carbide growth, decreasing the precipitation temperature and reducing the amount of primary carbide. When the rare earth content is excessive, rare earth atoms are enriched at the front of solid–liquid interface, resulting in forming large-scale rare earth inclusion. Therefore, in order to control the precipitation of primary carbide during ESR process, the rare earth content in the ESR ingot should be controlled. Liu et al. [27] and Gao et al. [28] demonstrated that appropriate rare earth content can refine microstructure and reduce alloy element segregation.

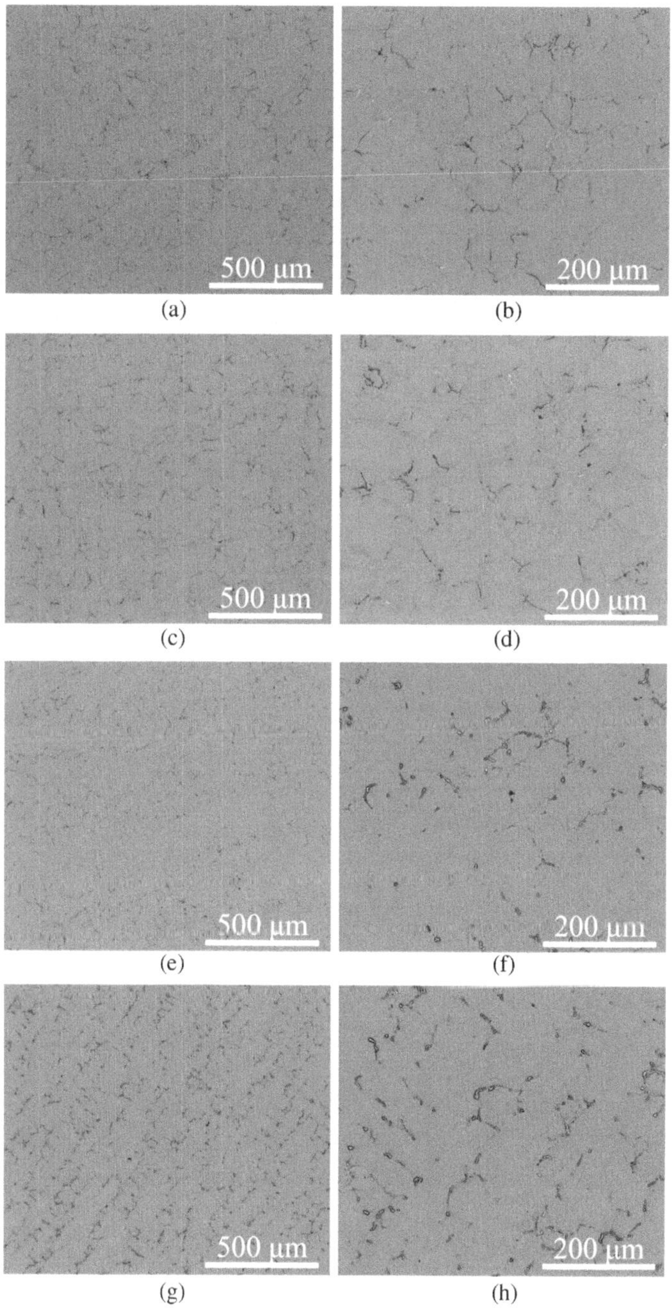

Fig. 6.8 SEM micrograph of carbides in BSE mode in ESR ingots: **a**, **b** 0RE; **c**, **d** 0.0005%%RE; **e**, **f** 0.0060%%RE; **g**, **h** 0.0086%%RE

Table 6.5 Basic parameters and characteristic parameters of carbides

Sample	Basic parameters of quantitative metallography							Characteristic parameters of carbides	
	Statistic of carbides parameters					Size of photo			
	N	A, μm^2	$\overline{D}$, μm	L', μm	W', μm	L, μm	W, μm	V_V, %	N_V
Y0	956	16,890	6.32	12.71	2.89	595.92	507.29	5.56	0.031
Y1	1117	15,651	5.78	8.32	2.36	595.92	507.29	5.18	0.028
Y2	1306	6807	3.19	4.01	2.02	595.92	507.29	2.25	0.015
Y3	1562	13,741	4.37	6.04	3.78	595.92	507.29	4.55	0.027

6.2.2 Effect of Rare Earth Inclusion on the Carbide

The chemical composition of 8Cr13MoV in ESR ingot is shown in Table 6.6, and the Ce content is 0.0097%.

The typical inclusions observed in original electrode and electroslag remelting after ESR with Ce addition were shown in Fig. 6.9.

As shown in Fig. 6.9, the Ce-contained inclusion in electrode is large and its distribution is not uniform. After ESR process, Ce combined with Al and $CeAlO_3$ inclusion with a smaller size was formed. Moreover, $CeAlO_3$ can be the heterogeneous nucleation for (Ti, V) (C, N) and M_7C_3. The inclusions and carbides in ESR ingot were shown in Fig. 6.10.

As shown in Fig. 6.10, the typical inclusion is large Al_2O_3 without Ce addition. After adding Ce, Al_2O_3 inclusion was transformed to $CeAlO_3$ inclusion, and the average size of inclusion was decreased from 5.1 μm to 1.5 μm. In addition, the complex inclusion, $CeAlO_3$ as a core surrounded by (Ti, V) (C, N) and M_7C_3, was observed. The observed results indicated Ce can modify Al_2O_3 to $CeAlO_3$ and reduce the inclusion size. As the heterogenous nucleation sites, $CeAlO_3$ inclusion can refine primary carbides.

The calculation model of two-dimensional lattice disregistry was shown in following equation:

$$\delta_{(hkl)n}^{(hkl)_s} = \sum_{i=1}^{3} \frac{\frac{\left|d_{[uvw]_s^i} \cos\theta - d_{[uvw]_n^i}\right|}{d_{[uvw]_n^i}}}{3} \times 100\% \tag{6.4}$$

where $(hkl)_s$ is a low-index plane of the substrate, $(hkl)_n$ is a low-index plane in the nucleated solid. $[uvw]_s$ is a low index direction in $(hkl)_s$, $[uvw]_n$ is a low-index direction in $(hkl)_n$. $d_{[uvw]s}$ is the interatomic spacing alone $[uvw]_s$, $d_{[uvw]n}$ is the interatomic spacing alone $[uvw]_n$, and θ is the angle between the $[uvw]_s$ and $[uvw]_n$.

Because of the similar crystal structure of (Ti,V) (C,N) and TiN, the misfit between $CeAlO_3$, Al_2O_3 and (Ti,V) (C,N) is analyzed by the misfit degree between $CeAlO_3$, Al_2O_3 and TiN, as shown in Fig. 6.11.

It can be seen from Fig. 6.11, the misfit value between (001) of $CeAlO_3$ and (110) of TiN is 2.63%. The misfit value between (0001) of Al_2O_3 and (110) of TiN is 5.35%. After adding Ce, Al_2O_3 inclusion can be modified to $CeAlO_3$. Compared to Al_2O_3, The misfit between $CeAlO_3$ and TiN is smaller, indicating $CeAlO_3$ is more likely to be heterogeneous nucleation core of TiN.

TiN and M_7C_3 are cubic FCC and hexagonal respectively. The misfit between (001) of TiN and (0001) of M_7C_3 is 4.7%, indicating TiN also can promote the formation of M_7C_3 and refine M_7C_3 carbide.

Table 6.6 Chemical composition of 8Cr13MoV steel (wt%)

C	Si	Cr	Mo	Mn	V	Ni	S	N	Ce
0.8	0.28	14.02	0.391	0.45	0.453	0.065	0.0043	0.011	0.0097

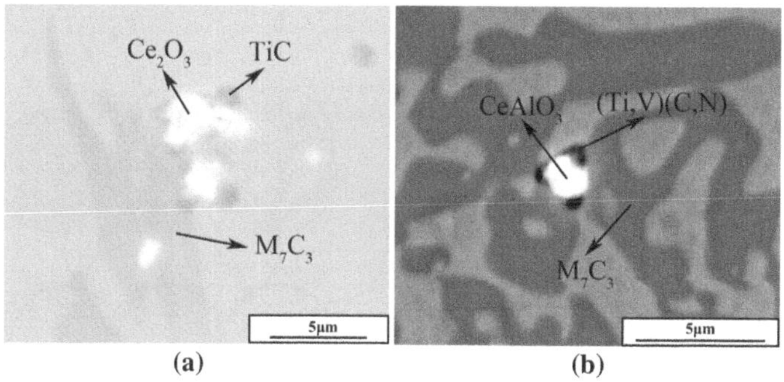

Fig. 6.9 Typical inclusions in electrode and ESR ingot with Ce addition. **a** Electrode; **b** ESR ingot

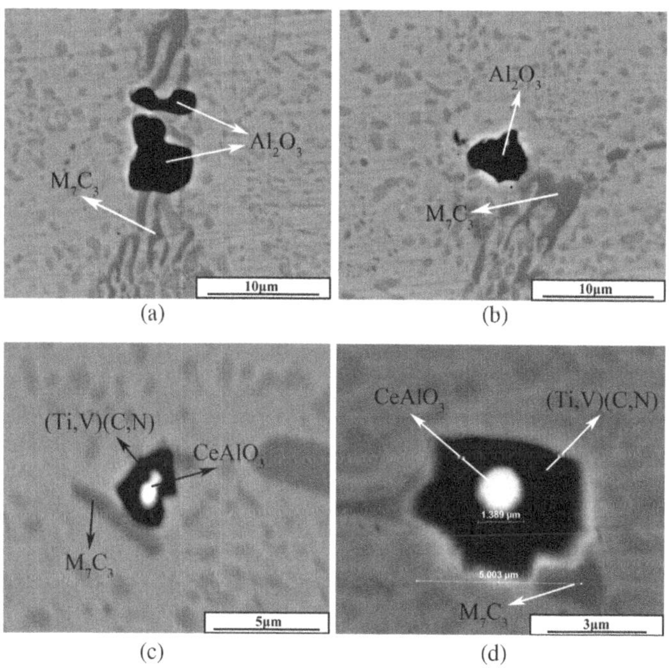

Fig. 6.10 Typical inclusions and carbides in ESR ingots. **a** and **b** Ce-undoped; **c** and **d** Ce content is 0.0097%

Al_2O_3 inclusion with high hardness can be modified into $CeAlO_3$ inclusion, with the reduction of inclusion size. On the one hand, the harm of Al_2O_3 inclusion to the subsequent rolling process is avoided. On the other hand, $CeAlO_3$ inclusion can promote and refine primary carbides as heterogeneous nucleation sites.

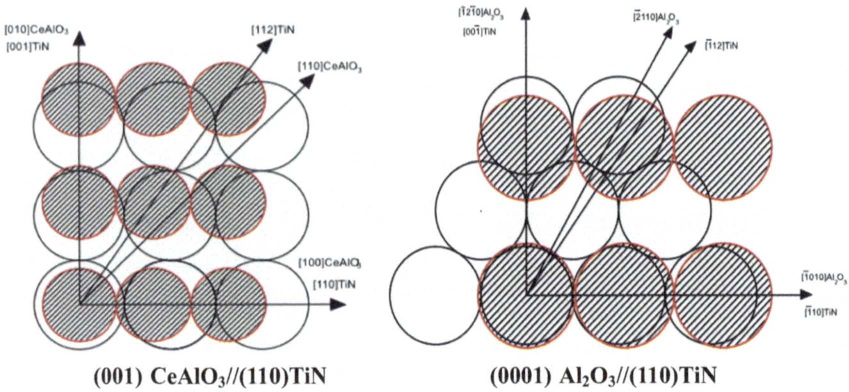

Fig. 6.11 Orientation relationship between $CeAlO_3$, Al_2O_3 and TiN

6.3 Effect of Heat Treatment on Carbides in Rare Earth Austenitic Hot-Work Die Steel

6.3.1 Microstructure of Rare Earth Microalloyed Austenitic Hot-Work Die Steel

After annealing treatment, the ESR ingots with different rare earth contents were performed through solid solution and aging heat treatment. The curve of heat treatment process is shown in Fig. 6.12.

The micrographs of the morphologies of austenite grain after heat treatment with different contents of yttrium were shown in Fig. 6.13.

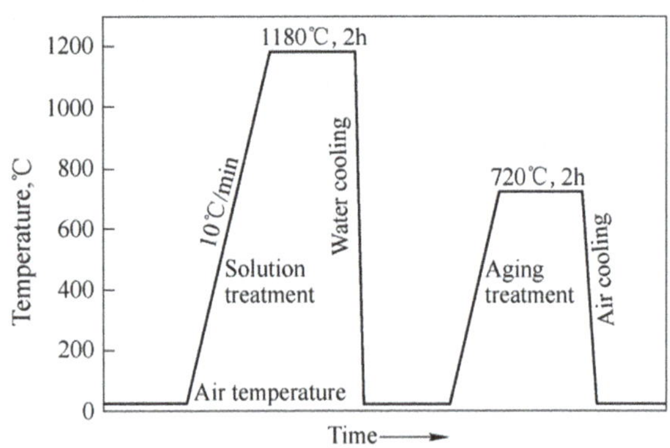

Fig. 6.12 Curve of heat treatment process of austenitic hot-work die steel

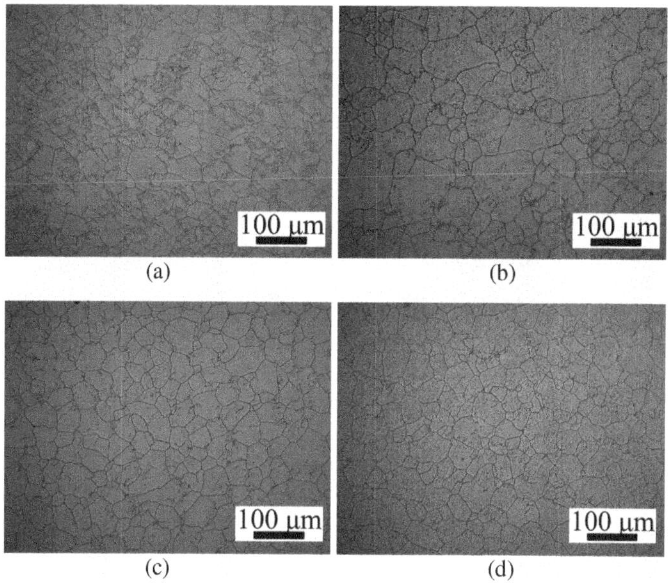

Fig. 6.13 OM micrograph of the morphology of austenite grain after heat treatment with different contents of yttrium: **a** 0RE; **b** 0.0005%RE; **c** 0.0060%RE; **d** 0.0086%RE

It can be seen from Fig. 6.13 that with increasing rare earth content, the average size of austenite grains gradually decreased. The distribution of austenite grains in Fig. 6.13 was counted using intercept method, and the result was shown in Fig. 6.14.

The sizes of austenite grains decreased with increasing addition of yttrium. The sizes distribution of austenite grains present uniform when the addition of yttrium reached to 0.0060%. In addition, the twin boundaries obviously increased resulted by the addition of yttrium. The grain boundaries areas gradually increased and the sizes of grain obviously decreased.

Generally, the larger the size difference between solvent atom and solute atom is, the stronger the effect of the delay of austenite recrystallization by solute atom is. Because of a large size difference between yttrium and iron atom, the rare earth atoms tend to segregate on the austenite grain boundary. The interaction between rare earth and grain boundary affects the grain boundary migration and the crystal growth rate in recrystallization process. Therefore, solute rare earth atoms have a strong drag effect on recrystallization, refining recrystallization grain size and raising recrystallization temperature obviously.

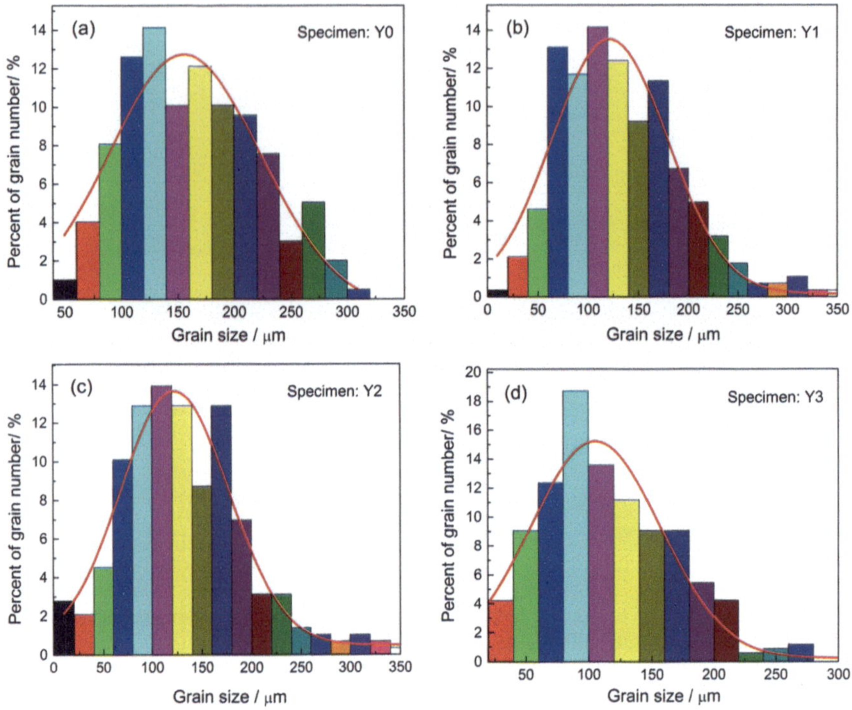

Fig. 6.14 Size distribution of austenite grain of specimens after heat treatment with different contents of yttrium: **a** 0 RE; **b** 0.0005%RE; **c** 0.0060%RE; **d** 0.0086%RE

6.3.2 Effect of Rare Earth on Grain Boundary in Austenitic Hot-Work Die Steel

The effect of rare earth on austenite grains in austenitic hot-work die steel after heat treatment was shown in Fig. 6.15.

As shown in Fig. 6.15, a lot of twin structures exists in austenite grain. With increasing rare earth content, the area ratio of twin grain boundary increases gradually. Meanwhile, the austenite grain is divided into smaller grains.

The effect of rare earth on the distribution of twin grain boundary in austenitic hot-work die steel after heat treatment is shown in Fig. 6.16, the twin grain boundary in austenite grains is described by blue line.

As seen from Fig. 6.16, with the increase of rare earth content, the proportion of twin grain boundary in austenite grain increases gradually. The austenite grain is divided into smaller grains by twin grain boundary.

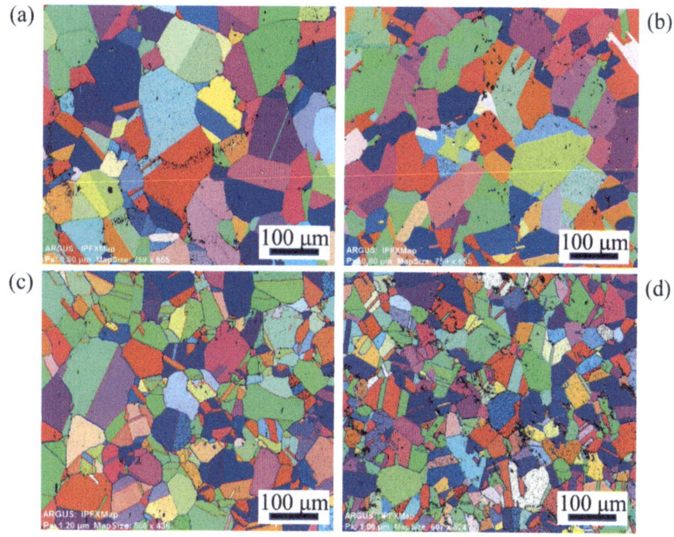

Fig. 6.15 Inverse pole figure map of specimens after heat treatment with different contents of yttrium (EBSD): **a** 0RE; **b** 0.0005%RE; **c** 0.0060%RE; **d** 0.0086%RE

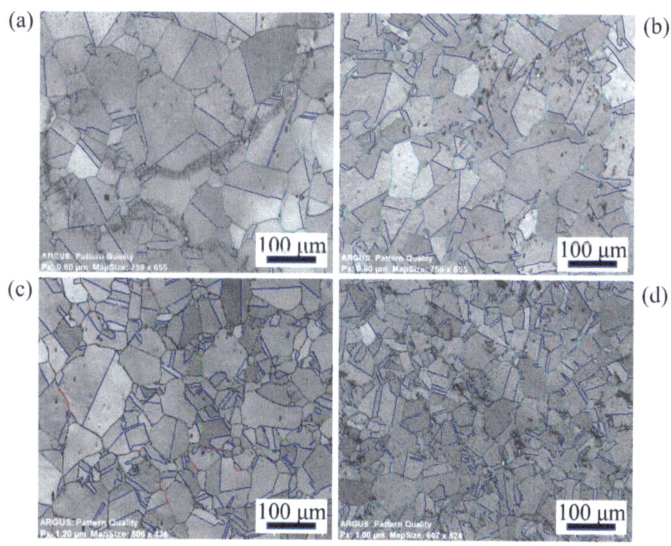

Fig. 6.16 Grain boundary microstructure of specimens after heat treatment with different contents of yttrium (EBSD): **a** 0RE; **b** 0.0050%RE; **c** 0.0060%RE; **d** 0.0086%RE

The formation of twins in austenitic steel is mainly affected by the stacking fault energy. Adding rare earth can effectively reduce the stacking fault energy, increase the proportion of twin grain boundary after high temperature deformation and refine austenite grains. In addition, rare earth can promote the recrystallization behavior of austenitic steel and increase the proportion of fine austenite grains.

Grain boundary orientation difference distribution in austenitic hot-work die steel after heat treatment with different rare earth contents is shown in Fig. 6.17.

As seen from Fig. 6.17, with the increase of rare earth content, the proportion of small angle grain boundary in austenitic hot-word die steel decreases gradually, as the proportion of large angle grain boundary increases. Especially, the proportion of ∑3 (60°) twin grain boundary increases obviously.

Many researches indicated that [29, 30] the twin grain boundary formed during high temperature deformation in high manganese austenitic steel can play the role of ordinary austenite grain boundary. Therefore, the area ratio of twin grain boundary can be increased by adding rare earth. The austenite grains can be refined by using twin grain boundary, hindering the movement of dislocation and improving the strength and toughness of the matrix.

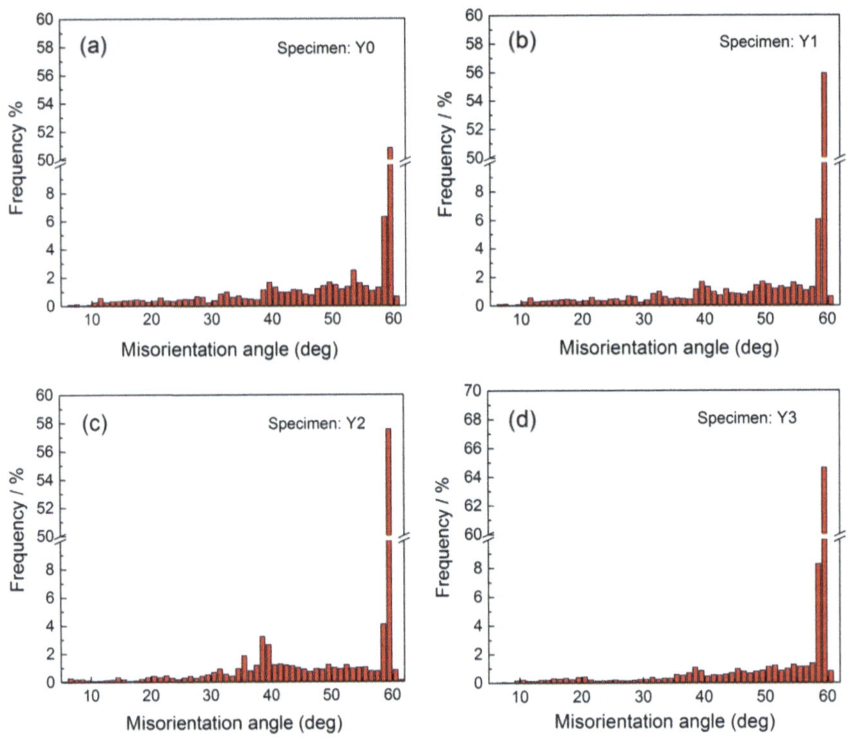

Fig. 6.17 Grain boundary orientation distribution in austenitic hot-work die steel: **a** 0RE; **b** 0.0050%RE; **c** 0.0060%RE; **d** 0.0086%RE

6.4 Effect of Rare Earth on Mechanical Properties of Austenitic Hot-Work Die Steel

The hardness and impact toughness of specimens after heat treatment with different contents of yttrium were shown in Fig. 6.18.

As shown in Fig. 6.18, with the increase of rare earth content, the hardness and toughness of austenitic hot-work die steel. When rare earth content is 0.0060%, the hardness and impact toughness are the highest, HRC 49.2 and 19.6 J respectively. When the rare earth content continues to increase to 0.0086%, the hardness and toughness decrease, especially the toughness decrease sharply.

The engineering tensile stress–strain curves of specimens after heat treatment with different contents of yttrium was shown in Fig. 6.19.

As seen from Fig. 6.19, with the increase of rare earth content, the tensile strength increases gradually. When rare earth content is 0.0060%, the tensile strength reaches the maximum. Further increasing rare earth content to 0.0086%, the tensile strength decreases.

The typical fracture morphologies of specimens after heat treatment with different contents of yttrium were shown in Fig. 6.20.

As seen from Fig. 6.20, a few dimples exist on the impact fracture surface of RE-undoped steel and 0.0005% RE steel, which are brittle intergranular fracture. For the steel containing 0.0060% rare earth, lots of dimples exist and can be considered ductile fracture. Increasing rare earth content to 0.0086%, some intergranular fracture is observed, which belong to mixed fracture. In addition, some carbides can be observed on the grain boundary or in the dimples. The element mappings of fracture surface of steel containing 0.0060% rare earth is shown in Fig. 6.21. Compared with other samples, the carbides in the dimples are smaller and distributes more uniformly.

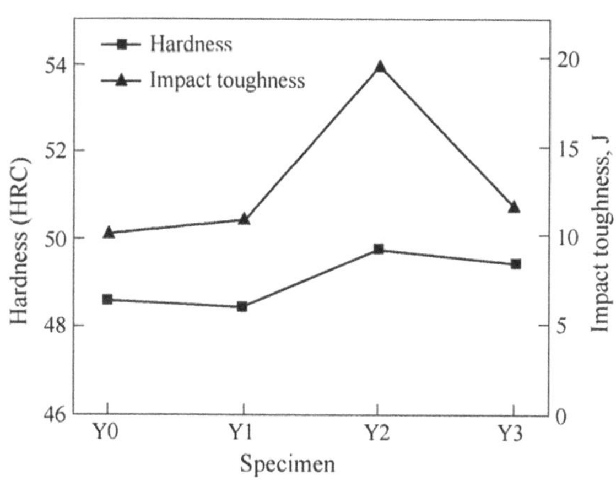

Fig. 6.18 Effect of rare earth on hardness and toughness of austenitic hot-work die steel

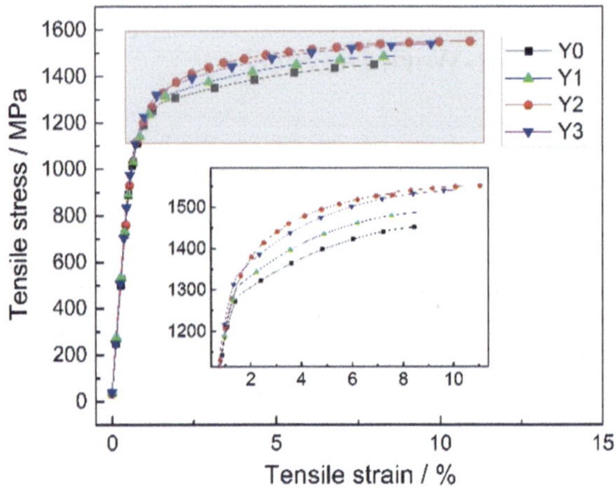

Fig. 6.19 Engineering tensile stress–strain curves of specimens after heat treatment with different contents of yttrium

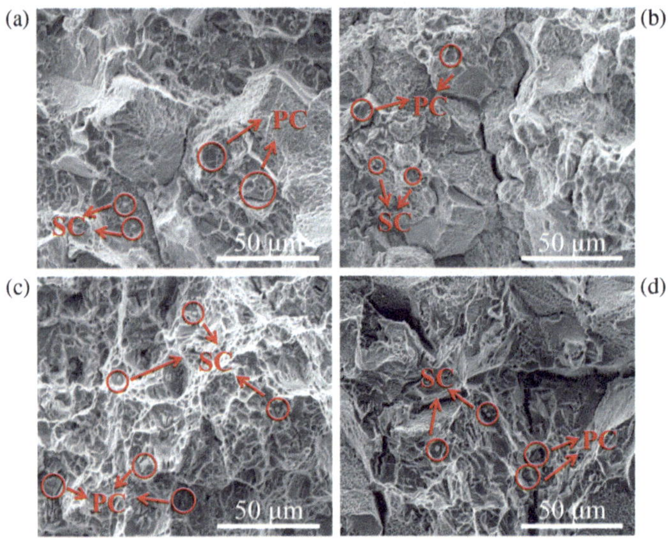

Fig. 6.20 The typical fracture morphologies of specimens after heat treatment with different contents of yttrium: **a** 0RE; **b** 0.0050%RE; **c** 0.0060%RE; **d** 0.0086%RE

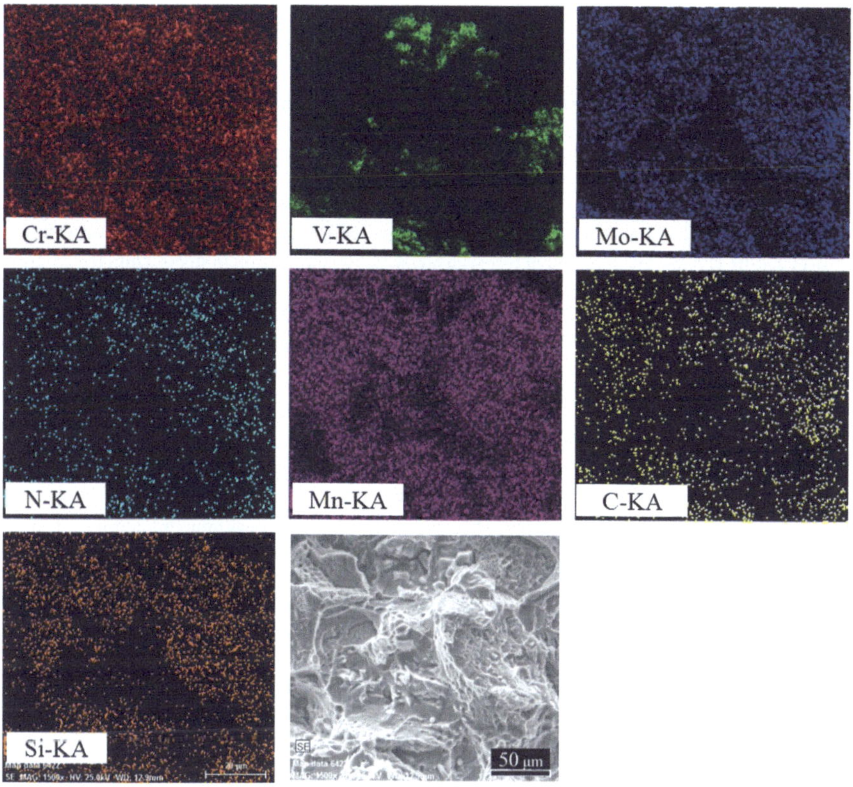

Fig. 6.21 Typical element mapping of fracture morphology of specimen Y2 after heat treatment

References

1. Yu SC, Wu SQ, Yang JQ et al (2004) Influence of rare earth on microstructure and mechanical properties of 5Cr21Mn9Ni4N steel. J Rare Earths 22(S1):122–125
2. Yang J, Hao FF, Li D et al (2012) Influence of rare earth on microstructure and mechanical properties of 5Cr21Mn9Ni4N steel. J Rare Earths 30(8):814–819
3. Lan J, He JJ, Ding WJJ et al (2001) Study on heterogeneous nuclei in cast H13 steel modified by rare earth. J Rare Earths 19(2):280–283
4. Lan J, He JJ, Ding WJ et al (2000) Influence of RE on solidification structure and impact toughness of cast H13 steel. Iron Steel 35(10):48–50
5. Xu GX (1995) Rare earths. Metallurgical Industry Press, Beijing, p 1995
6. Yang CD (2009) A study on processes and properties of Y-base rare earth inoculating high manganese steel. Dissertation, Kunming University of Science and Technology
7. Huo WX, Ren HP, Jin ZL et al (2012) Effects of microstructure and mechanical properties of high manganese steel containing different content of rare earth. Hot Working Technol 41(7):15–17
8. Hamidzadeh MA, Meratian M, Saatchi A (2013) Effect of cerium and lanthanum on the microstructure and mechanical properties of AISI D2 tool steel. Mater Sci Eng A 571:193–198

9. Hufenbach J, Helth A, Lee MH et al (2016) Effect of cerium addition on microstructure and mechanical properties of high-strength Fe85Cr4Mo8V2C1 cast steel. Mater Sci Eng A 674:366–374
10. Li YB, Wang FM, Li CR (2009) Effect of cerium on grain and carbide in low chromium ferritic stainless steels. J Chin Rare Earth Soc 27(1):123–127
11. Wang ZJ (2001) Modification of RE high manganese steel obviously prolong parts service life. N Technol N Proc 1:28–29
12. Zhang YY, Ma JR, Su C et al (2015) Grain boundary engineering process and morphology of carbides precipitated at grain boundaries in Inconel alloy 600. Shanghai Met 37(6):46–50
13. Li CL (2013) New development of research on rare earth application in steels. Chin Rare Earths 34(3):78–85
14. Lin Q, Ye W, Du YS et al (1992) Effects of rare earth metal on the properties of steels and optimization control. J Univ Sci Technol Beijing 14(2):225–231
15. Hu WH, Yuan Y, Liu X et al (2003) On the behavior of soluble aluminum in steel. Iron Steel 38(7):42–44
16. Zhang F, Lv XJ, Wang B et al (2011) Control of inclusions of non-oriented silicon steel sheets by RE treatment. Iron Steel Vanadium Titan 32(3):46–50
17. Yang JC, Liu X, Gao XZ et al (2007) Effect of rare earth element Ce on modification of inclusions in stainless steel 2Cr13. Spec Steel 28(3):30–31
18. Casper V, Grong O, Haakonsen F et al (2009) Progress in the development and use of grain refiner based on cerium sulfide or titanium compound for carbon steel. ISIJ Int 49(7):1046–1050
19. Song MM, Song B, Xin WB et al (2015) Effects of rare earth addition on microstructure of C-Mn steel. Ironmak Steelmak 42(8):594–599
20. Wen B, Song B (2012) In situ observation of the evolution of intragranular acicular ferrite at Ce-containing inclusions in 16Mn steel. Steel Res Int 83(5):487–495
21. Wang LM, Du T, Lu XLL et al (2001) Study of behaviors and application of micro-rare earth elements in steel. Chin Rare Earths 22(4):37–40
22. Deng XX, Wang XH, Jiang M et al (2012) Effect of inclusions on the formation of intragranular acicular ferrite in steel containing rare earth elements. J Univ Sci Technol Beijing 34(5):535–540
23. Pan N, Song B, Zhai QJ et al (2010) Effect of lattice disregistry on the heterogeneous nucleation catalysis of liquid steel. J Univ Sci Technol Beijing 32(2):179–182, 190
24. Pan N, Song B, Zhai QJ (2009) Catalysis on heterogeneous nucleation of solid compounds in liquid steel. Acta Metall Sin 45(12):1441–1445
25. Lin Q, Song B, Guo XM et al (2001) Effects of RE on microalloying in steel and application prospects. Chin Rare Earths 22(4):31–36
26. Wang MJ, Chen L, Wang ZX et al (2012) Effect of rare earth addition on continuous heating transformation of a high speed steel for rolls. J Rare Earths 30(1):84–89
27. Liu HH, Fu PX, Liu HW et al (2017) Carbides evolution and tensile property of 4Cr5MoSiV die steel with rare earth addition. Metals 7(10):436–449
28. Gao JZ, Fu PX, Liu HW et al (2015) Effect of rare earth on the microstructure and impact toughness of H13 steel. Metals 5(1):383–394
29. Rajib K, Lailesh K, Satyam S (2018) Grain boundary engineering of medium Mn TWIP steel: a novel method to enhance the mechanical properties. ISIJ Int 58(7):1324–1331
30. Choi WS, SandlÖbes S, Malyar NV et al (2018) On the nature of twin boundary-associated strengthening in Fe-Mn-C steel. Scr Mater 156:27–31

Chapter 7
Effect of Nitrogen on the Carbide in Steel

Abstract Nitrogen can stabilize the austenite structure, enlarge austenite phase zone, and increase the precipitation temperature of carbides in special steels. In this chapter, the effect of nitrogen on the microstructure, carbide and mechanical properties of high-Mn austenitic die steel is studied. It is found that nitrogen has no effect on the types and morphology of carbides, but changes the composition of carbides. There are more primary precipitates in high-Mn austenitic die steel after adding nitrogen, and the size of primary precipitates increases, which could be related to the increase of precipitation temperature of primary precipitates with nitrogen addition in the steel. After aging treatment, the secondary precipitates in high-Mn austenitic steel with nitrogen addition are only spherical or near spherical V(C, N) with 100–200 nm in size and uniform distribution in the steel, whereas the secondary precipitates are only V_8C_7, some of which are 0.5–1 μm in size and long strip in morphology, except a small amount of near spherical precipitates of 100–200 nm in size. The secondary precipitates in high-Mn nitrogen-containing austenitic steel are larger in volume fraction and smaller in size in comparison with that in high-Mn austenitic steel without nitrogen addition.

Keywords Nitrogen · Microstructure · Precipitates · Heat treatment · Mechanical property

Traditional hot work die steels are mostly martensitic steels, which has high hardness and high wear resistance, such as H13 [1]. However, the recovery of martensite matrix and carbide coarsening after service temperature exceeds 650 °C leads to rapid decline in strength and die failure, which limits their service temperature [2]. It would be natural to assume that die materials—which do not possess the mentioned barrier within the service temperature range and have a strength exceeding that of highly heat-resistant martensitic steel at a temperature exceeding 700 °C—should have an austenitic structure or belong to high- temperature alloys. Nitrogen is added to the austenite hot work die steel as an alloying element, which can stabilize the austenite structure and expand the austenite phase zone [3]. At the same time, the formation of VN plays a role in strengthening precipitation [4–6]. After aging and solution heat treatment, the precipitated phase of steel is finer and more uniformly dispersed, and the thermal stability at high temperature is improved [7]. The addition

J. Li and C. Shi, *Carbide in Special Steel*, Engineering Materials,
https://doi.org/10.1007/978-981-16-1456-9_7

of nitrogen in steel results in the formation of nitrides or carbonides, which can increase the precipitation temperature of the precipitated phase. It can be used as nucleating particles in the initial solidification process of liquid steel and hinder the migration and expansion of grain boundary, thus refining the initial austenite grain size.

7.1 Thermodynamic Analysis of the Effect of Nitrogen on the Precipitation Phase of Austenite Hot Work Die Steel

7.1.1 Effect of Nitrogen on Precipitation Temperature of Precipitation Phase

The chemical composition of austenitic steel with and without nitrogen is shown in Table 7.1. The solubility of nitrogen in austenitic steel is relatively high, up to 0.4%. The experimental steel was melted in a vacuum-induction furnace. Nitrogen-containing ferrochromium alloy was added in the smelting process to obtain nitrogen-containing steel (HMSA-N). The experimental steel without nitrogen (HMSA) is affected by nitrogen in the atmosphere during the smelting process, so that there is a very small amount of nitrogen in the experimental steel.

The addition of nitrogen to the austenite hot work die steel can not only stabilizes austenite, but also precipitates high temperature resistant carbonitrides (M(C, N)) to improve its high temperature strength. The thermodynamic software Thermo-Calc was used to calculate the precipitation phase and temperature of the high-manganese austenitic hot-working die steel (HMAS-N) with 0.15% nitrogen and high-manganese austenite with 0.0014% nitrogen (HMAS), as shown in Fig. 7.1.

The precipitation phase types of high manganese austenitic hot die steel (HMAS-N) and (HMAS) are the same, which are austenitic, MC or M(C, N), M_2C, $M_{23}C_6$, M_6C and M_7C_3 phases. However, the precipitation and disappearance temperatures of each phase are quite different, and the critical temperatures of each equilibrium precipitation phase are shown in Table 7.2.

For high manganese austenitic hot die steel HMAS-N containing 0.15% nitrogen, austenite began to form at 1394.79 °C. With the continuous decrease of temperature, carbonitrite (M(C, N)), M_2C, M_7C_3 and $M_{23}C_6$ phases were precipitated successively, and their initial precipitation temperatures were 1389.85 °C, 930 °C,

Table 7.1 Chemical composition of tested steel (wt%)

Steel	C	Mn	Cr	Mo	V	Si	P	S	N	Fe
HMAS-N	0.56	14.5	3.192	1.641	1.723	0.52	0.015	0.029	0.15	Bal.
HMAS	0.59	14.9	3.53	1.55	1.726	0.54	0.013	0.026	0.0014	Bal.

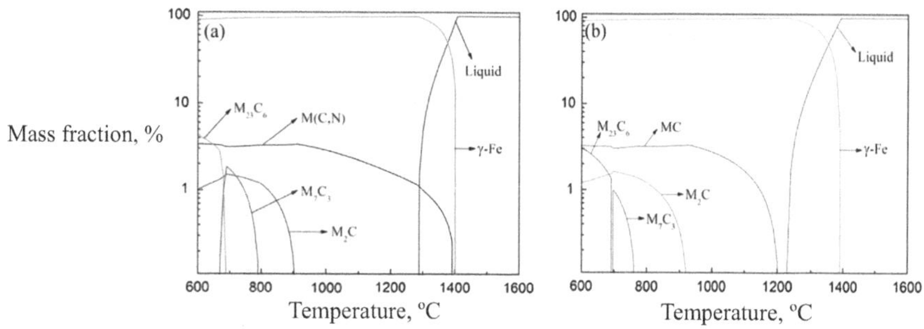

Fig. 7.1 Equilibrium phase precipitation in austenite hot work die steel steel calculated using Thermo-Calc: **a** HMAS-N; **b** HMAS

Table 7.2 Transformation temperatures of precipitates in steel (°C)

Steel	Precipitation temperature (T_S)					Disappearance temperature (T_f)		
	MC	M_2C	$M_{23}C_6$	M_6C	M_7C_3	M_2C	$M_{23}C_6$	M_7C_3
HMAS-N	1390	930	670	453	810	453	437	652
HMAS	1200	1200	699	453	77.0	453	399	699

810 °C and 679.5 °C, respectively. With the decrease of temperature, M_7C_3 phase continuously transforms to $M_{23}C_6$ phase, until the temperature drops to 660 °C, the transformation is completed and M_7C_3 phase disappears.

For the hot Austenite mold steel HMAS with nitrogen 0.0014%, austenite began to form at 1390 °C, and HMAS-N formed austenite at 1394.79 °C, so the nitrogen content had little effect on the austenite precipitation temperature. When the temperature drops to 1200 °C, MC phase begins to precipitate out. For HMAS-N, the precipitation temperature of M(C, N) was 1394.79 °C, and the temperature difference between the two steels was close to 200 °C, indicating that the addition of nitrogen element increased the precipitation temperature of MC phase. The precipitation temperature of M_2C, $M_{23}C_6$ and M6C in the two steels is almost the same, indicating that nitrogen has no influence on the precipitation temperature of these phases. The precipitation temperature of M_7C_3 in HMAS was 770, 40 °C lower than that of M_7C_3 in HMAS-N, and the temperature range of the precipitation phase in HMAS was narrower.

The results of thermodynamic calculation show that the precipitation temperature of MC phase can be significantly increased by adding an appropriate amount of nitrogen, thus improving the stability of V(C, N). The fine and dispersed MC phase will play a role in reducing the coarsening of austenite grains by pinning grain boundaries during solidification and solid solution heat treatment, thus improving the strength and toughness of steel. However, if a large amount of V(C, N) of large size is precipitated at the grain boundary, it is easy to become the crack source and significantly reduce the toughness of hot die steel [8]. Therefore, it is necessary to reduce the size of V(C, N) through process control in the forging process. The precipitated phase is dissolved in the austenite matrix as much as possible through the solid solution heat treatment, and then the fine and dispersed secondary carbide is precipitated out through the aging heat treatment, so as to improve the strength and toughness of the steel.

7.1.2 Effect of Nitrogen on the Composition of Precipitation Phase of Austenite Hot Work Die Steel

The effect of nitrogen on the element composition of the precipitated phase is shown in Fig. 7.2.

It can be seen from Fig. 7.2a that the main constituent elements of the MC-type precipitation phase are V, N, and C, so the MC phase is mainly V(C, N). V(C, N) is the high temperature precipitation phase. When the precipitation begins at 1390 °C, the content of N is the highest and the content of C is very low. With the decrease of temperature, the content of N in MC phase decreases continuously, while the content of C increases continuously. The content of V in the MC phase is almost unchanged in the temperature range of 800–1300 °C, which shows that the MC phase mainly exists in the form of N-rich VC_xN_{1-x} in the high temperature section. With the decrease of temperature, N element in VN is gradually replaced by C element. Therefore, MC phase mainly exists in the form of C-rich VC_xN_{1-x} at low temperature, and a small amount of Cr and Mo elements are also dissolved in M(C, N) phase.

According to Fig. 7.2b, M_2C is a close packed hexagonal crystal structure, which can be inferred as Mo_2C-type alloy carbide [9]. The main elements of M2C precipitate are Mo, V, C and a small amount of Cr and N.

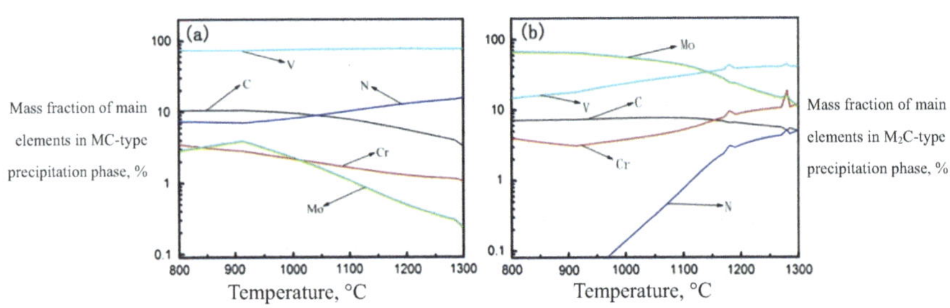

Fig. 7.2 Alloy element composition of precipitated phase: **a** MC; **b** M_2C

7.2 Effect of Nitrogen on the Microstructure and Precipitation Phase of Annealed ESR Ingot

7.2.1 Effect of Nitrogen on the Dendrites of Annealed ESR Ingot

Samples were taken at the edge, 1/2 radius and center of the ESR ingot, and then observed by optical microscope after corrosion, as shown in Fig. 7.3. At the same time, 20 data were randomly measured for the two steels. The change of dendrite spacing with sampling position is shown in Fig. 7.4.

According to Figs. 7.3 and 7.4, the primary dendrite spacing from the edge of the ESR ingot to the central position of the ESR ingot increases gradually. Nitrogen-containing austenitic hot-working die steel has a larger primary dendritic spacing at each position of the ESR ingot than that of non-nitrogen-free austenitic hot-working die steel. This can be explained by Kurz-Fisher theory [10]. The primary dendrite spacing can be expressed by the following equation:

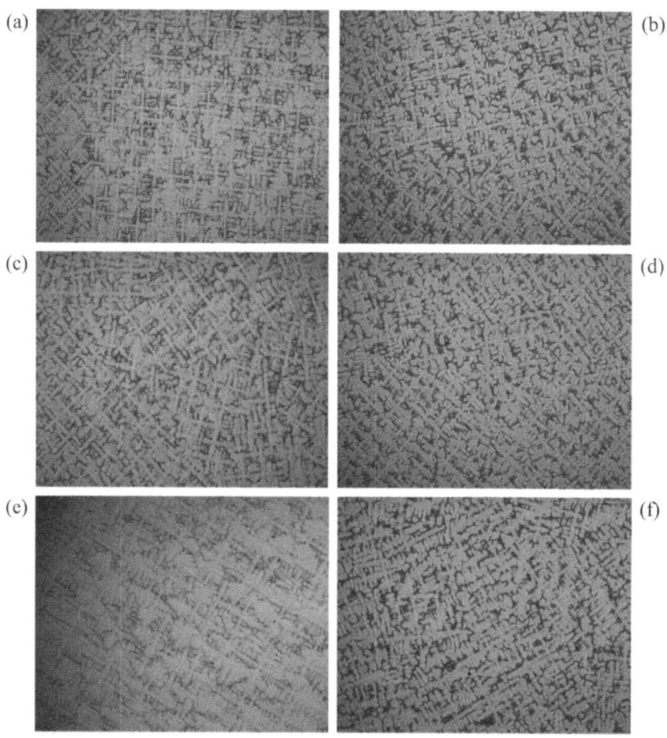

Fig. 7.3 Solidification structure of different positions of ESR ingot: **a**, **c** and **e** are the edge, 1/2 radius and center of HMAS respectively; **b**, **d** and **f** are the edge, 1/2 radius and center of HMAS-N respectively

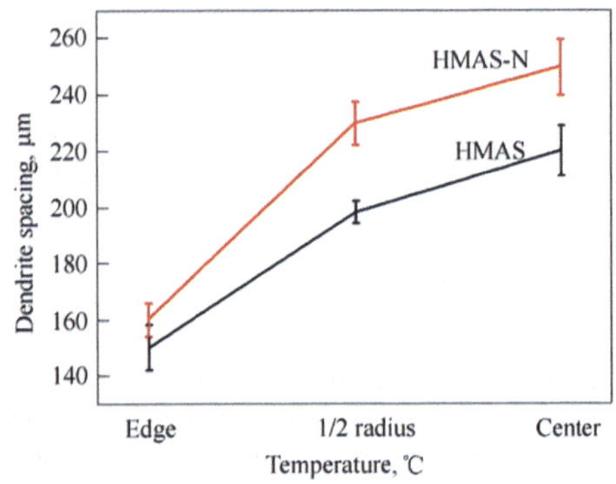

Fig. 7.4 Dendrite spacing of the two steels changes with position

$$\lambda_1 = \frac{4.3(\Delta T_0 D\Gamma)^{0.25}}{(KR)^{0.25}G^{0.25}} \tag{7.1}$$

where $\Delta T_0 = mC_0(1-k)/k$, represents the solidification temperature range; D is the solute diffusion coefficient; Gibbs coefficient Γ is the ratio of solid-liquid surface energy to the melting entropy; K is the equilibrium solute distribution coefficient; R is the solidification rate; G is the temperature gradient.

In the case where other variables are the same,

$$\Delta T_0 = mC_0(1-k)/k, \tag{7.2}$$

where m is the liquidus slope; C_0 is the initial solid solution of the solute element in the liquid.

According to Eq. (7.2), the primary dendrite spacing equation can be simplified to an equation related to the amount of solid solution:

$$\lambda_1 = A(C_0)^{0.25} \tag{7.3}$$

where the constant A in the equation represents other factors besides the solid solution amount C_0.

It can be seen that the dendrite spacing increases with the increase of nitrogen content [11]. When other variables are the same, the primary dendrite spacing is proportional to the original composition of the alloy. For the edge, 1/2 position, and center of HMAS, the primary dendrite spacing is 150 ± 8, 198 ± 4, and 220 ± 9 μm; For the edge, 1/2 position, and center of HMAS-N, the primary dendrite spacing is 160 ± 6, 230 ± 8, 250 ± 10 μm.

7.2.2 Effect of Nitrogen on Precipitation Phase of ESR Ingots

It can be seen from Fig. 7.1 and Table 7.2 that M(C, N), M_2C, $M_{23}C_6$ and M_7C_3 precipitate in sequence with the decrease of temperature. The precipitation temperature of M(C, N) is above1300 °C. This indicates that M(C, N) is a highly stable phase and remains undissolved during heat treatment process.

To predict phase precipitation during liquid steel solidification in a practical ESR refining process, the Scheil–Gulliver model included in Thermo-Calc software was employed to calculate the nonequilibrium phase precipitation in HMAS-N, as shown in Fig. 7.5. M(C, N) precipitates from liquid steel directly. As the austenite continues to precipitate from liquid steel, carbon and alloying element contents keep increasing. Primary carbides M_2C precipitate directly from liquid steel when the solid fraction of liquid steel exceeds 0.93.

The microstructure, carbide and carbide composition analysis of HMAS and HMAS-N ESR ingots after annealing are shown in Fig. 7.6.

Figure 7.6a and b are the microstructures of HMAS and HMAS-N, respectively. It can be seen that there are two kinds of precipitated phase with large amount. One is an irregular multilateral strip with a size of 10–50 μm, and the other is lamellar fish-skeleton-like with a size of 5–25 μm, mostly distributed in grain boundary and a small part in matrix, as shown in Fig. 7.6c and d.

Figure 7.6e and f are the elements mapping scanning images corresponding to Fig. 7.6c and d. They show that for the austenite hot work die steel HMAS ESR ingot containing 0.0014% nitrogen, vanadium, chromium and carbon are distributed on the strip precipitates. It indicates that the strip precipitates are mainly V alloy carbides, and Mo, Cr and carbon are distributed in the fishbone frame precipitation phase, i.e. the fish skeleton precipitate phase is Mo alloy carbide, and contains a certain amount. In addition to vanadium, chromium and carbon distributed in the strip like

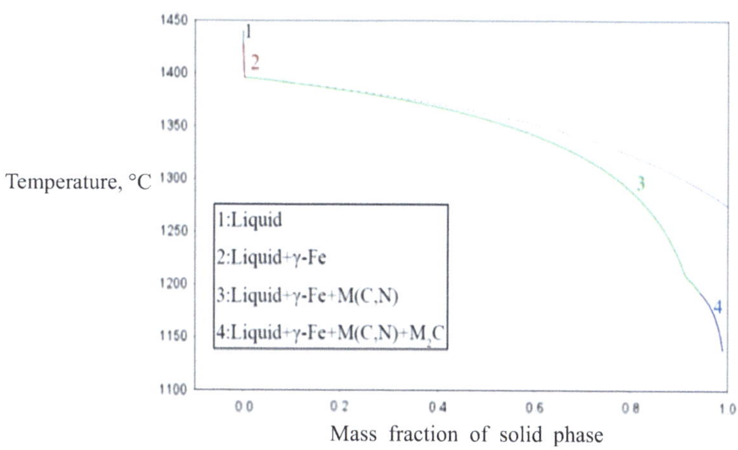

Fig. 7.5 Nonequilibrium phase precipitation of HMAS-N steel calculated using Thermo-Calc

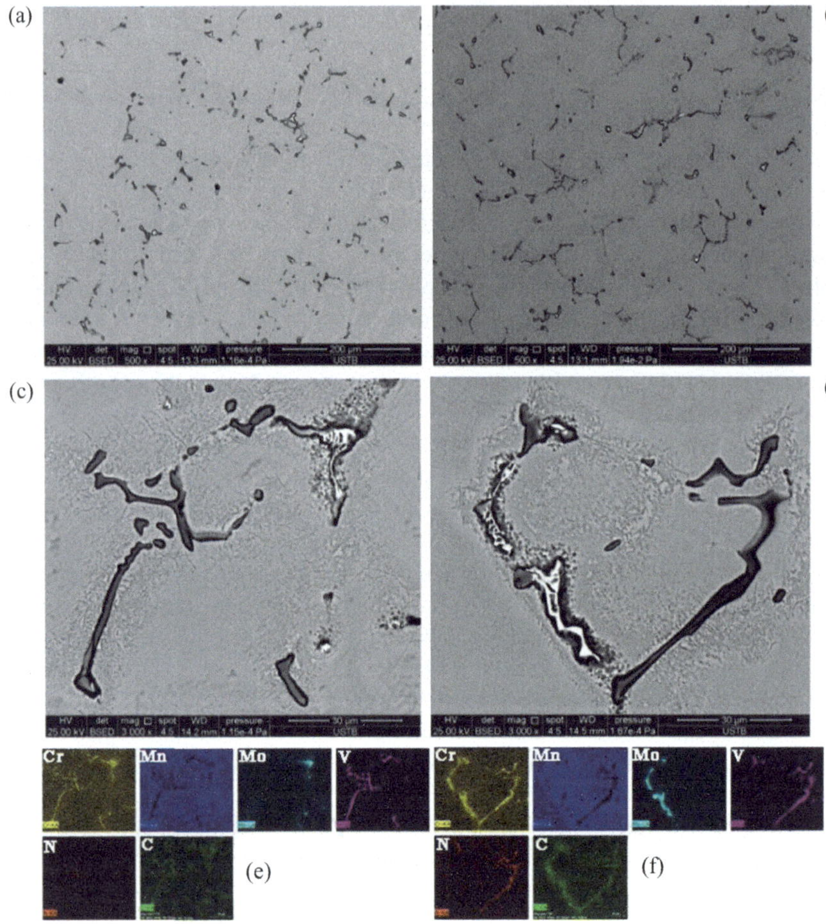

Fig. 7.6 Scanning electron microscopy (SEM) images of ESR ingot microstructure: **a**, **c** HMAS; **b**, **d** HMAS-N; **e** and **f** are scan photographs of **c** and **d**, respectively

precipitates, there are more nitrogen elements in HMAS ESR ingot of austenitic hot work die steel. HMAS containing 0.0014% nitrogen, that is, vanadium carbonitride and molybdenum alloy carbide.

Electrolytically extract the precipitate from steel in an organic solution to observe the three-dimensional morphology of the precipitated phase, as shown in Fig. 7.7.

From the analysis of Fig. 7.7c and d, it can be seen that the three-dimensional morphology of the irregular long strip or short rod precipitate should be irregular multilateral strip or slice shape with severe angles; rod-shaped or lamellar fish-skeleton-like precipitates are typical eutectic carbide morphologies. Table 7.3 shows the energy spectrum analysis of different types of carbides in Fig. 7.7b.

As shown in Table 7.3, M(C, N) (namely V(C, N)) contains a certain amount of the elements Cr and Mo. In addition, M_2C is Mo_2C, which contains more Cr and Mo

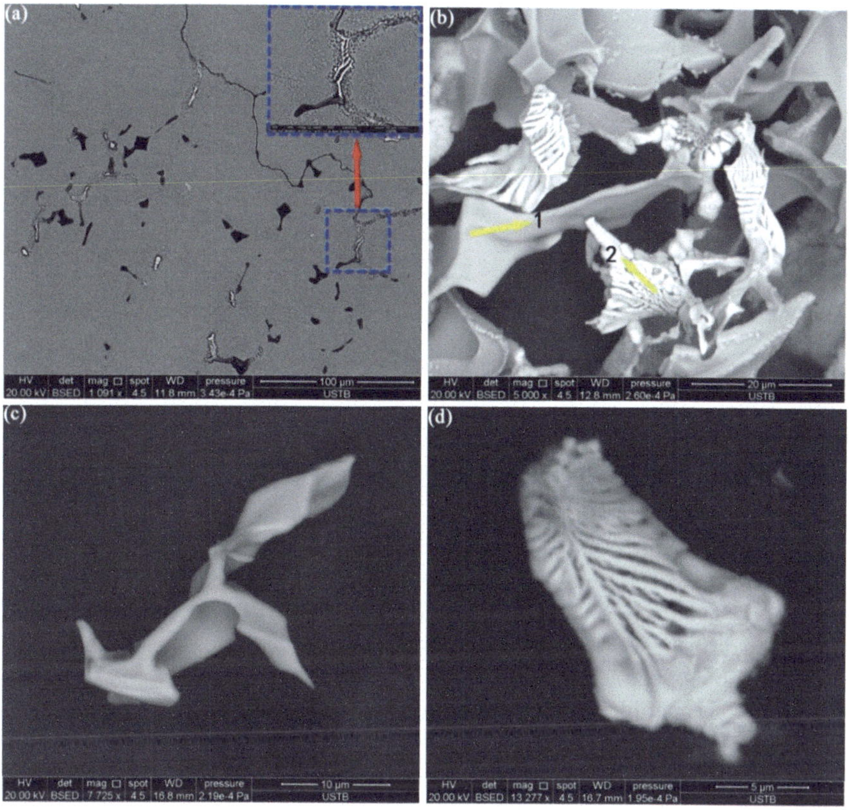

Fig. 7.7 SEM micrographs showing the three-dimensional morphology of carbides or carbonitrides extracted from ESR ingot

Table 7.3 Energy-dispersive X-ray spectrometer EDS analyzed results of carbides or carbonitrides (wt%)

Point	Element						
	C	V	Mo	Cr	N	Fe	Mn
1	14.68	41.60	0.64	1.68	33.11	6.31	1.98
2	56.69	6.88	17.35	7.94	0.05	5.55	5.54

atoms. The morphology of V(C, N) is an irregular multilateral strip or slice shape with severe angles. The morphology of Mo_2C is rod-shaped or lamellar fish-skeleton-like.

XRD technology was used to analyze the powder collected by electrolytic extraction, and the results are shown in Fig. 7.8. The precipitated phase of HMAS is V_8C_7 and Mo_2C, and the precipitated phase of HMAS-N is V(C, N) and Mo_2C.

There are a lot of vanadium nitrides in the ESR ingots of HMAS. This may be because the crystal lattice of nitride and solid solution matches better in structure

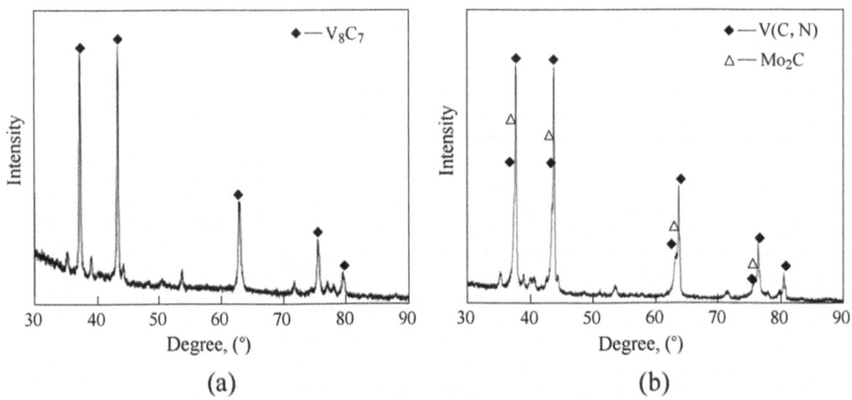

Fig. 7.8 XRD patterns of precipitates powder: **a** HMAS; **b** HMAS-N

and has a stronger binding force with the matrix than that of carbides. The diffusion coefficient of nitrogen is higher than that of carbon. In the aging process, the supersaturated nitrogen is preferred to be concentrated near the grain boundary and other defects, making nitrides more nucleated than carbides. Due to the high nitrogen content, the diffusion coefficient of chromium in austenite decreases, which hinders the precipitation of carbide [12].

Nitrogen has no effect on the morphology of the precipitated phase of austenitic hot die steel. However, compared with HMAS, there are more primary precipitation phases in HMAS-N, and the size of primary precipitation phase is larger. This may be related to the addition of N, which is beneficial to increase the precipitation temperature of MC. Both HMAS and HMAS-N have large primary precipitation phases, which need to be forged at high temperature to make large carbides or carbonitride more uniformly dispersed.

7.3 Effect of Heat Treatment Process on the Structure and Precipitates in Nitrogen-Containing Austenitic Die Steel

7.3.1 Effect of Solution Heat Treatment on Carbides in Steel

Solid solution heat treatment (SSHT) was carried out to dissolve coarse primary precipitates in the matrix, which were then reprecipitated from supersaturated solid solution during the subsequent aging process with the purpose of precipitation strengthening. If the solution temperature is too low, large primary precipitates are difficult to dissolve into matrix. However, if the solid solution temperature is too high,

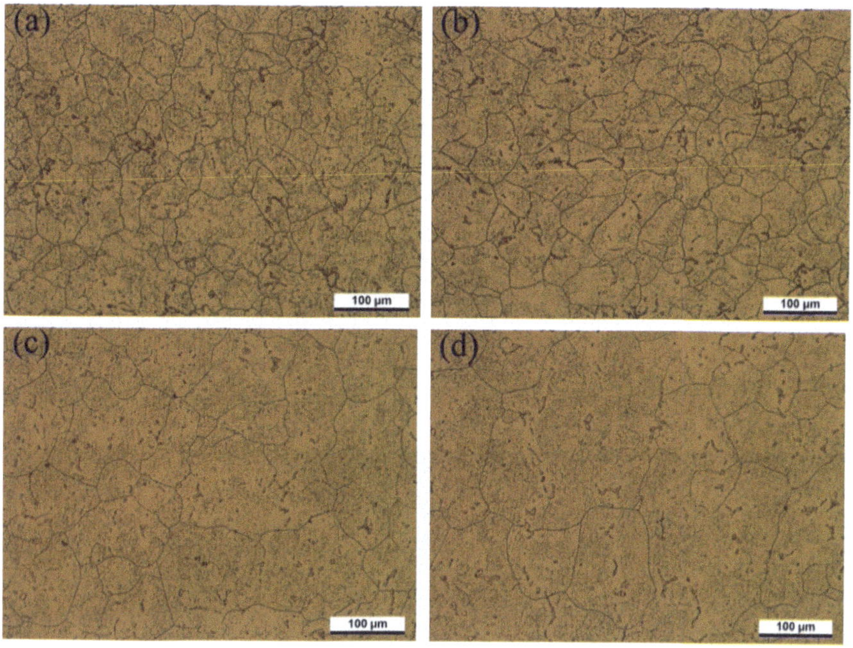

Fig. 7.9 Effect of solution temperature and solution time on microstructure evolution: **a** 1170 °C, 0.5 h; **b** 1170 °C, 2 h; **c** 1200 °C, 2 h; **d** 1230 °C, 3 h

austenitic grain size (AGS) will grow up excessively. The AGS tremendously influences diffusive and diffusionless phase transformations, precipitation, and mechanical properties such as strength, hardness, toughness, and ductility [13]. Hence, the SSHT process is quite important for microstructures and mechanical properties.

The microstructure evolution at different solution temperatures and solution times is shown in Fig. 7.9.

As shown in Fig. 7.9a and b, the precipitates were not completely dissolved at 1170 °C, which is consistent with thermodynamic calculation using Thermo-Calc software. Thermodynamic calculation illustrates that precipitation temperature of the primary V(C, N) reaches up to 1390 °C, as shown in Table 7.2. N addition in steel increased the melting point and stability of carbonitrides. Meanwhile, the pinning effect of undissolved V(C, N) [14] inhibits the growth of austenite grains [15].

More carbides, carbonitrides, and alloying elements dissolved into the austenite matrix with increasing solid solution time at 1170 °C. Austenitic grain-coarsening speed is accelerated, and undissolved V(C, N) pinning effect was weakened when the solution temperature exceeds 1200 °C. Under higher temperature conditions, the precipitates are more likely to dissolve and coarsen [16, 17], which weakens the solute drag effect of alloying elements at high temperatures [18]. The increase of the solution temperature has a great drive on the grain growth. Therefore, austenite grain coarsening is obvious when the solution heat treatment is carried out at 1230 °C.

7.3.2 Effect of Aging Heat Treatment on Carbides in Steel

To remove residual stress caused by SSHT and dispersed small secondary phase particles, high-temperature aging heat treatment was carried out [19, 20]. TEM and diffraction spot calibration results are shown in Figs. 7.10 and 7.11. The result shows that the precipitated phase is V_8C_7.

It can be seen from Fig. 7.10 that there are large secondary precipitate phases in the austenitic hot die steel without nitrogen after the optimal heat treatment process. The size of secondary precipitate phases is 0.5–1 μm. Most of the morphology is elongated, while a small number of nano-sized precipitate phases are near spherical.

After pre-aging at 650 °C for 1 h and then aging at 780 °C for 1 h, the structure of the nitrogen-containing austenite hot work die steel is shown in Fig. 7.12.

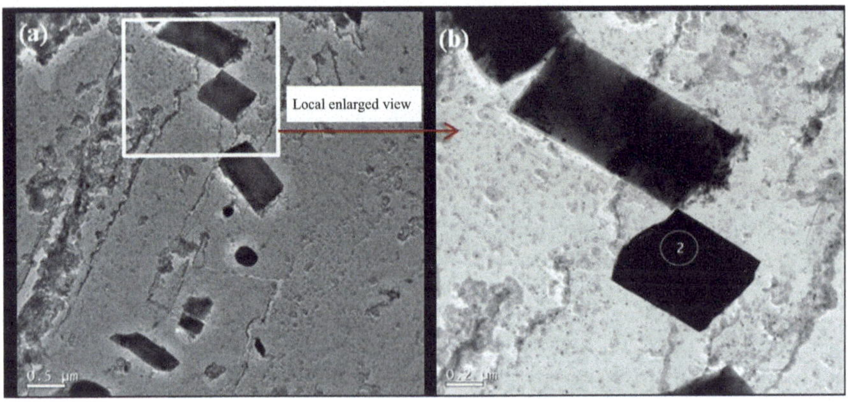

Fig. 7.10 TEM results of HMAS: **a** and **b** is (**a**) local enlarged view

Fig. 7.11 Diffraction spot calibration result

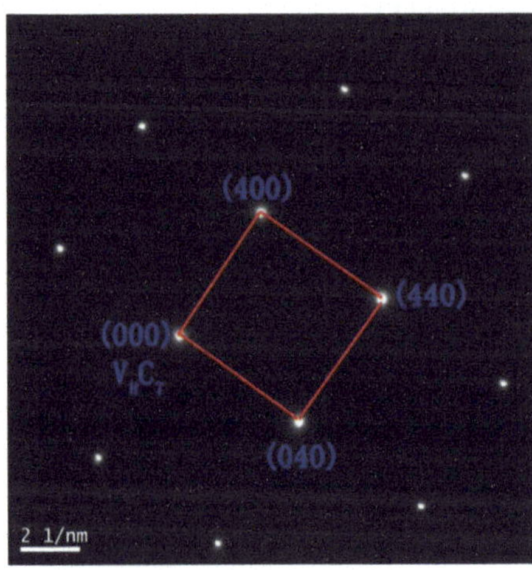

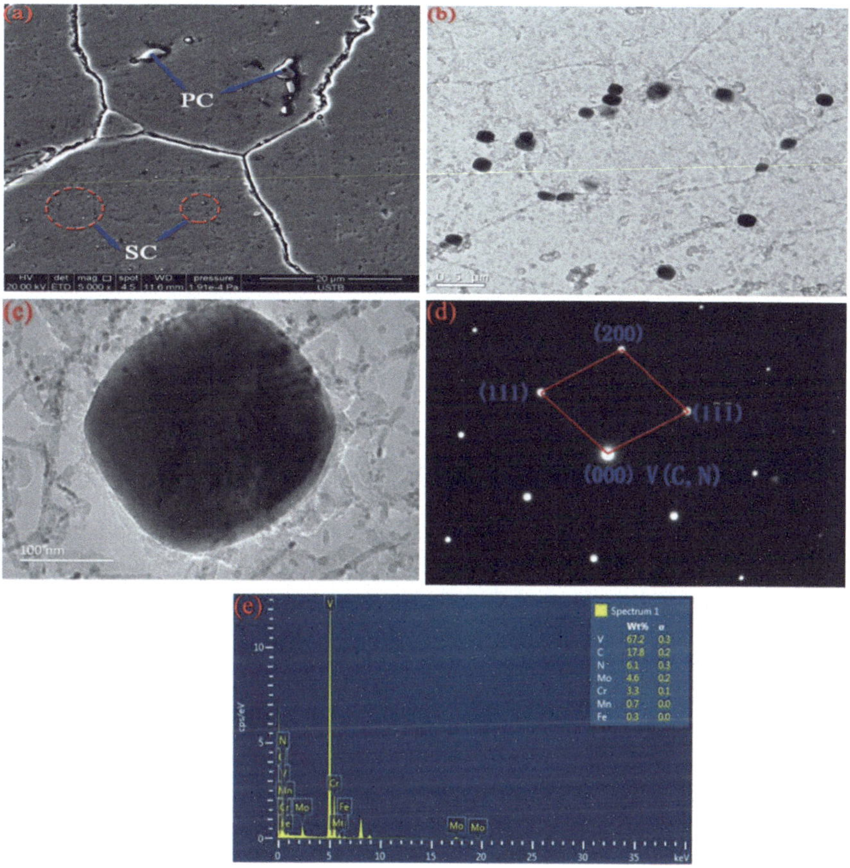

Fig. 7.12 Microscopic observation results of HMAS-N after the best heat treatment process: **a** SEM; **b–d** TEM and diffraction spot calibration results; **e** TEM energy spectrum analysis

It can be seen from Fig. 7.12a that there are both undissolved primary V(C, N) at the micron level and a large number of secondary precipitation phases at the nanometer level in the steel. Figure 7.11b shows the TEM image and diffraction pattern calibration results of precipitated phase, and Fig. 7.12c shows the TEM energy spectrum results of precipitated phase in Fig. 7.12d. It can be seen that the secondary precipitated phase is uniformly distributed in steel, with a size of about 100–200 nm and a shape of spherical or near-spherical morphology. This is because there is the following parallel orientation relationship between austenite and precipitated microalloy carbides [21]: (100) V(C, N) //(100)γ, (010) V(C, N) //(010)γ. Under the control of this parallel orientation relationship, the interface energy between austenite and precipitated phase plays a significant role, and the mismatch in all directions is the same, so the V(C, N) precipitated in austenite crystal should be spherical. The results of diffraction pattern calibration based on the nano-scale precipitated phases show that the precipitated phases are all V(C, N). V(C, N) has a high hardness (above HV1200), has a face-centered cubic structure, and has a coherent relation with austenite matrix, which is not easy to grow under a certain temperature. In general, dispersion strengthening of MC carbide can be maintained up to 650–700 °C [22]. The addition of N further increases the stability of the secondary precipitated phase, reduces the coarsening rate, and keeps the dispersion strengthening at a higher temperature. So as to ensure the material has a higher thermal stability at high temperature and meet the requirements of mold steel at a higher temperature.

In summary, most of the smaller particles of austenitic hot-work die steel with a nitrogen content of 0.15% (HAMS-N) are spherical or close to spherical, with uniform distribution. In addition, the volume fraction is significantly higher than the austenitic hot work die steel with a nitrogen content of 0.0014% (HMAS). It shows that the N element is beneficial to the precipitation of the second phase and makes the second precipitation phase smaller. For austenitic hot work die steel with a nitrogen content of 0.0014% (HMAS), the larger size of the precipitated phase is likely to become the source of cracks, resulting in a decrease in the toughness of the steel; For austenite hot-working die steel with a nitrogen content of 0.15% (HMAS-N), the small amount of dispersed secondary precipitated phase particles will increase the wear resistance of the die steel. At the same time, a large number of secondary precipitated phase particles will also reduce the solid solution alloy of the steel matrix, make the steel matrix softer, and make the steel have higher toughness, which is consistent with the measured value of actual impact toughness.

7.3.3 Effect of Heat Treatment on the Mechanical Properties of Carbides of N-Containing Austenitic Die Steel

The Rockwell hardness and impact energy of specimens aged with different aging heat treatment processes are shown in Fig. 7.13. The following two kinds of aging heat treatments (single-stage aging and two-stage aging) were carried out. Single-stage aging and two-stage aging schemes are shown in Table 7.4 and Table 7.5, respectively.

As for single-stage aging, shown in Fig. 7.13a, when the aging temperature was at 720 °C, significant secondary hardening occurred, and the hardness reached the maximum value of 47.24 HRC. It can be attributed to the enrichment and segregation of precipitate-forming elements at 720 °C, such as V, Mo, Cr, C, and N. As the aging temperature continued to increase, the hardness decreased to 45.04 HRC at 760 °C. As for double-stage aging, shown in Fig. 7.13a, the hardness increases with the re-aging temperature increasing from 700 to 760 °C. When the two-stage aging temperature was at 760 °C, the hardness reached the peak value of 46.06 HRC. As the re-aging temperature continued to increase, the hardness decreased and dropped to 43.84 HRC at 800 °C. Compared to that of single-stage, the hardness value of two-stage

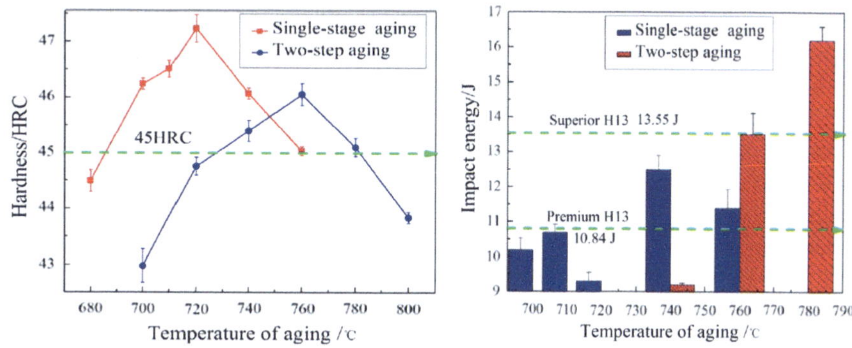

(a) **Rockwell hardness** (b) **Impact energy**

Fig. 7.13 Rockwell hardness and impact energy of specimens aged with different aging heat treatment processes

Table 7.4 Single-stage aging heat treatment process

The specimen number	Aging temperature (°C)	Aging time
S-680	680	2 h
S-700	700	
S-710	710	
S-720	720	
S-740	740	
S-760	760	

Table 7.5 Two-stage aging heat treatment process

The specimen number	Pre-aging temperature (°C)	Pre-aging time	Re-aging temperature (°C)	Re aging time
T-700	650	1 h	700	1 h
T-720			720	
T-740			740	
T-760			760	
T-780			780	
T-800			800	

aging is slightly less than that of single-stage aging at the same aging temperature. In addition, the temperature for maximum hardness delayed. To ensure that the steel has higher hardness and impact toughness to restrain crack initiation and growth in the process of service, the Charpy impact tests were carried out for the specimens with Rockwell hardness values >45 HRC after aging heat treatment. The impact energies after different aging heat treatments are shown in Fig. 7.13b. For single-aging at

720 °C, the hardness value reached the maximum value of 47.24 HRC, whereas the impact energy reached the minimum value at 9.2 J. The impact energy reached the maximum value of 12.5 J and the corresponding hardness value was 46.05 HRC at 740 °C. For two-stage aging, the impact energy gradually increased and reached the maximum value of 16.2 J at 780 °C with increasing temperature from 740 to 780 °C. As for two-stage aging heat treatment, impact energy greatly improved despite the hardness decreasing by 1–2 HRC compared with that of single-stage aging.

The maximum impact energy value of 16.2 J is not only higher than premium H13 (10.84 J), but also almost exceeds superior H13 (13.55 J) [23]. The newly prepared HMAS-N, in terms of the toughness, is considerably outstanding when taking H13 mechanical properties into consideration. Compared with the maximum impact energy of single-stage aging, two-stage aging impact energy increased 29.6%. Despite the lowest hardness value of 45.1 HRC at 780 °C re-aging temperature, it still meets the service requirements of hot work steel. The SEM fractograph of specimens after the Charpy test for a fixed solid solution temperature (1170 °C) and various aging processes are shown in Fig. 7.14.

Both quasi-cleavage facets and ductile dimples were observed, which indicated the mixed mechanism of fracture. The specimen S-720 showed in Fig. 7.14a is mainly brittle-fractured surface zones characterized by transgranular quasi-cleavage and intergranular fracture. It can be noticed that quasi-cleavage fracture facets encircled by ridges are much smaller than the austenite grain (58.4 μm), and this kind of morphology forms when a crack propagates along the primary precipitates. The size of intergranular fracture grains (about 60 μm) was similar to the size of austenite grains (58.4 μm). This implies that intergranular fracture along the boundaries of austenite grains operated partially for specimen S-720. As shown in Fig. 7.14b, quasi-cleavage facets show massive primary precipitates. It indicated that quasi-cleavage nucleated along the precipitates. As shown in Fig. 7.14c, some portions of the fracture surface exhibited ductile-fractured surface with fine dimples. The dimple percentage increases with increasing aging temperature (from 720 to 760 °C) for single-stage aging treatment. The specimens S-740 and S-760 showed mainly ductile-fractured areas with fines dimples and some quasi-cleavage features It is clear that the fracture surface of specimen S-760 contained microcracks and some spherical secondary precipitates precipitated from the aging process, as shown in Fig. 7.14e. Figure 7.15 shows fractograph EDS element distribution mappings of specimen S-760. The EDS results of the precipitates shown in Fig. 7.14b and e are presented in Table 7.6.

Figure 7.15 showed that elements N, V, C were enriched in the same area, confirming that undissolved coarse primary carbonitrides are V(C, N), which also contains a small amount of Cr. Spherical secondary carbides are Mo-rich carbides. As shown in Fig. 7.14f, specimen T-780 presented more ductile dimples and fewer quasi-cleavage facets.

The undissolved primary precipitates, austenite grain size, strength of the matrix, secondary precipitates, and the strength of grain boundary have great effects on the impact energy [24]. Aging heat treatment has little effect on primary precipitates and austenite grain size. For specimen S-720, the massive undissolved primary V(C, N) had a higher hardness value (47.2 HRC) and is not easy to deform, which is

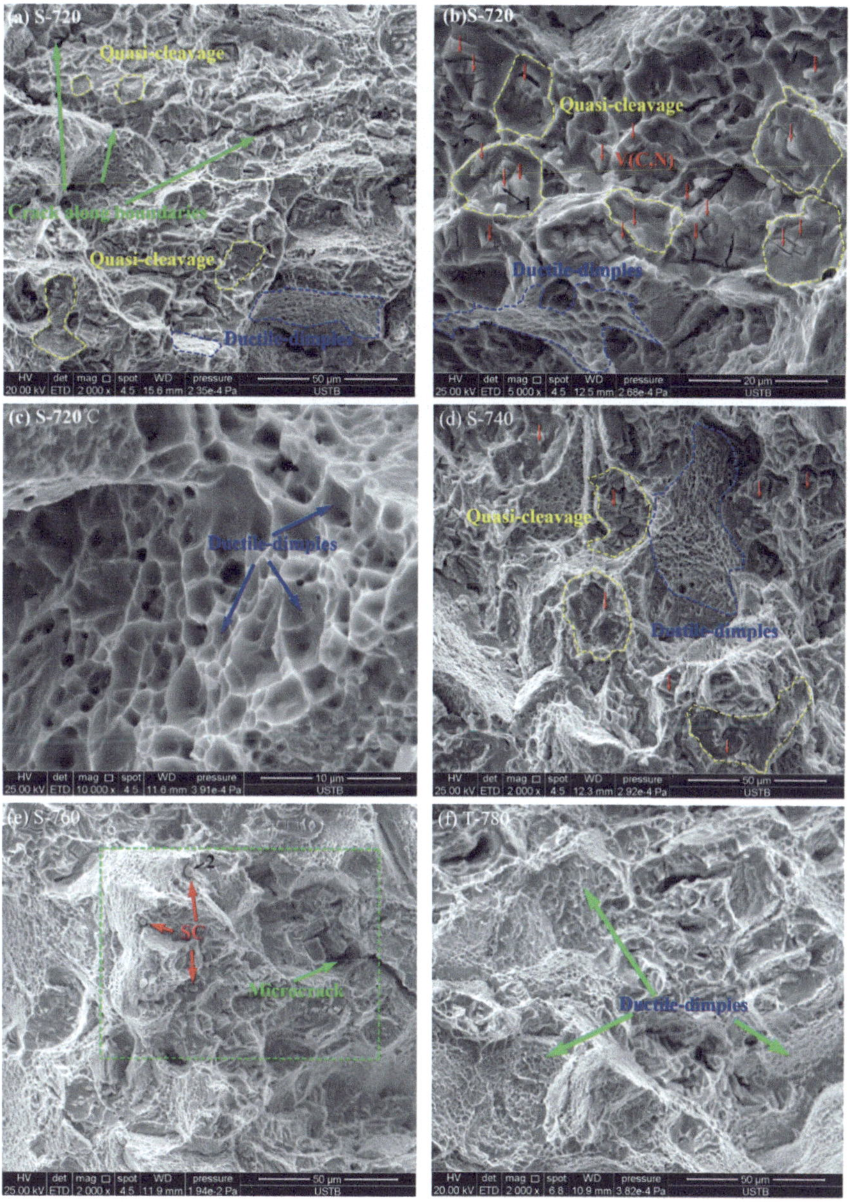

Fig. 7.14 SEM fractographs of specimens after different aging heat treatment processes: **a**–**c** S-720; **d** S-740; **e** S-760; **f** T-780. **b** and **c** are highly magnified images taken at the region enriched with quasi-cleavage facets and dimples of specimen S-720, respectively. SC represents secondary precipitates

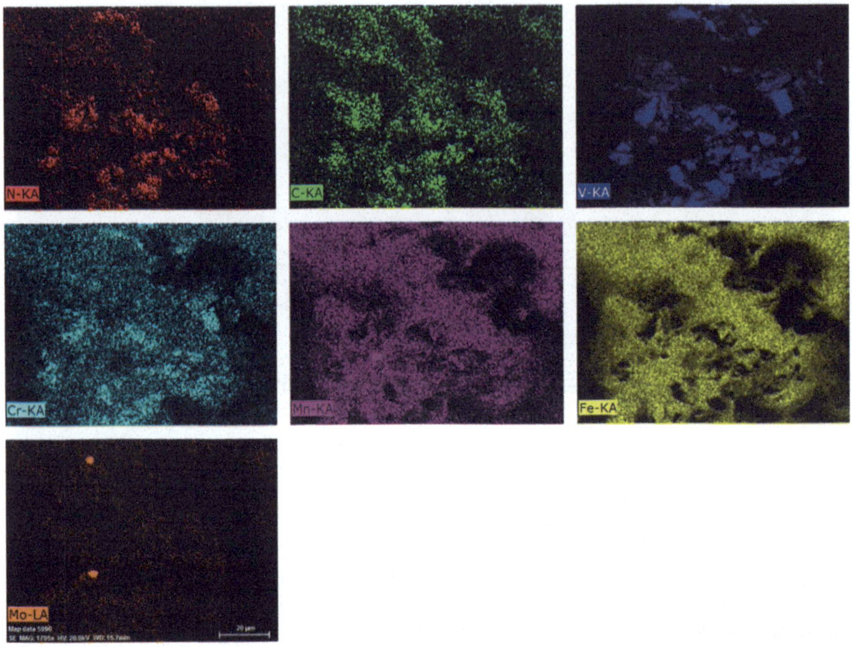

Fig. 7.15 Energy-dispersive X-ray spectrometer (EDS) element mappings of S-760 fractograph

Table 7.6 EDS results for precipitates (wt%)

Point	Element						
	C	V	Mo	Cr	N	Fe	Mn
1	28.48	40.50	0.24	0.92	27.57	1.72	0.57
2	37.07	0.26	16.57	1.65	0.00	43.62	0.83

responsible for embrittlement for the lower impact energy (9.3 J). A lower binding force among precipitates and steel matrix resulted from the existence of primary and secondary precipitates. It is much easier to generate cracks at the regions where primary V(C,N) are enriched. With increasing aging temperature from 720 to 740 °C for single-stage aging, the amount of secondary precipitates increased, which meant the interstitial atoms dissolved in the matrix precipitated and the degree of supersaturation decreased, resulting in the softening of the steel matrix [25]. The increasing number of secondary precipitates has a negative effect on impact energy. The softer the base metal, the larger the plastic district around the crack tip when the stress concentration is high enough and plastic deformation occurs. A larger plastic district will result in greater crack propagation and larger impact energy [26]. For specimen S-720 and S-740, the base metal plays the main role on impact energy compared to the negative effect of secondary precipitates. Hence, impact energy increased from 9.3 to 12.5 J with increasing temperature, from 720 to 740 °C, for single-stage aging temperature. When increasing temperature to 760 °C for the single-stage aging, the amount of secondary precipitates increases and the coarsening of the austenite grain boundary has a major effect on the impact energy, which led to impact energy decreasing from 12.5 to 11.4 J.

For two-stage aging, the low-temperature pre-aging of the two-stage aging process is equivalent to that of the nucleation stage, and the high-temperature aging is the stabilization stage [27]. The first stage in aging of the alloy is to promote the formation of the GP (Guinier-Preston) zone, and then to obtain a large number of dispersed GP areas, and to make the GP areas extend to a certain scale without dissolution in the second stage of aging. In the second high-temperature aging process, the aging precipitation sequence of the alloy changes from GP zone to secondary precipitates, and the alloy enters the aging stage. This improves the working capacity of tools for hot deformation. The two-stage aging may be more conducive to distributed homogeneous precipitation of secondary precipitates and reducing growth of secondary precipitates, which will weaken the reduction of the impact energy compared to single-stage aging. As a result, the impact energy increases from 9.2 to 16.2 J with a slight decrease of the strength when increasing the re-aging temperature from 740 to 780 °C.

References

1. Ji TY, Wu XC (2013) Microstructure and properties of H13 modified hot work die steel. JISRI 25(5):31–38
2. Jiang QC, Sui HL, Guan QF (2004) Thermal fatigue behavior of new type high-Cr cast hot wore die steel. ISIJ Int 44(6):1103–1107
3. Lu SY, Zhang TK, Kang XF (1995) Stainless steel. Atomic Energy Press, BeiJing
4. Grabovskii VY, Kanyka VI (2001) Austenitic die steels and alloys for hot deformation of metals. Met Sci Heat Treat 43:402–405
5. Jiang H, Wu XC, Shi NN (2012) Effect of nitrogen on the microstructure and properties of austenitic hot work die steel. Mater Mech Eng 36(1):58–61
6. He YL et al (2011) Thermodynamic and kinctic calculation of precipitation behavior of Cr-rich carbide. T Mater Heat Treat 32(1):134–137
7. Zhang Y, Li J, Shi CB (2017) Effect of heat treatment on the microstructure and mechanical properties of nitrogen-alloyed high-Mn austenitic hot work die steel. Metals 7(3):94
8. Medvedeva A et al (2009) High-temperature properties and microstructural stability of hot-work tool steels. Mater Sci Eng, A 523(1):39–46
9. Wu HL et al (2011) Effect of Nb on MX precipitated Phase in K2 spring steel. J Univ Sci Technol B 33(8):927–935
10. Kurz W, Fisher DJ (1984) Fundamentals of solidification. Trans Tech Publications, New York
11. Ye W, Zhang X, Yang GT (2012) Effect of carbon and nitrogen on the macroscopic solidification structure of ferritic stainless steel ingot. Spec Cast Nonferrous Alloys 32(9):801–803
12. Gao FB (2014) Effect of nitrogen on the structure and properties of 201 stainless steel. Dissertation, Inner Mongolia University of Science & Technology
13. Lee SJ, Lee YK (2008) Prediction of austenite grain growth during austenitization of low alloy steels. Mater Des 29(9):1840–1844
14. Speer JG, Michael JR, Hansen SS (1987) Carbonitride precipitation in niobium/vanadium microalloyed steels. Metall Trans A 18A(2):211–222

15. Matsuo S, Ando T, Grant NJ (2000) Grain refinement and stabilization in spray-formed AISI 1020 steel. Mater Sci Eng, A 288(1):34–41
16. Lin YL, Lin CC, Tsai TH et al (2010) Microstructure and mechanical properties of 0.63C-12.7Cr martensitic stainless steel during various tempering treatments. Mater Manuf Processes 25:246–248
17. Dutra JC, Siciliano F, Padilha AF (2002) Interaction between second-phase particle dissolution and abnormal grain growth in an austenitic stainless steel. Mater Res 5(3):379–384
18. Leyson GPM, Curtin WA, Hector L et al (2010) Quantitative prediction of solute strengthening in aluminium alloys. Nat Mater 9(9):750–755
19. Lee SJ, Jung YS, Baik S et al (2014) The effect of nitrogen on the stacking fault energy in Fe–15Mn–2Cr–0.6C–xN twinning-induced plasticity steels. Scripta Mater 92:23–26
20. Jack DH, Jack KH (1973) Carbides and nitrides in steel. Mater Sci Eng 11(1):1–27
21. Gong WM, Yang CF, Zhang YQ et al (2004) Precipitation kinetics of V(C, N) in austenite for low carbon steel microalloyed with vanadium and nitrogen. J Iron Steel Res 16(6):41–46
22. Teng ZK, Liu CT, Ghosh G (2010) Effects of Al on the microstructure and ductility of NiAl-strengthened ferritic steels at room temperature. Intermetallics 18(8):1437–1443
23. North America Die Casting Association. NADCA Die Material Committee (2003) NADCA recommended procedures for H13 tool steel. NADCA, p 207
24. Mesquita RA, Barbosa CA, Morales EV et al (2011) Effect of silicon on carbide precipitation after tempering of H11 hot work steels. Metall Trans A 42(2):461–472
25. Yan P, Liu ZD, Bao HS et al (2014) Effect of tempering temperature on the toughness of 9Cr–3W–3Co martensitic heat resistant steel. Mater Des 54(2):874–879
26. Ding RL (2009) Metal materials and heat treatment. Machinery Industry Press, Beijing
27. Werenskiold JC, Deschamps A, Brechet Y (2000) Characterization and modeling of precipitation kinetics in an Al–Zn–Mg alloy. Mater Sci Eng A 293(1):267–274

Chapter 8
Feasibility Analysis of Titanium on Carbide Control in High Carbon High Alloy Steel

Abstract In recent years, control of carbides in high carbon and high chromium cast iron and die steel by alloying element Ti is a research hotspot. This chapter takes high carbon martensitic steel 8Cr13MoV as an example to ascertain the feasibility of Ti-alloying 8Cr13MoV steel. The results show that TiC precipitates first in the liquid phase, which can act as nucleation core to refine the microstructure and eutectic carbide M_7C_3. Ti inhibits the precipitation of M_7C_3 from austenite and promotes the transformation of M_7C_3 to $M_{23}C_6$ during cooling. With the increase of Ti content in the steel, the amount of Cr-rich carbides decreases, which improves the hardenability of the steel, increases the content of martensite, refines the martensite structure, and greatly improves the tensile strength of the as-cast structure. Appropriate amount of Ti in steel can reduce the total amount of carbides and disperse relatively of carbides in the steel, which is beneficial to improve the mechanical properties of the steel after heat treatment. Excessive Ti content in steel results in the generation of a large number of TiC and inhibits the spheroidizing growth of secondary carbides during annealing, which deteriorates the subsequent processing properties of the steel.

Keywords Titanium · Carbides · Solidification microstructure · Forged microstructure

During the solidification process, the segregation of alloying elements will lead to the formation of a large number of eutectic carbides. This kind of carbide will become the core of microcrack initiation [1], and will take away the chromium element in the matrix, resulting in the decrease of corrosion resistance. It is very limited to adjust the solidification process parameters to improve the segregation. Therefore, many researches have been carried out to improve the properties of materials by modifying the carbide structure, including reducing the volume fraction and size of carbides, shortening the distance between carbides and changing the morphology of carbides. Common methods include increasing nucleation or inhibiting eutectic carbide growth [2–12].

In recent years, it is a research hotspot to control carbides in high carbon and high chromium cast iron and die steel by using alloy element Ti [2, 5, 13–15]. The results show that Ti is a strong carbide forming element, which can effectively improve the segregation of carbon, refine eutectic carbide, reduce the volume fraction of carbide,

J. Li and C. Shi, *Carbide in Special Steel*, Engineering Materials,
https://doi.org/10.1007/978-981-16-1456-9_8

and improve the impact toughness and wear resistance of the material. High carbon martensitic stainless steel and high chromium cast iron have the same eutectic type of carbides. Therefore, it is necessary to study the effect of Ti on carbide in tool and die steel. Taking the high carbon martensitic steel 8Cr13MoV as an example, the feasibility of Ti controlling carbide in high carbon alloy steel is analyzed.

8.1 Effect of Titanium on the Carbides in Ingot

8.1.1 Effect of Titanium on the Type of Carbides

The chemical composition of 8Cr13MoV ingots with different Ti contents are shown in Table 8.1.

Carbides extracted from steel matrix were analyzed by XRD and results are shown in Fig. 8.1. Carbides in sample No. 1 were confirmed to be M_7C_3. The type of carbide changed after titanium addition because titanium combined carbon and nitrogen to form TiC and Ti(C, N). Meanwhile, M_7C_3 changed to $M_{23}C_6$ after titanium addition. This result was consistent with the thermodynamic calculation which indicated that titanium could enhance the complete transition from M_7C_3 to $M_{23}C_6$.

8.1.2 Effect of Titanium on the Composition of Carbides

The SEM images of ESR ingot microstructure are shown in Fig. 8.2. It can be seen that the addition of Ti can refine the grain size of steel, but it is not obvious. Further adding Ti will reduce the grain size. The size of carbides in Ti bearing steel is large and mainly distributed on the grain boundary, which is a typical feature of eutectic carbides [16]. These carbides presented two kinds of morphology in sample Nos. 2 and 3, i.e., single black carbides and complex carbides with black and gray particles. The size of complex carbide decreases with the increase of Ti content in the ingot.

The EDS results of carbides in Fig. 8.2 are shown in Table 8.2. According to EDS results, the complex carbides were confirmed to be TiC as nucleus of M_7C_3

Table 8.1 Chemical composition of 8Cr13MoV steel (wt%)

Sample no.	C	Cr	Mo	Mn	Si	V	Ni	Ti
1	0.78	13.6	0.20	0.50	0.44	0.14	0.16	0.043
2	0.79	14.12	0.20	0.48	0.44	0.20	0.20	0.50
3	0.77	13.44	0.21	0.44	0.33	0.16	0.16	0.77
4	0.78	14.05	0.21	0.42	0.35	0.16	0.16	1.20

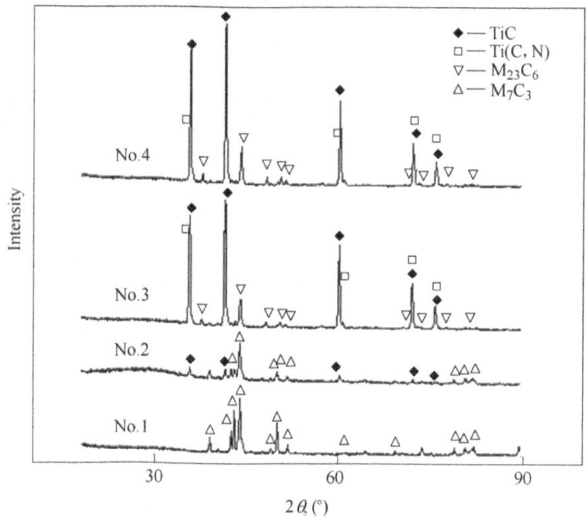

Fig. 8.1 XRD results for samples with different titanium content

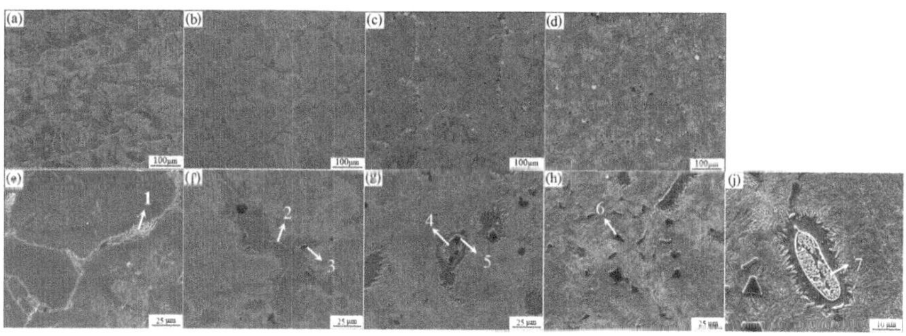

Fig. 8.2 SEM images of ESR ingot microstructure: **a** and **e** No. 1; **b** and **f** No. 2; **c** and **g** No. 3; **d**, **h** and **j** No. 4

Table 8.2 EDS results for carbides (wt%)

Point	Element				
	C	Cr	Fe	Ti	Mo
1	13.22	51.89	31.25	–	1.45
2	13.45	50.22	31.88	2.12	2.33
3	26.33	–	1.18	70.10	2.38
4	13.81	53.38	26.94	1.05	2.48
5	23.86	4.43	8.89	51.76	7.79
6	20.28	2.91	12.2	61.47	3.14
7	8.68	42.53	47.90	–	1.40

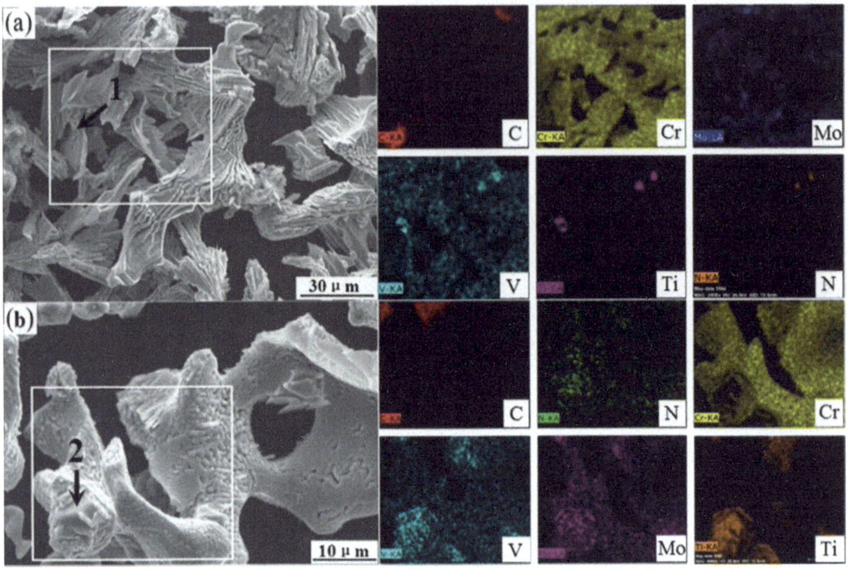

Fig. 8.3 EDS element mappings of carbides: **a** No. 1; **b** No. 3

and $M_{23}C_6$ carbide. Carbides in sample No. 4 have three kinds of morphologies, which included single granular, irregular flake and rod-like with interior tiny laminar structure.

The EDS element mappings of carbides are shown in Fig. 8.3. It was found by EDS shown in Fig. 8.3a that the enrichment of titanium, vanadium and nitrogen in the area where (Ti, V)N precipitated. (Ti, V)N has a high melting point and acts as the core of eutectic carbides after precipitating from liquid steel. It was proved that titanium was prior to combine with nitrogen when the content of titanium was low. In sample No. 2, it was clear that nitrogen, vanadium, molybdenum and titanium are enriched in the same area.

8.1.3 Effect of Titanium on the Morphology of Carbides

The SEM images of carbides extracted from steel are shown in Fig. 8.4. In sample with 0.043%Ti, the size of carbides was large, and the maximum size is over 50 μm. The carbides were typical eutectic carbides and its morphology was skeleton shape made of a cluster of long strip carbide. In sample with 0.50%Ti, there are two kinds of carbides were observed in the sample with 0.50%Ti. The morphology of M_7C_3 carbide is the same as that in the sample with 0.043%Ti. The overall size of M_7C_3 carbide is small, part of which is attached to TiC, as shown in Fig. 8.4f, and the others are the independent TiC particle. In sample with 0.77%Ti, the size of carbides was relatively small. Most of carbides were complex structure of TiC combining with

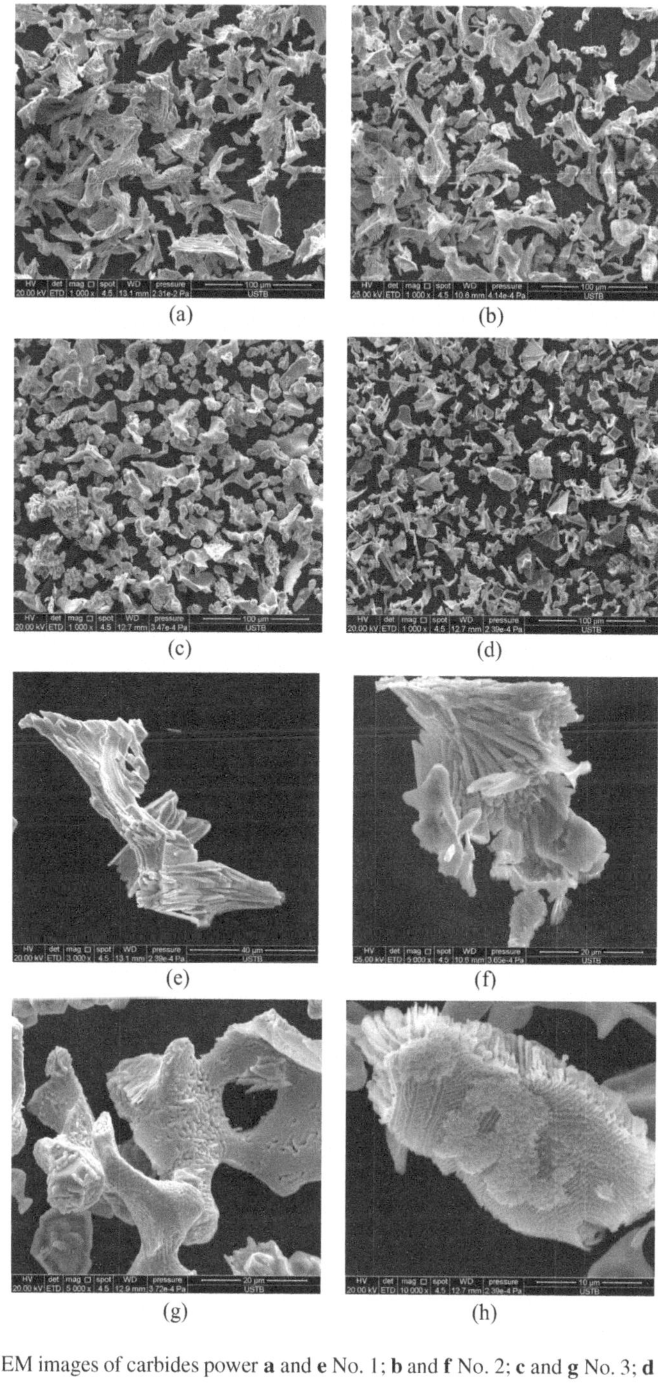

Fig. 8.4 SEM images of carbides power **a** and **e** No. 1; **b** and **f** No. 2; **c** and **g** No. 3; **d** and **h** No. 4

$M_{23}C_6$. Others were pure TiC, as shown in Fig. 8.4g. The size of complex carbides was large, and their morphology was similar with that of eutectic carbides, but different from that in sample with 0.043%Ti. It was deduced from the morphology that the large carbides $M_{23}C_6$ were transformed from eutectic carbides M_7C_3 rather than that precipitated directly from austenite. In sample with 1.20%Ti, the morphology of most carbides was small irregular flake TiC and Ti(C, N). Only a small amount of carbide $M_{23}C_6$ was regular shape as shown in Fig. 8.4f, which corresponded to the rod-like carbides shown in Fig. 8.2j. The small carbides $M_{23}C_6$ in sample 1.20%Ti were different from that in sample with 0.77%Ti. These carbides were transformed from carbides M_7C_3 precipitated from austenite.

8.2 Effect of Titanium on Forged Microstructure and Annealed Organization

8.2.1 Influence of Titanium on Forged Microstructure

The SEM images of forged ESR ingot microstructure were shown in Fig. 8.5. The eutectic carbides in sample with 0.043%Ti and 0.50%Ti were broken. Meanwhile, the small secondary carbides in sample No.1 precipitated around the eutectic carbides. No small secondary carbides precipitated around the broken eutectic carbides as

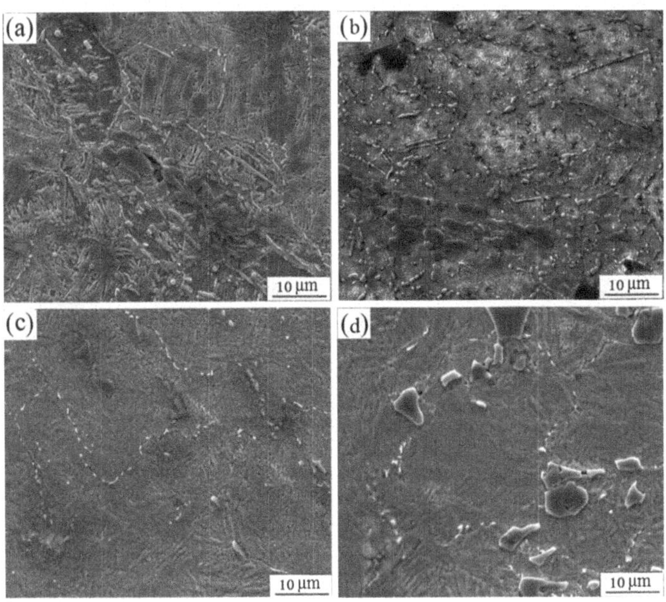

Fig. 8.5 SEM images of specimens after forging: **a** No. 1; **b** No. 2; **c** No. 3; **d** No. 4

shown in sample 0.77 and 1.20%Ti. It was found that secondary carbides precipitated along the grain boundaries in these samples. These secondary carbides were much less after added titanium in the steel. The precipitated carbides in sample 1.20%Ti along grain boundaries were granule shape.

8.2.2 *Influence of Titanium on Spheroidizing Annealed Microstructure*

The SEM images of spheroidizing annealed microstructure were shown in Fig. 8.6. The amount of carbides decreased with adding Ti in the steel, which was 1.27 particles per μm^2, 1.18 particles per μm^2, 1.06 particles per μm^2 and 0.93 particles per μm^2, respectively. The martensite structure was different before annealing so that the most carbides of sample with 0.043%Ti were chain and had certain directivity. During annealing, the carbides would precipitate along the acicular martensite and become long strip shape easily [17]. The precipitated carbides have certain directivity after annealed if separated incompletely during annealing. In summary, titanium addition could inhibit the precipitation of the secondary carbides greatly during heat treatment,

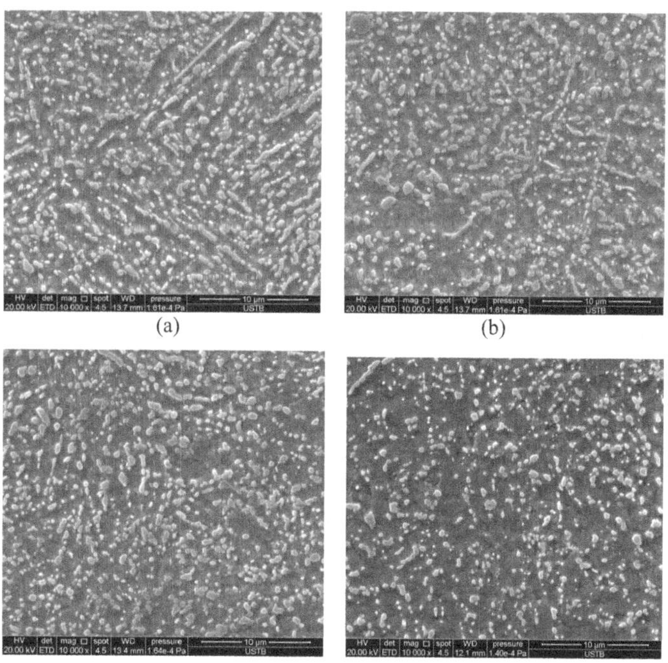

Fig. 8.6 SEM images of specimens after spheroidizing annealing: **a** No. 1; **b** No. 2; **c** No. 3; **d** No. 4

and this effect increases with increasing titanium content. The growth of carbide in the spheroidizing annealing process relies on the diffusion of carbide forming elements. Ti is strong carbide forming elements. With the increase of Ti content, the content of TiC increases, which restrains the diffusion of carbon element and causes the reduction of the precipitation of secondary carbides in spheroidizing annealing process.

8.3 Feasibility Analysis of Ti Treatment on High Carbon Alloy Steel

8.3.1 Effect Mechanism of Ti on Carbides in High Carbon Alloy Steel

The effect of Ti on the equilibrium phase during the cooling process of 8Cr13MoV was calculated by Thermo-Calc, and results were shown in Fig. 8.7. In this phase diagram, the content of other alloying elements was fixed, and Ti content and temperature were treated as variable.

As shown in Fig. 8.7, when the content of titanium reached up to 0.1%, TiC would precipitate at the temperature of 1450 °C. The precipitation temperature of TiC increased with the increasing content of titanium. When the content of titanium reached up to 0.5%, TiC would precipitate at the temperature of 1600 °C. Therefore, eutectic carbide TiC could act as the nucleation core of high-temperature ferrite or austenite, which could contribute to grain refinement [18]. It can be concluded that the grain refinement is due to the precipitation of TiC in the liquid phase, which provides a large number of nucleation cores for δ-Fe. Some studies have shown that [3], adding Ti to high carbon and high chromium cast iron can not only refine the grain, but also effectively reduce the segregation of C and Cr elements, which is consistent with the reduction of M_7C_3-type eutectic carbides. The disregistry between

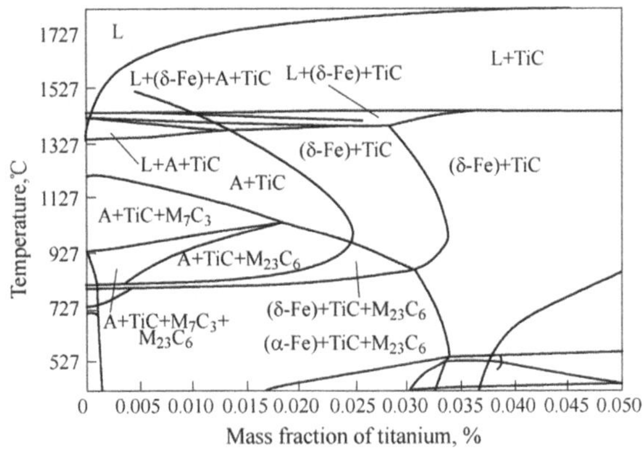

Fig. 8.7 Equilibrium phase formation of 8Cr13MoV steel calculated using Thermo-Calc

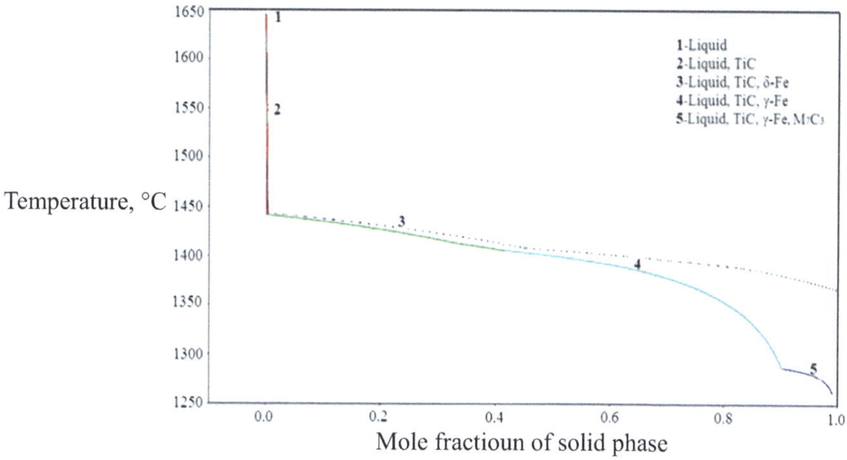

Fig. 8.8 Non-equilibrium phase precipitation in steel calculated using Thermo-Calc

TiC and M_7C_3-type carbides calculated by the formula of mismatch degree proposed by Bramfitt [18, 19] is 4.7%. It is proved that TiC is an effective nucleation for M_7C_3-type carbides and plays an important role in the refinement of M_7C_3-type carbides. The calculation results of non-equilibrium solidification carried out by using Scheil module in Thermo-Calc are shown in Fig. 8.8. It can be seen from Fig. 8.8 that M_7C_3-type carbides begins to precipitate in the liquid phase when the molar mass fraction of solid phase reaches about 90%. At this time, the solid phase is ferrite and TiC.

In the equilibrium solidification, M_7C_3-type carbides precipitates from austenite., it can be seen from Fig. 8.7 that the precipitation temperature range of M_7C_3-type carbides decreases with the increase of Ti content, indicating that Ti element has a restraining effect on the precipitation of M_7C_3-type carbides. In addition, increasing Ti content increased the transition temperature of M_7C_3-type carbides to $M_{23}C_6$-type carbides and promoted the transformation of M_7C_3-type carbides to $M_{23}C_6$-type carbides. When the content of Ti in ESR ingot reaches 0.77%, $M_{23}C_6$-type carbides will be formed during solidification, but due to the effect of Ti element, M_7C_3-type carbides will be transformed into $M_{23}C_6$-type carbides finally.

The M_7C_3-type carbides with large size precipitated in high chromium cast iron will seriously reduce its wear resistance and toughness. The researchers found that adding Ti element could refine M_7C_3-type carbides to improve the service performance of high chromium cast iron. The effect mechanism of Ti on carbides in high carbon alloy steel in this chapter is similar to that reported by other scholars [20–22]. As a strong carbide forming element, Ti element is easy to react with C to form TiC in liquid alloy. The TiC particles precipitated first in the liquid alloy and acted as the heterogeneous nucleation core of M_7C_3-type carbides, which plays the role of refining carbides.

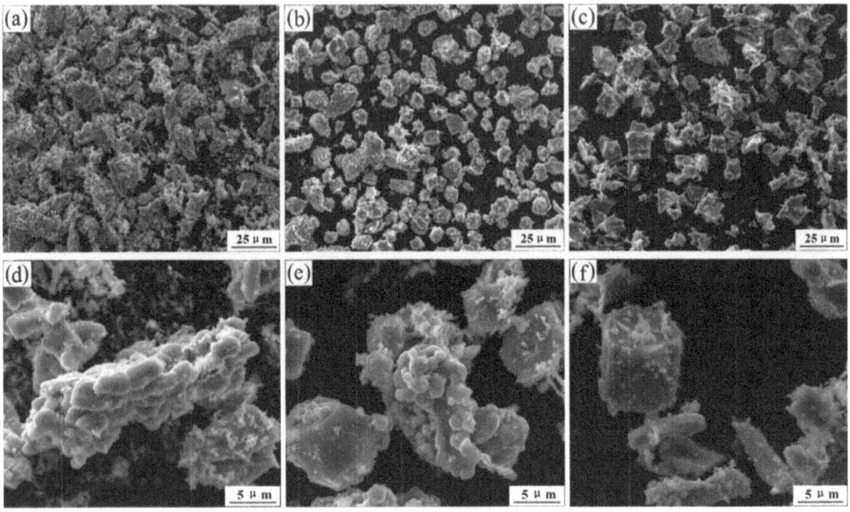

Fig. 8.9 Morphology of carbide after forging and spheroidizing annealing **a** and **d** No. 1, **b** and **e** No. 3; **c** and **f** No. 4

8.3.2 Feasibility of Ti Treatment on Carbide Control in High Carbon Alloy Steel

In the process of heat treatment, $M_{23}C_6$-type carbides is easier to be dissolved into the matrix than M_7C_3-type carbides. The transformation of M_7C_3-type carbides to $M_{23}C_6$-type carbides is beneficial to carbides control in heat treatment process. To further verify whether the lager sized $M_{23}C_6$-type carbides in the ingot can be dissolved into the matrix, the ESR ingot is forged, spheroidized and annealed. The SEM images of carbides extracted by anode electrolysis are shown in Fig. 8.9.

It can be seen from Fig. 8.9 that a large number of fine secondary carbides are precipitated in the micstructure after spheroidizing treatment and attached to the large particle carbides. It can be seen from Fig. 8.9b that the large complex carbide in the ingot with 0.77%Ti disappears, which proves that the $M_{23}C_6$-type carbides wrapped in the outer layer of TiC has basically dissolved. The $M_{23}C_6$-type carbides wrapped in the TiC, which were not completely, has been spheroidized, as shown in Fig. 8.9d. According to the size of carbide, the size of carbides in the sample with 0.043%Ti is the largest, and the size of carbides in the sample with 0.77 and 1.20%Ti have no difference. Part of carbides in the sample with 1.20%Ti are larger than that in the ingot with 0.77%Ti. It can be seen from Fig. 8.9c and d that there are 5–10 μm TiC particles in the sample with 0.77 and 1.20%Ti after spheroidizing annealing, which indicates that large sized TiC particles precipitated in the liquid alloy still exist in steel after heat treatment.

The measured results of mechanical properties after forging and spheroidization annealing are shown in Table 8.3.

Table 8.3 Mechanical properties of steel samples before and after heat treatment

Sample no.	Hardness		Tensile strength, MPa		Elongation after fracture, %	
	Before (HRC)	After (HRB)	Before	After	Before	After
1	52.9	95.2	635.95	739.53	1	17.85
2	53.2	95.0	889.11	740.19	1	17.88
3	56.7	93.6	974.49	714.01	1	20.95
4	57.6	94.1	1315.66	694.39	1	20.41

It can be seen from Table 8.3 that the sample with 0.77%Ti after forging and spheroidizing annealing has the lowest hardness and the highest elongation after fracture. In the process of heat treatment, the larger sized $M_{23}C_6$-type carbides continue to dissolve into matrix of sample with 0.77%Ti, which is more suitable for subsequent machining.

The tensile fracture after spheroidizing annealing is shown in Fig. 8.10. There were large carbides at the fracture surface of sample with 0.043%Ti, which are labeled in Fig. 8.10a. There are no large carbides in the fracture of sample with 0.77%Ti. TiC particles uniformly distributed on the fracture surface of samples 0.77 and 1.20%Ti. The number of TiC in sample with1.20%Ti was larger than that in sample with 0.77%Ti. Due to the existence of a large amount of TiC, the tensile strength and elongation after fracture decreased.

Figure 8.11 shows Images TiC carbides in 8Cr13MoV steel with 0.77%Ti. It can be seen from Fig. 8.11 that TiC can not be completely broken and dissolved in subsequent forging, hot rolling and annealing processes, and it largerly with the size of about 5–10 μm.

The high chromium cast iron is generally cast and formed without large deformation process such as rolling. However, the high carbon martensitic stainless steel 8Cr13MoV, used for mading high grade knives and scissors, needs to undergo forging, hot rolling, annealing and cold rolling in the production process, and are finally prepared into all kinds of knives and scissors. In the processes of deformation, such as rolling, these micron-level TiC will generate stress concentration and

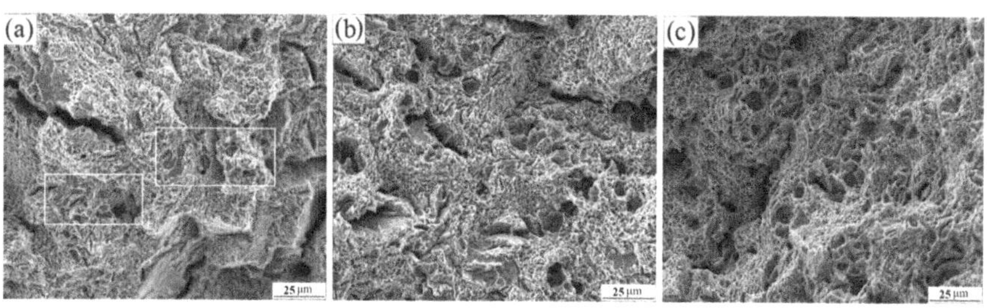

Fig. 8.10 SEM images of tensile fracture after forging and spheroidizing annealing: **a** No. 1; **b** No. 3; **c** No. 4

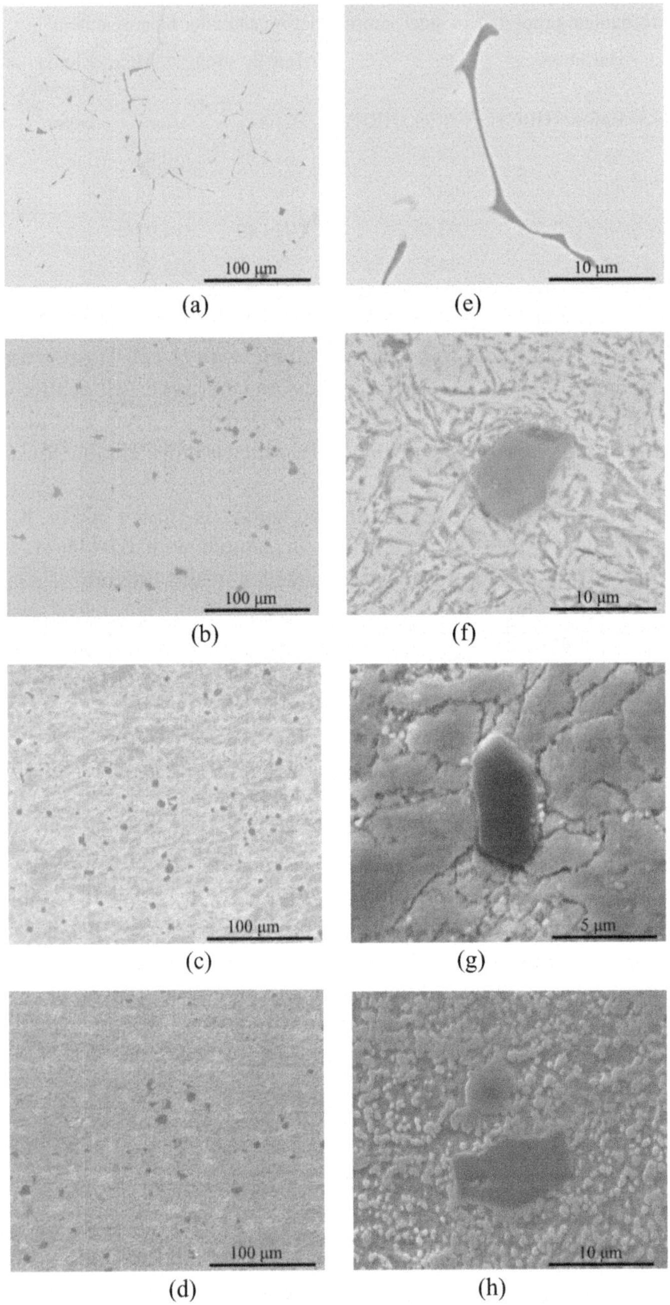

Fig. 8.11 Images TiC carbides in 8Cr13MoV steel with 0.77%Ti: **a** and **e** Ingot; **b** and **f** Forging; **c** and **g** Heat rolling; **d** and **h** Spheroidizing annealing

become the source of cracks, causing brittle fracture of the material and seriously affecting the service life of the material. It can be seen from Fig. 8.10b and c that a large number of micron-level TiC are observed in the dimple of the tensile fracture of the sample with 0.77%Ti and 1.20%Ti indicating that the fracture was mainly caused by the rupture between the thick TiC and the matrix. Moreover, the thickness of the blade of the knife and scissor products is within 5 μm. The TiC on the blade is easy to fall off, causing a gap in the blade and seriously affecting the service life of the product. Therefore, TiC particles precipitation in liquid alloy at high temperature cannot be completely eliminated in the process of heat treatment and deformation. For high-carbon alloy steel produced by rolling and other deformation processes, it is difficult to effectively control the precipitation and growth of TiC in the liquid alloy, it is not suitable for titanium alloying treatment.

References

1. Karagöz S, Fischmeister HF (1998) Cutting performance and microstructure of high speed steels: contributions of matrix strengthening and undissolved carbides. Metall Mater Trans A 29(1):205–216
2. Wu XJ, Xing JD, Fu HG et al (2007) Effect of titanium on the morphology of primary M7C3 carbides in hypereutectic high chromium white iron. Mater Sci Eng A 457(1):180–185
3. Bedolla-Jacuinde A, Correa R, Mejía I et al (2007) The effect of titanium on the wear behaviour of a 16%Cr white cast iron under pure sliding. Wear 263(1):808–820
4. Bedolla-Jacuinde A, Correa R, Quezada JG et al (2005) Effect of titanium on the as-cast microstructure of a 16%chromium white iron. Mater Sci Eng, A 398(1–2):297–308
5. Arikan MM, Çimenoğlu H, Kayali ES (2001) The effect of titanium on the abrasion resistance of 15Cr–3Mo white cast iron. Wear 247(2):231–235
6. Chen HX, Chang ZH, Lu JC et al (1993) Effect of niobium on wear resistance of 15%Cr white cast iron. Wear 166(2):197–201
7. Ma Q, Wang CC, Shoji H (1995) Modification of hypoeutectic low alloy white cast irons. J Mater Sci 31(7):1865–1871
8. Bratberg J, Frisk K (2004) An experimental and theoretical analysis of the phase equilibria in the Fe-Cr-V-C system. Metall Mater Trans A 35(12):3649–3663
9. Lewellyn RJ, Yick SK, Dolman KF (2004) Scouring erosion resistance of metallic materials used in slurry pump service. Wear 256(6):592–599
10. Wang MJ, Mu SM, Sun FF et al (2007) Influence of rare earth elements on microstructure and mechanical properties of cast high-speed steel rolls. J Rare Earths 25(4):490–494
11. Fu HG, F DM, Zou DN et al (2004) Structures and properties of high-carbon high speed steel by RE-Mg-Ti compound modification. J Wuhan Univ Technol 19(2):48–51
12. Duan JT, Jiang ZQ, Fu HG (2007) Effect of RE-Mg complex modifier on structure and performance of high speed steel roll. J Rare Earths 25(7):259–263
13. Bu FZ, Wang YB, Zheng LH et al (2018) Precipitation behavior of nano-carbides during tempering of Ti-Mo microalloyed steel. JISRI 30(11):928–934
14. Cho KS, Sang IK, Park SS et al (2016) Effect of Ti addition on carbide modification and the microscopic simulation of impact toughness in high-carbon Cr-V tool steels. Metall Mater Trans A 47(1):26–32
15. Mirzaee M, Momeni A, Keshmiri H et al (2014) Effect of titanium and niobium on modifying the microstructure of cast K100 tool steel. Metall Mater Trans B 45(6):2304–2314
16. Yu WT, Li J, Shi CB et al (2016) Effect of Spheroidizing annealing on microstructure and mechanical properties of high-carbon martensitic stainless steel 8Cr13MoV. J Mater Eng Perform 26(2):1–10

17. Bjärbo A, Hättestrand M (2000) Complex carbide growth, dissolution, and coarsening in a modified 12 pct chromium steel-an experimental and theoretical study. Metall Mater Trans A 32(1):19–27
18. Bramfitt BL (1970) The effect of carbide and nitride additions on the heterogeneous nucleation behavior of liquid iron. Metall Mater Trans B 1(7):1987–1995
19. Razavinejad R, Firoozi S, Mirbagheri SMH (2012) Effect of titanium addition on as cast structure and macrosegregation of high-carbon high-chromium steel. Steel Res Int 83(9):861–869
20. Huang ZF, Xing JD, Zhi XH et al (2011) Effect of Ti addition on morphology and size of primary M7C3 type carbide in hypereutectic high chromium cast iron. Mater Sci Technol 27(1):426–430
21. Wu X, Xing J, Fu H et al (2007) Effect of titanium on the morphology of primary M7C3 carbides in hypereutectic high chromium white iron. Mater Sci Eng A 457(1–2):180–185
22. Chung RJ, Tang X, Li DY et al (2009) Effects of titanium addition on microstructure and wear resistance of hypereutectic high chromium cast iron Fe–25wt.%Cr–4wt.%C. Wear 267(1):356–361

The manufacturer's authorised representative in the EU is Springer
Nature Customer Service Centre GmbH, Europaplatz 3, 69115 Heidelberg,
Germany. If you have any concerns regarding our products, please
contact ProductSafety@springernature.com

Printed and bound by CPI Group (UK) Ltd, Croydon, CR0 4YY
30/09/2026
02225506-0008